Virology

a practical approach

Edited by

B W J Mahy

The Animal Virus Research Institute, Pirbright,
Woking, Surrey GU24 0NF, UK

Oxford University Press, Walton Street, Oxford OX2 6DP

First published 1985
Reprinted 1986, 1987, 1991

British Library Cataloguing in Publication Data

Virology : a practical approach.—(Practical approach series)
1. Viruses
I. Mahy, B.W.J. II. Series
576′.6484 QR301

ISBN 0-904147-78-9

Oxford, England.

Preface

Virology has long been an esoteric branch of microbiology, fascinating to those pursuing an understanding of the nature of viruses, but a closed book to most biologists. The study of viruses began because they are agents of the most important diseases of plants and animals, as well as man. Despite the elimination of smallpox, many virus diseases such as influenza, foot-and-mouth disease, herpes simplex or rabies are only poorly controlled, and new virus diseases such as Lassa fever, African swine fever or acquired immune deficiency syndrome (AIDS) pose problems which require urgent research effort by virologists. At the same time virology has entered a new phase of growth as a result of the contributions it has made, and is continuing to make, to molecular and cellular biology. Virologists have always exploited technical advances in other branches of science, especially those developed for the study of biological macromolecules, but only recently have molecular and cellular biologists begun to use viruses as tools to probe complex questions concerning cellular structure and function.

It could be argued that the greatest contribution of virology to science and mankind has been the provision of reverse transcriptase, which forms the cornerstone of modern genetic engineering. But many of the current concepts in molecular and cellular biology (such as introns, splicing or oncogenes) arose directly from studies of virus structure and function. It is certain that the study of viruses will continue to expand and the techniques peculiar to virology will need to be used by an ever greater number of scientists.

This book has been compiled with both aspects of virology in mind. It provides practical recipes and protocols for handling most of the animal viruses of current interest either as disease agents or as probes of cell function. Within the space available, it is not possible to give complete coverage of all the virus groups, and I take responsibility for the selection of topics. However, the retroviruses are an important omission: they were deliberately not included since they have recently been covered extensively by two excellent volumes from the Cold Spring Harbor Laboratory.

Instead of devoting each chapter to a different technique, such as assay, radiolabelling, or purification, the chapters have, for the most part, been arranged in virus families. This is because the methods used are frequently peculiar to a particular group of viruses, and whilst some general principles recur, the reader bent on working with a particular virus (e.g., polio) should find all the standard techniques set out in a single chapter. A Dictionary of Virology by K.E.K.Rowson, T.A.L.Rees and B.W.J.Mahy is a useful volume to guide the non-virologist through the nomenclature and jargon of the trade. Recourse may also need to be made to a volume on animal cell and tissue culture methods which is available in the Practical Approach series.

It is a pleasure to thank all the authors for their diligence and co-operation, as well as the series editors and staff of IRL Press for their speedy response to requests, which made the editorial work most enjoyable. I am also deeply grateful to Chris Smale for artistic advice, and to my secretaries, Mary Wright in Cambridge and Hazel West in Pirbright for their forebearance.

B.W.J.Mahy

Contributors

T.Barrett
Division of Virology, Department of Pathology, University of Cambridge, Laboratories Block, Addenbrooke's Hospital, Hills Road, Cambridge CB2 2QQ, UK

P.Beard
Institute of Experimental Cancer Research, Chemin des Boveresses, CH-1066 Epalinges, Switzerland

J.C.S.Clegg
Centre for Applied Microbiological Research, Porton Down, Salisbury SP4 0JG, UK

E.A.Gould
Arbovirus Unit, London School of Hygiene and Tropical Medicine, Winches Farm Field Station, 395 Hatfield Road, St. Albans, Herts., UK

R.Hull
John Innes Institute, Colney Lane, Norwich NR4 7UH, UK

S.C.Inglis
Division of Virology, Department of Pathology, University of Cambridge, Laboratories Block, Addenbrooke's Hospital, Hills Road, Cambridge CB2 2QQ, UK

R.A.Killington
Department of Microbiology, University of Leeds, Leeds LS2 9JT, UK

M.McCrae
Department of Biological Sciences, University of Warwick, Coventry CV4 7AL, UK

P.D.Minor
National Institute of Biological Standards and Control, Holly Hill, Hampstead, London NW3 6RB, UK

P.Morgan-Capner
Department of Virology, Preston Infirmary, Preston PR1 6PS, UK

J.R.Pattison
Department of Medical Microbiology, University College and The Middlesex School of Medicine, Faculty of Clinical Sciences, University Street, London WC1E 6JJ, UK

K.L.Powell
Department of Microbiology, University of Leeds, Leeds LS2 9JT, UK

B.Precious
National Institute for Medical Research, The Ridgeway, Mill Hill, London NW7 1AA, UK

C.R.Pringle
Department of Biological Sciences, University of Warwick, Coventry CV4 7AL, UK

W.C.Russell
National Institute for Medical Research, The Ridgeway, Mill Hill, London NW7 1AA, UK

H.Türler
Department of Molecular Biology, University of Geneva, 30 quai Ernest-Ansermet, CH-1211 Geneva 4, Switzerland

W.H.Wunner
The Wistar Institute of Anatomy and Biology, 36th Street at Spruce, Philadelphia, PA 19104, USA

Contents

Abbreviations

BBS	borate-buffered saline
BMV	bovine mammilitis virus
BSA	bovine serum albumin
CCV	channel catfish virus
CEF	chick embryo fibroblasts
CFT	complement fixation test
CMC	carboxymethylcellulose
CMV	cytomegalovirus
cpe	cytopathic effect
CS	calf serum
CVS	challenge virus standard
DAPI	4′,6-diamidino-2-phenylindole
DEAE	diethylaminoethyl
DEP	diethyl pyrocarbonate
DGV	dextrose-gelatin-veronal
DI	defective-interfering
DMSO	dimethylsulphoxide
EDTA	ethylenediamine tetraacetic acid
EHV-1	equine herpes virus type 1
EID_{50}	50% egg infectious dose
ELISA	enzyme-linked immunosorbent assay
FA	fluorescent antibody
FCS	foetal calf serum
Fetr	feline rhinotracheitis virus
FITC	fluorescein isothiocyanate
HA	haemagglutination
HAI	haemagglutination-inhibition
HAU	haemagglutinating units
HEL cells	human embryonic lung cells
Hepes	N-2-hydroxyethylpiperazine-N′-2-ethanesulphonic acid
HFF	human foreskin fibroblasts
HL	haemolysin
HLI	haemolysin-inhibition
HSV	herpes simplex virus
HVS	herpesvirus saimiri
LD_{50}	lethal dose at the 50% end point
MEM	minimal essential medium
m.o.i.	multiplicity of infection
NCS	newborn calf serum
NIEP	non-infectious enveloped particles
NP-40	Nonidet P-40
OMK cells	Owl Monkey kidney cells
OSC	optimum sensitising concentration
PBS	phosphate-buffered saline

PD_{50}	50% protective dose
PEG	polyethylene glycol
PMSF	phenylmethylsulphonyl fluoride
PRV	pseudorabies virus
PTA	phosphotungstic acid
PVM	murine pneumonia virus
rbc	red blood cells
RIA	radioimmunoassay
RIP	radioimmunoprecipitation
RS virus	respiratory syncytial virus
RSB	reticulocyte standard buffer
RuBPC	ribulose bisphosphate carboxylase
SDS	sodium dodecyl sulphate
SNI	serum neutralisation index
SV40	simian virus 40
TCA	trichloroacetic acid
$TCID_{50}$	dilution of virus required to infect 50% of cultures
TD	Tris-Dulbecco
TMV	tobacco mosaic virus
TPB	tryptose phosphate broth
UA	uranyl acetate
VBS	veronal-buffered saline
VSV	vesicular stomatitis virus

CHAPTER 1

Purification, Biophysical and Biochemical Characterisation of Viruses with Especial Reference to Plant Viruses

ROGER HULL

1. INTRODUCTION

Among the various features used in a descriptive characterisation of a virus are the biophysical and biochemical properties of the particles. Before these properties of a virus can be determined the particles have to be purified. The optimum purification procedure differs from virus to virus and it would not be realistic to list all the various procedures in this chapter. The reader is referred to other chapters in this book for methods of purifying viruses of animals. In this chapter I will discuss the principles behind designing purification procedures for plant viruses and give some examples. Once the particles are purified, the methodology for determining the biophysical and biochemical properties is the same for most, if not all, viruses be they from animals, plants or bacteria. This chapter will be confined to the methods for obtaining these measurements from the simpler viruses comprising a protein coat surrounding the nucleic acid genome.

2. PURIFICATION OF PLANT VIRUSES

Purification procedures for many viruses are given in the CMI/AAB Descriptions of Plant Viruses (1) and in (2). However, although there is no universal technique, there are various basic features and useful facts which can be used in designing methods for virus purification.

2.1 **Host Plants**

Ideally one should have a systemic host in which to grow the virus and a local lesion host for assaying various stages of purification. Choice of a systemic host should take account of the amount of virus produced, the ease of extraction of the virus (e.g., is the plant horticulturally soft? does the plant contain unhelpful substances such as large amounts of polyphenol oxidase?), the ease of growing the plant and the possibility of contamination with other viruses. The local lesion host should be used to ascertain the time of maximum virus content in the propagation host and also to check various stages in virus purification.

In most cases, however, it may not be possible to satisfy many of the above criteria. Often one cannot find a suitable local lesion host. The technique of dot blotting (see Section 3.8) can be used to replace local lesion assays.

2.2 Extracting the Virus from the Plant

To extract virus particles from the plant, the cell walls have to be broken and the cell contents released. This is usually done in the presence of a buffer to control the pH and of additives to prevent the released enzyme activities from damaging the virus particles. The action of enzymes is also slowed by extracting in the cold. Unless otherwise stated all purification steps should be at 0–5°C.

The most frequently used type of instrument for breaking plant cells is the blender, either top- or bottom-driven. There are many types on the market and choice should be governed by the efficiency of breaking up the plant material. This can be affected by the speed at which the blades rotate, the size and angle of the blades and the shape of the vessel. Breakage of plant cells can be increased by prior freezing of the tissue, though in some cases this will reduce virus yield. There may be problems with blenders if the propagation host is very fibrous (a feature to be taken into account in host selection) or if the virus particles are long and flexuous; in the latter case particles of, say, closteroviruses are broken by blending (3). These problems can be overcome by using a sap-press or a pestle and mortar. For viruses which are limited to vascular tissue, for example, phloem-limited luteoviruses, the leaf tissue should be initially disrupted with cellulases and pectinases (4).

The choice of extraction buffer can greatly affect the outcome of purification attempts. Viruses with elongated particles (e.g., potyviruses) which tend to aggregate or to be absorbed onto cellular debris are best extracted in alkaline buffers (pH 8–9) of moderate ionic strength (e.g., 0.1 M); however some rod-shaped viruses [e.g. tobacco mosaic virus (TMV)] can be damaged by alkaline buffers. Acidic buffers of about pH 5 (e.g., 0.1 M sodium acetate adjusted to pH 4.8 with acetic acid) are useful in the extraction of many viruses with isometric particles; there is the added advantage of many of the host proteins being precipitated at this pH. Some viruses with isometric particles, for example, cucumber mosaic and tobacco ringspot viruses, may be precipitated at around pH 5 and so pHs closer to neutrality have to be used for them. On the other hand, particles of other viruses, for example, bromoviruses, swell at around pH 7 making the viral RNA susceptible to nucleases and so, in their case, the use of lower pHs has an additional advantage.

The most common additives are reducing agents to prevent the action of polyphenol oxidases 'tanning' the viral coat protein. Commonly used reducing agents are 10 mM ascorbic acid (the pH of the buffer may need to be re-adjusted), 20 mM sodium ascorbate, 0.5% 2-mercaptoethanol, 20–40 mM sodium sulphite, 10 mM sodium thioglycollate or 10–20 mM sodium diethyldithiocarbamate; for hosts with a high 'tanning' level, hide powder can be used (5). Chelating compounds are also used to reduce enzymic activity and to dissociate ribosomes; 5–50 mM sodium ethylene diamine tetraacetic acid (EDTA) is the most commonly used. However, consideration must be taken of the possible involvement of divalent cations in virus particle stabilisation, e.g., sobemoviruses. The particles of some viruses, for example, caulimoviruses, are contained within proteinaceous inclusion bodies from which they have to be released. With

Table 1. Removal of Major Contaminants.

Contaminant	*Treatment*[a]
Ribosomes	10 mM Na_2 EDTA, 0.5 M NaCl, 8% butanol, chloroform, heat, bentonite.
RuBPC	8% butanol, butanol/chloroform, pH 4.9, heat, bentonite.
Phytoferritin	10 mM Na_2 EDTA, pH 4.9, 8% butanol, butanol/chloroform, heat, bentonite.
Membranes and organelles	Organic solvents, non-ionic detergents.

[a]These treatments are alternatives.
EDTA or NaCl are added to extract to disrupt ribosomes or phytoferritin so that they will not sediment with virus.
8% butanol is added dropwise to stirred extract and denatured products removed by low speed centrifugation.
Organic solvents, chloroform, 1:1 mixture of butanol* + chloroform*, acetone*, ether*, or freon: equal volume mixed with extract, the emulsion broken by centrifugation and the aqueous phase recovered.
Non-ionic detergents. Make extract 1 – 5% with Triton X-100 or Nonidet P-40. Heat. Extract heated to 50 – 60°C for 10 min and denatured proteins removed by low speed centrifugation.
Bentonite: Mg bentonite used as in (6).
*Safety note. These are either very volatile and flammable even at low temperatures or may be carcinogenic.

caulimoviruses this can be effected by overnight treatment with 2.5 – 5% Triton X-100 in 1 – 2 M urea.

2.3 Clarifying the Extracts

To purify the virus particles they have to be separated from all the other constituents of the host cytoplasm. These include organelles, membranes, ribosomes and proteins, especially ribulose bisphosphate carboxylase (RuBPC) and phytoferritin. *Table 1* lists agents which may be effective in removing the various contaminants. In developing a purification procedure for a new virus the effects of these various clarification methods have to be tested, preferably using local lesion or dot-blot assay (see Section 3.8). The low-speed centrifugation for removing precipitated contaminants or for breaking organic solvent emulsions is usually at 10 000 *g* for 10 min.

2.4 Concentrating the Virus

This is usually performed by precipitation using polyethylene glycol (PEG) or by high-speed centrifugation. Typical conditions for PEG precipitation are: for rod-shaped viruses 2.5% w/v PEG, 0.1 M NaCl, for isometric viruses 10% w/v PEG, 0.1 M NaCl; it is important to have sufficient salt present to effect the precipitation. Usually the PEG and salt are stirred into the virus solution for about 1 h and the virus precipitate recovered by centrifugation at 10 000 *g* for 10 min. For high-speed centrifugation most typical isometric viruses (80 – 130S) are sedimented at 70 000 *g* for 3 h, most rod-shaped viruses (130 – 180S) at 50 000 *g* for 2 h.

Table 2. Sucrose Rate-zonal Density Gradients.

A. To obtain approximately linear gradients by diffusion the following quantities of sucrose should be layered and allowed to diffuse at 4°C overnight.

Tube size (inches)	Rotor	*ml of sucrose*			
		40%[a]	30%	20%	10%
3 x 1	SW25	7	7	7	4
3.5 x 1	SW27	8	8	8	6
3.5 x 9/16	SW41	3	3	3	2
2 x 0.5	SW50.1	1	1	1	1

[a]Percentages give as w/v.

B. Approximate centrifuging times for different viruses in sucrose density gradients[b]

Rotor	Speed (r.p.m.)	*Time for different viruses*		
		BMV[c]	SBMV	TMV
SW27	24 000	4.0 h	3.0 h	2.5 h
SW41	36 000	2.5 h	2.0 h	1.5 h
SW50.1	45 000	1.5 h	1.25 h	1.0 h

[b]10–40% sucrose gradients in appropriate buffer containing 0.1 M salt and run at 4°C.
[c]BMV = brome mosaic virus, S_{20w} = 87; SBMV = southern bean mosaic virus, S_{20w} = 115; TMV = tobacco mosaic virus, S_{20w} = 194.

2.5 Further Purification

After clarification and concentration, the virus is usually not pure enough for most biochemical or biophysical techniques. Further purification can be effected by either rate-zonal or equilibrium centrifugation. Details of the theory and much of the practice of centrifugation can be found elsewhere (7). As far as plant viruses are concerned the following techniques can be used for further purification.

2.5.1 *Rate-zonal Centrifugation*

This is usually in gradients of sucrose dissolved in the buffer used for virus purification. Borate buffer should not be used for sucrose gradients. For most simple viruses (those comprising just coat protein and nucleic acid) gradients of 10–40% sucrose are used. These can be made by:

(i) Using a gradient maker which will mix the input 10% and 40% sucrose solution appropriately (see ref. 7).

(ii) Layering 10, 20, 30 and 40% sucrose solutions in the amounts shown in *Table 2* and allowing these to diffuse at 4°C overnight, or

(iii) Freezing a 25% sucrose solution in the centrifuge tubes and then thawing slowly.

The virus preparation is layered on top of the gradient, care being taken not to overload. Large tubes (e.g., SW27) should not be loaded with more than 5 mg virus nor small tubes (e.g., SW50) with more than 1 mg virus. Centrifugation is in swinging bucket rotors and guidelines to the times and speeds are given in *Table 2*.

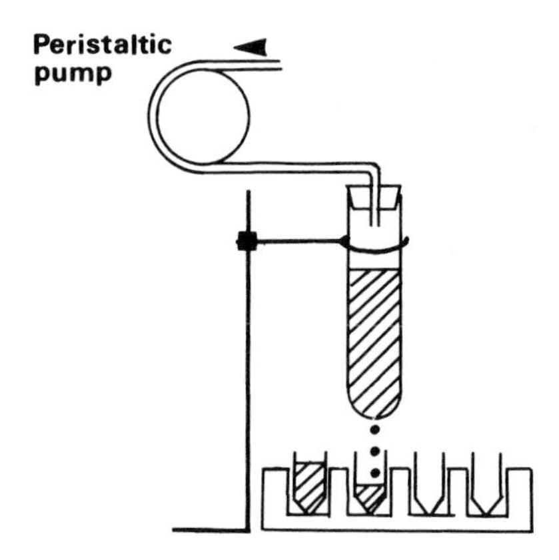

Figure 1. Simple apparatus for fractionating gradients. The apparatus is set up as shown in the diagram, care being taken to ensure that joints are airtight; the bung in the top of the gradient tube should be coated lightly with Vaseline. Vaseline should also be spread very lightly over the bottom of the tube, before puncturing with a needle, to ensure even drop size. The flow of air into the tube is controlled by the rate of the peristaltic pump; dripping can be stopped by switching the pump off.

Virus bands can be visualised by top light and can be removed from the top of the tube using a syringe with a needle bent at right angles near the tip, through the side of the tube with a syringe and needle (use Vaseline or Sellotape to prevent leakage) or by dripping from the bottom of the tube. In the latter case a simple apparatus such as that illustrated in *Figure 1* can be used to control the dripping. More sophisticated apparatus such as the Isco Density Gradient Fractionator, in which the contents of the gradient are displaced upwards through a spectrophotometer flow cell and a record obtained of the optical density, can be used. A rule of thumb in determining the method for obtaining the virus band is that it should not pass the region of the tube where contaminants have banded, that is, if the contaminants, for example, RuBPC, are above the virus band the gradient should not be displaced upwards nor should the band be taken by syringe through the top of the tube. The sucrose can be removed by dialysis against the appropriate buffer or by precipitating the virus with PEG.

2.5.2 *Isopycnic Centrifugation*

Isopycnic centrifugation for purification of plant viruses is usually in either caesium chloride (CsCl) or caesium sulphate (Cs_2SO_4). The particles of some viruses, for example, cucumber mosaic, or alfalfa mosaic, are disrupted by CsCl; they are however stable in Cs_2SO_4. Particles of other viruses, for example, bromoviruses, are unstable in CsCl at pHs above 7; therefore they should be in pH 5 buffer. If possible CsCl should be used since the Cl^- ion is chaotropic and will remove plant proteins etc. associating non-structurally with the virus particles.

Suggestions on the starting densities of CsCl and Cs_2SO_4 gradients for typical viruses are given in *Table 3*. The virus should be suspended in an appropriate buffer and CsCl or Cs_2SO_4 added to give the correct starting density; virus concentra-

Table 3. Isopycnic Density Gradient Centrifugation.

	Approximate banding densities (g/cm³)	
	CsCl	*Cs_2SO_4*
Protein	1.28 (29% w/v)[a]	1.24 (24.5% w/v)
Virus containing 5% RNA	1.32 (32% w/v)	1.27 (27% w/v)
Virus containing 20% RNA	1.36 (35.5% w/v)	1.30 (29.5% w/v)
Virus containing 35% RNA	1.42 (39.5% w/v)	1.33 (32% w/v)
Virus containing 40% RNA	1.50 (44.5% w/v)	1.38 (35.5% w/v)
RNA	>1.90	1.63 (49.5% w/v)
DNA	1.70 (55.5% w/v)	1.42 (38% w/v)

[a]Approximate amount of Cs salt to achieve required density. Densities should be checked using a refractometer and the following relationships between refractive index and density.

			Density range (g/cm³)
CsCl	$\delta\ 25°$	$= 10.8601\ \pi_D^{25°} - 13.4974$	1.25 – 1.90
Cs_2SO_4	$\delta\ 25°$	$= 12.1200\ \pi_D^{25°} - 15.1662$	1.15 – 1.40
		$= 13.6986\ \pi_D^{25°} - 17.3233$	1.40 – 1.70

tion should not be greater than 1 mg/ml. If aluminium caps are used the centrifuge tubes should be only three-quarters filled with CsCl or Cs_2SO_4 solution, the cap put on while the tube is dry and the rest of the tube filled with paraffin oil. Aluminium tube caps can be damaged by caesium salts and may then fail during centrifugation. Isopycnic gradients are generated by centrifuging either in a Beckman R40 at 36 000 r.p.m. or an R50 at 40 000 r.p.m. for 18 h. Separations are better in angle rotors than in swinging bucket rotors for reasons given in reference 7.

Virus bands can be located by top illumination and removed with a syringe and needle through the side of the tube. The Cs^+ salt is then dialysed away.

2.6 What is Pure?

The degree to which a virus needs to be purified depends upon what biophysical or biochemical techniques are going to be applied to it. Clearly preparations used for analyses of viral nucleic acids or coat proteins would have to be much less contaminated with host material than those being used for electron microscopy or sedimentation studies.

2.7 Preserving Virus Preparations

Purified preparations of some viruses can be preserved frozen at −20°C though it should be noted that freezing can affect the structure and stability of some viruses. A useful preservative is chlorbutanol (1,1,1-trichloro-2-methyl-2-propanol) which does not affect either the biological properties (of non-enveloped viruses) or the u.v. absorption spectrum. A crystal or two of sodium azide will also prevent bacterial contamination though care must be taken in handling the preparations.

2.8 Specific Purification Procedures

Four purification procedures are listed below which incorporate various points noted above. Before acquiring and growing any of the viruses, advice should be obtained on any possible phytosanitary regulations.

2.8.1 *Brome Mosaic Virus*

This method can be used for the purification of viruses with isometric particles not precipitated at pH 5.

(i) Grow virus in barley for 10 – 12 days. Harvest the leaves. Carry out further operations at 4°C.
(ii) Blend in 2 ml 0.1 M sodium acetate (pH 4.8) per gram of leaves.
(iii) Centrifuge at 10 000 *g* for 10 min. Recover the supernatant.
(iv) Add PEG 6000 to give 10% and NaCl to give 0.1 M to the supernatant and stir for 1 h.
(v) Centrifuge at 10 000 *g* for 10 min, resuspend the pellet in one-tenth the original volume of 0.1 M sodium acetate pH 5.0.
(vi) Centrifuge at 10 000 *g* for 10 min, recover the supernatant.
(vii) Add CsCl to give a density of 1.36 g/cm^3 (see *Table 3*) and centrifuge at 36 000 r.p.m. for 18 h in a Beckman R40 rotor.
(viii) Recover the band from the gradient, dialyse against at least 100 volumes of 0.1 M sodium acetate (pH 5.0) for 3 – 4 h to remove CsCl.
(ix) Repeat steps iv and v. Resuspend in 1 ml of 0.1 M sodium acetate (pH 5.0) per 100 g of starting leaf material.

2.8.2 *Alfalfa Mosaic Virus*

This is a method which can be used for a virus which is precipitated at low pHs and which is salt sensitive.

(i) Propagate the virus in tobacco cv. Xanthi NN for 10 – 12 days. Harvest the leaves. Carry out all further steps at 4°C.
(ii) Blend each 100 g of leaves in 100 ml of 10 mM K_2H/KH_2PO_4 pH 7.1 to which 1.0 g of ascorbic acid and 3.5 ml of 50% K_2HPO_4 has been added.
(iii) Add an equal volume of a 1:1 mixture of chloroform + butanol; repeat the blending (do this step in a fume hood).
(iv) Centrifuge the mixture at 5000 *g* for 5 min. Recover the top aqueous phase.
(v) Centrifuge the aqueous phase at 27 000 r.p.m. for 3 h in a Beckman R30 rotor.
(vi) Resuspend the pellets in one-tenth the original volume of 10 mM phosphate buffer (pH 7.1).
(vii) Add Cs_2SO_4 to give a density of 1.30 g/cm^3 (see *Table 3*) and centrifuge at 36 000 r.p.m. for 18 h in a Beckman R40 rotor.
(viii) Recover the band from the gradient and dialyse against at least 100 volumes of 10 mM phosphate buffer (pH 7.1).
(ix) Repeat steps (v) and (vi) resuspending the pellet in 1 ml of 10 mM phosphate buffer (pH 7.1) per 100 g of starting leaf material.

2.8.3 *Tobacco Mosaic Virus*

This procedure is only for viruses with very stable particles. It does not require a high speed centrifuge.

(i) Propagate the virus in tobacco cv. White Burley for 2–3 weeks (N.B. This virus is very contagious and can be spread to other nearby hosts by simple contact of plants or by handling).
(ii) Blend leaves in 1 ml of 10 mM Tris-HCl (pH 7.4) per gram of leaves. Strain the sap through cheesecloth.
(iii) Heat the sap at 55°C for 10 min.
(iv) Centrifuge at 10 000 *g* for 10 min and recover the supernatant.
(v) Add PEG to a final concentration of 2.5% and NaCl to 0.1 M, stir for 1 h and centrifuge at 10 000 *g* for 10 min.
(vi) Resuspend the pellets in 10 mM Tris-HCl (pH 7.4) and centrifuge at 10 000 *g* for 10 min, recover the supernatant.
(vii) Repeat steps (v) and (vi) 2–3 times. Resuspend the final pellet in 1 ml of 10 mM Tris-HCl (7.4) per 10 g starting leaf material.

This gives a preparation of moderate purity. Further purification can be effected by rate-zonal or isopycnic centrifugation, techniques which necessitate high-speed centrifuges.

2.8.4 *Potato Virus X*

(i) Propagate the virus in tobacco for 2–3 weeks.
(ii) Blend each 1 g of leaves in 1 ml of 0.02 M sodium borate pH 8.2 containing 0.01 M of sodium ascorbate.
(iii) Strain through cheesecloth and add Triton X-100 to the sap to give 0.5%; stir for 1 h.
(iv) Centrifuge at 10 000 *g* for 10 min and recover the supernatant.
(v) Layer the extract over a cushion of 5 ml of 30% sucrose in 10 mM Tris-HCl, 1 mM EDTA (pH 8.0) in tubes of a Beckman R30 rotor; centrifuge at 27 000 r.p.m. for 3 h.
(vi) Resuspend the pellets in one-tenth the volume of 10 mM Tris-HCl, 1 mM EDTA (pH 8.0) overnight.
(vii) Centrifuge the extract at 10 000 r.p.m. for 10 min, recover the supernatant and centrifuge at 45 000 r.p.m. for 1.5 h in a Beckman R50 rotor.
(viii) Resuspend the pellets in 1 ml of 10 mM Tris-HCl, 1 mM EDTA (pH 8.0) per 50 g of starting leaf material.

Further purification, if necessary, is by isopycnic centrifugation.

3. BIOPHYSICAL CHARACTERISATION

3.1 **Reasons for Biophysical Characterisation**

The biophysical properties of viruses are studied to provide information on the size and shape of the virus particles, on their molecular weight and on their structure. In many cases the particles can be treated as simple macromolecules with regularly repeating surface structure. Some highly sophisticated techniques (e.g., X-ray crystallography) are now being used but I will just describe those which can be performed in a moderately well equipped laboratory.

3.2 **Ultraviolet Spectra**

The majority of simple viruses comprise mainly protein and nucleic acid each of which has a characteristic absorption spectrum in the ultraviolet (*Figure 2*). The absorption spectrum of the virus particles is a combination of both nucleic acid and protein. However, since the nucleic acid has a much higher specific absorption at its maximum (~260 nm) (RNA $E^{0.1\%}_{260\text{nm}} = 25$; DNA $E^{0.1\%}_{260\text{nm}} = 20$) than does protein ($E^{0.1\%}_{260\text{nm}}$ = about 0.5) the nucleic acid spectrum dominates. The specific absorptions of nucleic acids do not vary markedly with composition (*Table 4*) but those of proteins depend on the proportion of aromatic amino acids. Thus the spectral characteristics of a virus should not be used to determine the percentage nucleic acid.

The absorption spectrum can also be used to measure virus concentration. To find the extinction coefficient for an unknown virus, the following procedure should be followed:

(i) Dialyse the virus preparation extensively against a suitable buffer, e.g., 10 mM sodium acetate (pH 5.0) or 10 mM phosphate buffer (pH 7.1). Keep the dialysing solution.

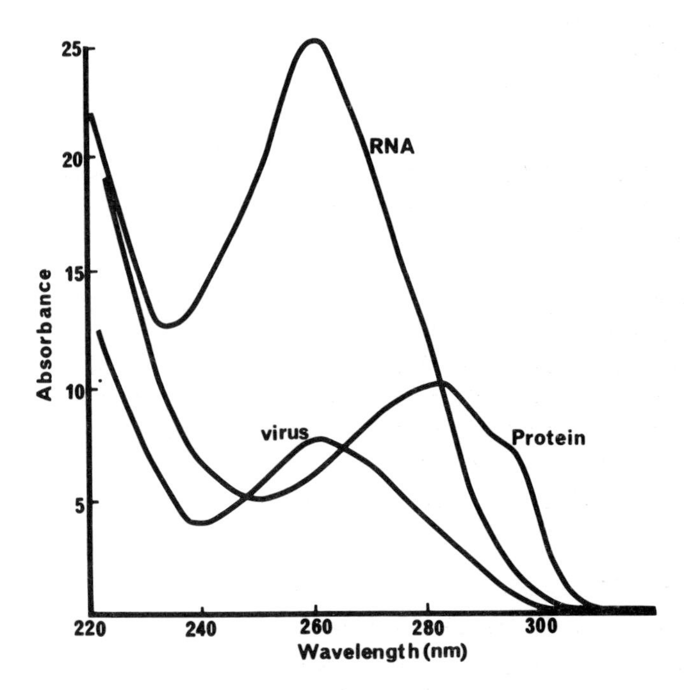

Figure 2. Comparative absorption spectra of RNA, a virus containing 20% RNA and a typical protein. The RNA and virus are at 1 mg/ml, the protein is at 10 mg/ml. Note the shoulder on the protein spectrum at 290 nm due to the absorbance by tryptophan.

Table 4. Specific Absorbance of Nucleotide Base Ratios and Specific Absorbance of Viral RNAs.

Nucleotide	E_{260}[a] x 10^{-3}	*Base ratios (%)*			
		Brome mosaic	*Alfalfa mosaic*	*Tobacco mosaic*	*Potato virus X*
Adenosine	14.2	27	26.5	29.8	32
Cytosine	6.8	21	21.8	18.5	24
Guanosine	11.8	28	24.4	25.3	22
Uridine	9.8	24	27.3	26.3	22
	E_{260}[b] x 10^{-2}	35.59	35.04	35.59	35.41

[a]Molar absorbance at 260 nm for nucleotide in acidic solution (pH 2).
[b]Absorbance of 1 mg/ml at 260 nm in acidic solution (pH 2).

(ii) Take a portion of the preparation, dilute as necessary and measure its u.v. absorption spectrum from 220 to 360 nm. Correct the 260 nm reading for light scattering if necessary (see below).

(iii) Measure several samples of known volume of the preparation and of the same volume of dialysing solution into pre-dried and pre-weighed vessels. Take the solution down to dryness either by freeze-drying or by placing first in an incubator at 50°C and, when apparently dry, in a vacuum desiccator.

(iv) Keep weighing the vessels at intervals until constant weights are achieved.

(v) The extinction coefficient can then be calculated from the E_{260} value and the weight of the virus preparation minus the weight of dialysing buffer.

To determine the correction needed for light scattering:

(i) Measure the absorbance at several wavelengths between 300 and 360 nm.

(ii) Plot the log wavelength/100 against log (100 x absorbance), extrapolate to 260 nm.

(iii) Take the antilog of the 260 nm extrapolate and divide by 100 to give the light scattering value.

(iv) Subtract the light scattering value at 260 nm from the A_{260nm} of the virus preparation to give the corrected value.

3.3 Sedimentation Properties

From studies on the sedimentation of viruses, information can be obtained on the number of sedimenting components the preparation comprised and on their sedimentation coefficients. The sedimentation properties of the particles in a virus preparation can be determined either in rate-zonal sucrose gradients or in the analytical centrifuge.

For details of both systems the reader is referred to reference (7). Further information on the use of the analytical ultracentrifuge and a simple graphical method for estimating sedimentation coefficients are given in reference (8). Since the subject is covered so comprehensively in these two references, one being a companion volume to this, I will just describe a simple method for estimating sedimentation coefficients.

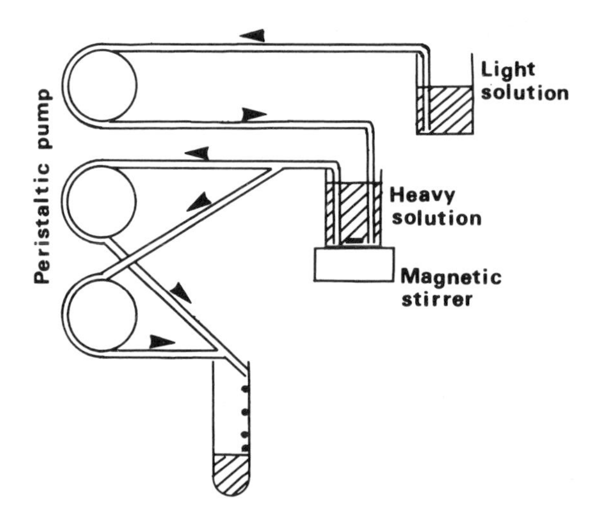

Figure 3. Simple apparatus for making a linear gradient. The three-track peristaltic pump is set up as shown in the diagram. The tubing from the heavy solution to the gradient centrifuge tube is put into the solution when the tubing from the light solution is full. Different types of gradients can be made by varying the layout, e.g., an exponential gradient can be made by using just two tracks of the peristaltic pump.

Gradients of 10–40% sucrose are constructed as described in Section 2.5.1. The best gradients for this purpose are those made using a gradient maker which can be either as described in reference 7 or as in *Figure 3*. Marker viruses (*Table 2B*) are chosen so that their sedimentation coefficients span that expected for the unknown virus and their nucleic acid contents are similar to that of the unknown. The latter feature will compensate for any changes in sedimentation coefficient due to the increasing density of the sucrose. The markers and the unknown virus are centrifuged into a gradient with sister gradients containing either markers or unknown virus so that bands can be identified. The sedimentation coefficient of the unknown virus can then be determined from the distance it has moved into the gradient relative to the distances that the markers move. This can be done rather crudely by visualizing the bands using top illumination and measuring distances from the meniscus. Alternatively the gradient can be passed through a spectrophotometer flow cell or can be fractionated and the A_{260nm} of each fraction measured. This will give a much more accurate determination of the migration of the markers and unknown into the gradient column. If very small amounts of viruses have to be used, the fractions containing the markers and unknown can be detected by 'dot-blots' (see Section 3.8).

3.4 Diffusion Coefficients

Although the diffusion coefficient of a virus particle is inversely related to its Stoke's radius, it cannot be determined from electron microscope measurements of the particle's dimensions. This is because the particles in the electron microscope are not hydrated; in solution they are and this affects their apparent radius.

The only practical way to measure the diffusion coefficient of a virus is by creating a sharp boundary between the virus solution and the solute and then measuring the spread of the boundary. This can be done in the analytical ultracentrifuge and the methods are fully described in reference 8.

3.5 Molecular Weight Estimations

There are two commonly used approaches for estimating the molecular weight of the particles of a given virus, from hydrodynamic data and from data on the coat protein and nucleic acid.

The method involving hydrodynamic data is based on the Svedberg equation:

$$M = \frac{RTs}{D(1 - \bar{\nu}\delta)}$$

in which the symbols have the following meanings:

M = weight of 1 mol of virus (molecular weight)
R = gas constant per mol (8.314 J/mol/K)
s = sedimentation coefficient (S_{20w} in Svedbergs)
D = diffusion coefficient (cm^2/sec)
$\bar{\nu}$ = partial specific volume (see below)
δ = solvent density (see below)
T = absolute temperature (°Kelvin).

The partial specific volume can be measured directly in a pycnometer or estimated indirectly from the proportion of nucleic acid and protein in the particles. Thus, if the percent nucleic acid is known (see Section 4.3), the partial specific volume can be calculated using $\bar{\nu}$ for RNA = 0.55 cm^3/g and for protein 0.74 cm^3/g (if the composition of the coat protein is known $\bar{\nu}$ can be calculated more precisely from the weighted sum of the $\bar{\nu}$s of the individual amino acids).

Since diffusion and sedimentation coefficients are usually measured at 20°C and corrected for salt effects, δ can be taken as 0.9982 and T as 293.3°K.

The second method for estimating molecular weight involves determining the percentage of the particle which is nucleic acid (see Section 4.3), the molecular weight of the nucleic acid and the number of nucleic acid molecules per particle. In addition or alternatively, the percentage of the particle which is coat protein and the number of subunits and their molecular weight can be used. If the nucleic acid data are used in this calculation and the molecular weight of the coat protein subunit is known, the number of subunits per particle can be estimated. This can be of assistance in interpreting observations on structure by electron microscopy.

The major difference between the two methods is that the hydrodynamic approach measures the hydrated particle, whereas no account of hydration is taken in the other approach. Thus, for some purposes the hydrodynamic approach is the only one. However, if one wishes to derive other data such as number of protein subunits or nucleic acid molecules per particle from the molecular weight estimate, the second approach is the best.

3.6 Isopycnic Banding Density

Factors to be considered in isopycnic banding of viruses in CsCl or Cs_2SO_4 (the most commonly used salts) are given in Section 2.5.2. Details of other materials for making isopycnic gradients can be found in reference 7. As with the determination of sedimentation coefficients, banding densities can be measured in the analytical or in the preparative ultracentrifuge. For use of the analytical ultracentrifuge the reader is referred to reference 8. To determine the banding density in the preparative ultracentrifuge, the virus particles are centrifuged to equilibrium in tubes in an angle rotor, the gradients are fractionated, the densities of fractions are measured and the fraction(s) containing the virus particles identified. Fractionation of the gradient can be effected using equipment such as that illustrated in *Figure 1*. Other methods of harvesting gradients are described in Chapter 2, Section 5.4. The densities of fractions can be measured using a refractometer and the relationships between density and refractive index given in *Table 3* and reference 7 or by weighing known volumes. The fractions containing the virus can be identified by measuring the A_{260nm} or by the dot-blot method (see Section 3.8). The latter is of considerable use with gradient materials, for example, metrizamide, which have a high u.v. absorbance.

The use of banding densities to estimate the proportion of protein and nucleic acid in virus particles can lead to erroneous results. There can be differences in the binding or penetration of Cs^+ ions into the particles and also in the effect of the gradient materials on the hydration of the particles.

3.7 **Electron Microscopy**

A detailed description of methods for preparing samples for and using an electron microscope is beyond the scope of this chapter. However, as electron microscopy is one of the most useful methods for determining the shape and structure of virus particles, some hints are listed below.

3.7.1 *Negative Stains*

The main method for visualising particles is be negative staining. While PTA (2% phosphotungstate acid adjusted to pH 6.5 – 7.0 with NaOH or KOH) is suitable for many viruses, the particles of some viruses are destroyed or damaged in this stain; these viruses are often those which are salt labile. An alternative stain is UA (2% aqueous unbuffered uranyl acetate pH $\leq$4.2) but this can damage some other viruses. Other stains which are less damaging include ammonium molybdate (pH 6.5), sodium tungstate adjusted to pH 6.5 with formic acid and unbuffered methylamine tungstate (pH 6.4). By trying a variety of stains one may be found which best reveals the capsid structure.

3.7.2 *Magnification Markers*

The magnification of electron micrographs should be carefully calibrated, preferably by an internal marker incorporated onto the grid. The magnification given on the electron microscope is not always reliable and there can even be variations in magnification across the grid. One of the best internal markers is the crystal lattice of catalase (17.2 nm) (ref. 9); particles of TMV which have a modal length of 300 nm and a helix pitch of 2.3 nm can also be used.

3.8 **Dot-blot Detection of Viruses**

This method involves the immobilisation of the nucleic acid of the virus onto nitrocellulose and its detection by hybridisation to a probe of complementary homologous nucleic acid. There is no need to extract the nucleic acid from the virus particles prior to dot-blotting. Dot-blots can be made from fractions from sucrose or caesium gradients without removing the gradient medium.

3.8.1 *Immobilisation of Nucleic Acid from Single-stranded RNA or DNA Viruses*

(i) Mark a sheet of nitrocellulose (Schleicher and Schull BA85) with soft (4B) pencil so that spots can be identified. Disposable gloves should be worn when handling nitrocellulose.
(ii) Wet the nitrocellulose in distilled water and then soak it in 20 x SSC (1 x SSC = 0.1 M NaCl, 0.015 M sodium citrate) for 5 – 10 min.
(iii) Blot off excess SSC from the nitrocellulose.
(iv) If required make dilutions of virus preparations or fractions in 50 mM Na_2H/NaH_2PO_4 (pH 7.0).
(v) Place 5 μl spots of virus preparations or fractions at the appropriate places on the nitrocellulose. Allow the spot to soak in.
(vi) Cover the nitrocellulose with Whatman 3MM paper and bake it at 80°C *in vacuo* for at least 2 h. If a vacuum oven is not available, air dry the nitrocellulose, then dry it under a heat lamp and bake it at 80°C for 2 h.

3.8.2 *Immobilisation of Nucleic Acid from Double-stranded RNA or DNA Viruses*

(i) Prepare nitrocellulose as in stages i – iii above.
(ii) Take 8 μl aliquots of virus preparation or fractions and add 2 μl of 0.5 M NaOH. Leave at room temperature for 10 min for DNA and 5 min for RNA.
(iii) Add 2 μl of 3 M sodium acetate (pH 5.5); make dilutions in 1 x SSC if required.
(iv) Apply 5 μl spots to nitrocellulose and bake as described in stages (v) and (vi) above.

N.B. if the DNA is supercoiled it should be depurinated by adding 5 μl of 0.5 M HCl to 5 μl aliquots prior to the addition of the NaOH (in this case 2 μl of 3 M NaOH).

3.8.3 *Preparation of Probes*

Probes are usually of double-stranded DNA (e.g., a clone) or of complementary DNA (cDNA) made by reverse transcribing RNA. There are many techniques for making probes and, for details, the reader is referred to reference (10). Below are given two methods which are commonly used.

Nick translation of double-stranded DNA

(i) Stock solutions

10 x NT buffer 50 mM Tris-HCl (pH 7.8)
5 mM $MgCl_2$
10 mM 2-mercaptoethanol

Stock deoxyribonucleotides (dNTPs) 0.1 mM dATP, TTP, dCTP, dGTP. Store frozen until needed. Thaw for as short a period as possible.

(ii) DNase: use RNase-free grade, e.g., Sigma DN-EP. Store frozen as 2 mg/ml in 10 mM Tris-HCl (pH 7.8), 1 mM $MgCl_2$. Dilute 1/1000 just prior to use.

Take 1 μg of DNA, 2.5 μl of 10 x NT buffer, water to 24 μl, 1 μl of dilute DNase, incubate at 37°C for 15 min. Heat at 70°C for 10 min to inactivate the DNase.

(iii) Dry down 20 μCi of α-^{32}P-labelled dATP or dCTP in a vacuum desiccator.

(iv) Add DNA solution from (ii) to dried labelled dNTP and 0.5 μl of each of the other three stock dNTPs. Add 2 units of *Escherichia coli* DNA polymerase I and incubate at 15°C or at room temperature for 60 min. Stop the reaction by heating at 70°C for 10 min.

(v) The unincorporated dNTPs are removed by column chromatography. Suspend Sephadex G-100 in G-100 buffer (5 mM Tris-HCl, 1 mM EDTA adjusted to pH 7.9). Make the column in a 1 ml plastic disposable pipette by removing most of the cotton wool plug and pushing the rest to the bottom of the pipette. Fill the pipette with G100 buffer and remove air bubbles. Allow the solution to drip out, refilling with the Sephadex suspension until the column is packed. Wash the column with G-100 buffer for about 30 min. Add 2 μl of 0.06% orange G in 10% Ficoll and 0.5% SDS to the nick-translated DNA and load onto the column. The column can be monitored using a standard mini monitor Geiger counter, or fractions can be taken. The nick-translated DNA is in the exclusion volume, the unincorporated dNTPs and the orange G running together in the retained volume.

Reverse transcription of RNA. The use of random primers for reverse transcription gives a population of cDNAs which should be representative of all the RNA.

(i) 10 x RT buffer 0.5 M Tris HCl (pH 8.0)
0.1 M $MgCl_2$
1.4 M KCl

10 x ME 0.3 M 2-mercaptoethanol

Stock dNTPs 10 μl of 10 mM of each of the dNTPs which are not being used for label, plus 10 μl of 0.1 mM of unlabelled dNTP which is also being used for label.

Primer — Suspend calf thymus or salmon sperm DNA at 5 mg/ml in 10 mM Tris-HCl (pH 7.4), 10 mM $MgCl_2$. Add DNase I to give 70 μg/ml and incubate at 37°C for 2 h. Autoclave at 121°C for 10 min.

(ii) Take 1 μg of RNA, 20 μl of primer DNA, and water to give 30 μl. Heat to 100°C for 1 min and allow to cool to room temperature.

(iii) Dry down 20 μCi of [α-^{32}P]dATP or dCTP in a vacuum desiccator.

(iv) Add RNA solution from (ii) above to dried down radioactive label and then add:
 5 μl of 10 x buffer;
 5 μl of 10 x ME;
 10 μl of dNTPs;
 10 – 12 units of reverse transcriptase (from avian myeloblastosis virus).
 Incubate for 1.5 h at 37°C.

(v) Add 11 μl of 3 M NaOH, 10 mM of EDTA and heat at 70°C for 3 h or at 37°C overnight.

(vi) Add 11 μl of 3 M HCl.

(vii) Add 5 μl of orange G solution and pass down a column of Sephadex G50 as described in (v) of Section 3.8.3.

3.8.4 *Hybridisation to Nitrocellulose*

The nitrocellulose is pre-hybridised to block non-specific sites and then incubated under hybridisation conditions such that the probe can bind to homologous immobilised nucleic acid. There are basically two types of hybridisation conditions, 65°C in 3 x SSC or 42°C in 3 x SSC, 50% formamide. Hybridisation is faster under the former conditions but in some cases immobilised RNA may be damaged; in those cases the formamide conditions should be used.

(i)

20 x Denhardt's solution	0.4% bovine serum albumin 0.4% Ficoll 400 0.4% Polyvinyl pyrrolidone (mol. wt. 44 000) 5 mM EDTA (pH 7.8)
20 x SSC	3 M NaCl 0.3 M sodium citrate
DNA	10 mg/ml salmon sperm or calf thymus DNA in 5 mM EDTA pH 7.8, sonicated until non-viscous, heated and quick-cooled.
Formamide	Formamide deionised by stirring with 'Ambertite' monobed resin MB3 until the pH is 6.8 – 7.0. The resin is removed by filtering through miracloth or muslin. Use immediately or store at – 70°C.
	Pre-hybridisation and hybridisation solutions: (a) 4 x Denhardt's solution 3 x SSC 10 μg/ml DNA (b) As (a) but made up to 50% formamide.

(ii) Place a dry nitrocellulose filter in a thick gauge polythene bag and seal three edges close to the filter. Add 0.2 ml of pre-hybridisation solution per cm^2 of filter, remove air bubbles and seal the fourth side, allowing enough space for cutting and re-sealing. Incubate in a water bath at 65°C for solution (a) or 42°C for solution (b) for 2–5 h.

(iii) Make up the hybridisation solution which is fresh pre-hybridisation solution (0.1 ml/cm^2 filter) to which $1-5 \times 10^6$ c.p.m. of nick-translated probe or $0.5-2.5 \times 10^6$ c.p.m. of reverse-transcribed probe has been added. The probe should be heated to 100°C for 1–2 min and rapidly cooled on ice before adding to the solution. Cut open the bag, pour off the pre-hybridisation solution, add the hybridisation solution, remove air bubbles and re-seal. Incubate in the water bath at 65°C for solution (a) or 42°C for solution (b) for 18–24 h.

(iv) Open the bag and remove the filter (this is best done by cutting three sides of the bag and opening it out flat). Place the filter in a container and wash with 4 x 250 ml 2 x SSC, 0.1% sodium dodecylsulphate (SDS). All the washes for hybridisation in solution (a) should be at 65°C for 10–15 min each but for solution (b) the first should be at room temperature for 10 min, the others being at 65°C.

(v) Blot the filter dry on Whatman 3MM paper, mount it on a sheet of Whatman 3MM, cover with cling film and autoradiograph it (see reference 11).

4. BIOCHEMICAL ANALYSES

This section describes methods for determining the type, proportion and sizes of nucleic acid in the virus particles and also the number of species and molecular weights of the protein subunits making up the capsid. These data can be used in conjunction with the biophysical data obtained by the methods given in Section 3 to characterise the particle composition and structure.

4.1 **Extraction of Nucleic Acids**

Viruses vary in the ease with which the nucleic acid can be extracted. It is advantageous to relax the protein-protein interactions if possible before removing the coat protein. If this can be done it is usually effected by the use of alkaline pHs (but remember that RNA is hydrolysed at a significant rate by pHs above 8.5) and chelating agents such as EDTA. Below are listed three methods by which nucleic acid can be extracted from most viruses. The first two involve denaturing the coat protein and then removing it; the denaturing conditions also reduce nuclease action. The third method can be applied to those viruses with capsids made up of proteins that appear to resist denaturation, at least under mild conditions.

When handling nucleic acids, great care must be taken to avoid nucleases. All containers should be either disposable plasticware which should be autoclaved before use or good quality glassware which should be well washed with detergents, rinsed and baked at 200°C. If necessary, articles can be rinsed with

0.1% diethyl pyrocarbonate (DEP) before use. Solutions should be autoclaved if possible and should then be maintained under sterile conditions. It is useful to remember that DNase is inhibited by EDTA (1 – 10 mM). It is more difficult to inhibit RNase but SDS (0.1 – 0.5%) or DEP (0.1%) should control the inhibition. However, DEP may render the nucleic acid biologically inactive and SDS will also inhibit other enzymes (e.g., reverse transcriptase) or translation systems.

4.1.1 *Phenol-SDS Extraction*

(i) Buffer saturated phenol is prepared by adding 55 ml of buffer to 500 g of phenol crystals (which should be white) and warming the bottle in warm water. When all the crystals are in solution a further 110 ml of buffer is added. On cooling there should be an aqueous phase on top of the phenol. 8-Hydroxyquinoline (0.1%) and *m*-cresol (10%) can be added to prevent oxidation of the phenol and to inhibit impurities; the phenol solution should not be used if it is coloured pink or brown.

Great care must be exercised in handling phenol solutions as they can cause serious burns. Gloves should always be worn and eye protection used if tubes are being shaken. Never mouth pipette phenol solutions. Dispose of waste solutions in a safe manner and always rinse phenol-containing vessels before asking other people to wash them up. In the event of phenol splashes on the skin the affected area should be washed well with a solution comprising three parts of saturated solution of polyethylene glycol and one part of 95% ethanol which should be kept to hand. Medical advice should also be sought.

(ii) The virus suspension should be in the appropriate buffer with EDTA (1 – 10 mM) present if necessary. SDS is added to give 1% and the solution heated to 50°C for 5 min. The solution should then be clear.

(iii) An equal volume of buffer-saturated phenol is added and the tube shaken to give a good emulsion. Do not use Parafilm to cover the tube as this will dissolve; cling-film is satisfactory.

(iv) The two phases are separated by centrifuging at 5000 *g* for 10 min. The aqueous upper phase is recovered.

(v) Steps (iii) and (iv) are repeated at least once or until there is no more denatured protein on the interface.

(vi) The aqueous phase is washed twice with anhydrous diethyl ether, the top ether phase being discarded.

(vii) If the salt concentration in the nucleic acid suspension is below 0.1 M 1/20 volume of 3 M sodium acetate (pH 6) is added. The nucleic acid is then precipitated by adding 2.5 volumes of cold absolute ethanol and placing the tube at –20°C for 18 h or at –70°C for 1 h.

Notes: for efficient extraction of nucleic acid, the protein concentration of the virus suspension should not exceed 5 mg/ml. If the nucleic acid contains a poly(A) sequence, extraction should be in a 1:1 mixture of phenol and chloroform.

4.1.2 *SDS-sodium Perchlorate Extraction (12)*

This method avoids the use of phenol.

(i) To a solution of virus (buffer should not contain potassium) add 0.25 volume of 25% (w/v) SDS (a good grade of SDS will dissolve to give this concentration if heated gently) and heat to 60°C for 3 min in a polyallomer tube.
(ii) Add three volumes of 100% (w/v) $NaClO_4$ and shake well.
(iii) Centrifuge for 3 – 5 min in a bench centrifuge with swing-out buckets. This will give a protein-SDS raft overlaying the liquid phase.
(iv) Remove the liquid phase by puncturing the bottom of the tube.
(v) Precipitate the nucleic acid as in stage (viii) above.

4.1.3 *Protease-SDS Extraction*

(i) Prepare Pronase or Protease K solution at 10 mg/ml in 0.015 M sodium citrate, 0.15 M sodium chloride (SSC) and pre-incubate 37°C for 30 min to eliminate nucleases.
(ii) Incubate the virus solution in SSC with Pronase (1 – 2% by weight of virus) and 0.1 – 0.5% SDS at 37°C for 3 – 18 h.
(iii) Carry out a phenol or SDS/$NaClO_4$ extraction as described above.

4.2 **Determining the Type of Nucleic Acid**

The nucleic acid in the virus can be either DNA or RNA, double-stranded or single-stranded. There are two approaches to determining the type, either chemically or enzymatically.

4.2.1 *Chemical Determination of Nucleic Acid Type*

This approach is not widely used nowadays. However, in certain circumstances it may be necessary to adopt this method. For DNA detection the diphenylamine reaction (13) should be used. For RNA detection the orcinol reaction (14) or its modification (15) should be used.

4.2.2 *Enzymatic Determination of Nucleic Acid Type*

This approach should be performed in association with gel electrophoresis analysis of the nucleic acid (see Section 4.4).

RNase treatment

(i) To free RNase from DNase make up a 10 mg/ml solution of pancreatic RNase A in 10 mM Tris-HCl (pH 7.5), 15 mM NaCl, heat to 100°C for 15 min, cool slowly to room temperature, aliquot and store at –20°C.
(ii) Single-stranded RNA is digested in 2 x SSC (see Section 4.1.3) for 15 min at 25°C using 5 μg/ml RNase A. Double-stranded RNA is not digested under these conditions but is digested when the salt concentration is reduced to 0.1 x SSC.

DNase treatment

(i) The DNase should be free from RNase (see manufacturer's descriptions) and should be made up at 10 mg/ml in 10 mM Tris-HCl (pH 7.5), 2 mM $MgCl_2$.

(ii) To test for DNA, the nucleic acid should be treated in 10 mM Tris-HCl (pH 7.5), 2 mM $MgCl_2$ with 10 μg/ml of DNase at 37°C for 15 min. The reaction can be stopped by heating to 70°C for 5 min or by the addition of EDTA to give 5 mM.

4.3 Determination of Nucleic Acid Content

As noted in Section 3.5, the proportion of nucleic acid and protein in the virus particle is needed to estimate indirectly the partial specific volume of the particle. The use of either u.v. absorption spectra or isopycnic banding densities for estimation of the percentage of nucleic acid content can lead to fallacious results. There are four approaches which can be used for determining the RNA content, three direct and one indirect. DNA content can be estimated using the methods described in Sections 4.3.1, 4.3.2 and 4.3.4.

4.3.1 *Phosphorus Content*

The RNAs of viruses contain 8.1 – 9.6% phosphorus, the value varying according to base ratio. Methods for determining base ratio can be found in reference (14). However, for a rough assessment 9.3% can be taken.

(i) Dialyse the virus extensively against distilled water and make up to a known concentration (e.g., 5 mg/ml).

(ii) Pipette 0.5 ml of virus into each of the three tubes calibrated to take 4 ml.

(iii) Add 0.5 ml of 60% perchloric acid to each tube.

(iv) For the standard, dissolve 112.5 mg of oven-dry KH_2PO_4 in 500 ml of water (giving a solution containing 50 μg Pi/ml). Take, for example, 0.5, 0.3 and 0.2 ml portions of this into marked tubes (giving 25, 15 and 10 μg Pi/tube) make up to 0.5 ml if necessary and add 0.5 ml of 60% perchloric acid to each.

(v) Place all the tubes in a digestion rack or sand bath at 190 – 200°C for 20 – 30 min. Take care that the virus solutions do not contain large quantities of other organic materials, for example, glycerol, as these might explode.

(vi) Allow tubes to cool and add five drops of 30% phosphorus-free hydrogen peroxide to each.

(vii) Return the tubes to the digestion rack for 15 min.

(viii) Allow to cool, add 0.4 ml of 5% ammonium molybdate $[(NH_4)_6Mo_7O_{24}.4H_2O]$, mix by swirling and add 0.1 ml of an aqueous solution containing 0.2% 1,2,4-amino-naphthol sulphonic acid, 12% sodium bisulphite and 2.4% sodium sulphite. Make up to the 4 ml mark with water.

(ix) Read the blue colour at 700 mμ in a spectrophotometer against a blank prepared in the same way using 0.5 ml of water.

(x) Prepare a standard curve and use this to calculate the amount of phosphorus present in the virus.

4.3.2 *Diphenylamine or Orcinol Reaction*

The amount of DNA or RNA in a virus preparation can be estimated using the quantitative diphenylamine (for DNA) or orcinol (for RNA) reaction. Reference (13) should be consulted for the diphenylamine reactions and (14) and (15) for the orcinol reaction.

4.3.3 *Hydrolysis of RNA*

This method also requires knowledge of the base ratio of the RNA for accurate estimation; methods for base ratio determination are given in reference (14).

(i) To a known amount of virus in 0.01 M of sodium phosphate buffer (pH 7.1) add one-tenth volume of 11.9 M HCl and leave at room temperature in a stoppered container overnight.
(ii) Centrifuge at 3000 r.p.m. for 20 min to remove denatured protein and recover the supernatant containing hydrolysed nucleotides.
(iii) Using 1 M HCl as a diluent, measure the E_{260nm} of the supernatant.
(iv) Use the base ratio to calculate the specific absorbance of nucleotides (see *Table 4*) and thus estimate the amount of RNA in the known weight of virus.

4.3.4 *Indirect Method*

It is sometimes possible to estimate the percentage nucleic acid from knowing the number of protein subunits per particle (from structural studies), the molecular weight of the protein subunits and the molecular weight of the nucleic acid (from gel electrophoresis).

4.4 **Gel Electrophoresis of Nucleic Acids**

The number of nucleic acid species in a virus preparation and their molecular weights can be determined by gel electrophoresis. Gel electrophoresis can also be used in experiments to determine the type and structure of nucleic acid (see Section 4.2.2). The reader is referred to the companion volume (11) for details of gel electrophoresis of nucleic acids. Below are given some further techniques not noted in (11).

For accurate determination of the molecular weight(s) of single-stranded nucleic acid by gel electrophoresis, the unknown nucleic acid and the markers have to be denatured. This is to remove the secondary structure which varies between different nucleic acids and also according to the concentration of salts and divalent cations.

4.4.1 *Denaturation by Glyoxal*

(i) Deionise 40% glyoxal by stirring with the mixed bed resin AG 501-X8(D) (BioRad), checking the pH until it becomes steady. This can take several hours and several batches of resin. Filter off the resin using muslin or miracloth. Aliquot the deionised glyoxal and store at $-70°C$.

(ii) Make up denaturant (GFP)

Deionised glyoxal	0.894 ml
Deionised formamide (see Section 3.8.4)	3.89 ml
500 mM sodium phosphate pH 7.0	0.111 ml
Distilled water	0.195 ml
	5.0 ml

Aliquot and store at $-70°C$.

(iii) Take 2 μl of RNA or DNA (1 mg/ml), add 18 μl of GFP, heat to 55°C for 10 min and cool to room temperature. Add 3 μl of loading solution comprising 10% Ficoll, 0.06% bromophenol blue, 0.06% orange G and 0.5% SDS.

(iv) Make up agarose gel, for example, 1.2% for RNAs of 2000–5000 nucleotides in 25 mM Tris adjusted to pH 7.2 with acetic acid, 1 mM Na_2EDTA (pH 7.2), 5 mM sodium acetate. The samples are loaded and electrophoresed using the Tris-acetate running buffer until the orange G reaches the bottom of the gel.

(v) Staining. If the RNA is fully denatured it is not stained with an intercalating agent such as ethidium bromide; this is a good test for denaturation. The RNA can be de-glyoxylated by treating the gel with 50 mM sodium hydroxide for 10–15 min (excessive treatment will hydrolyse the RNA), neutralising in 200 mM sodium acetate (pH 5.0) and staining with 1 μg/ml ethidium bromide. Bands of RNA are then visualised by fluorescence under u.v. light. Alternatively a non-intercalating stain such as toluidine blue can be used. Gels are stained in 0.05% toluidine blue in 10 mM sodium phosphate pH 7.0 for 60 min and are then destained in phosphate buffer.

4.4.2 *Denaturation by Formaldehyde*

(i) 10 x running buffer in 200 mM Hepes. 10 mM Na_2EDTA adjusted to pH 7.8 with sodium hydroxide.

(ii) The sample is prepared by taking: 3 μl of 10 x running buffer; 15 μl of deionised formamide (see Section 3.8.4); 5 μl of 36–37% formaldehyde; 1 μg RNA, water and loading solution (see Section 4.4.1) to give 30 μl. Heat to 65°C and cool on ice.

(iii) Make up agarose gel, for example, 1.2% for RNAs of 2000–5000 nucleotides. Use 1.2 g of agarose; 10 ml of 10 x buffer; 16.7 ml of 36–37% formaldehyde; and 73.3 ml of water. Load samples and electrophorese using 1 x running buffer until orange G has reached the bottom of the gel. Stain as in Section 4.4.1.

4.4.3 *Denaturation with Formamide*

Formamide denaturation is easily reversible without damaging the RNA (e.g., by alkali). Thus, if the need is to recover the RNA for biological experiments this is a useful method. Denaturation is not as complete as with the above two systems.

(i) Stock solutions

(a) 50 x buffer Tris 21.7 g
$NaH_2PO_4.2H_2O$ 23.4 g
$Na_2EDTA.2H_2O$ 1.85 g
water to 1 litre
The pH should be 7.6–7.8

(b) Deionised formamide (see Section 3.8.4). It is important that the 'pH' of the formamide is around 7.0.

(c) Loading solution. 30% sucrose, 0.06% bromophenol blue, 0.06% orange G in formamide.

(ii) Prepare the samples by adding an equal volume of loading solution to the RNA solution. Heat to 90°C for 2 min and cool.

(iii) Make agarose gel in a 1:1 mixture of 1 x buffer and formamide. Pour the gel, allow to cool to room temperature and then put at 4°C until set (~20 min). Load the samples and electrophorese; the running buffer (the 1:1 mixture of 1 x buffer and formamide) needs circulating during electrophoresis.

(iv) After electrophoresis the gel should be washed for about 15 min in water and then stained with ethidium bromide.

4.5 Gel Electrophoresis of Proteins

As with nucleic acid, gel electrophoresis is the method of choice for determining the number of species and size of the protein coat of virus particles. The protein is denatured with SDS, thus eliminating most of the problems associated with the size-charge ratio of non-denatured proteins. For methods of denaturation and gel electrophoresis of proteins the reader is referred to the companion volume (16). Special attention should be paid to the methods for detecting atypical migration of SDS-denatured proteins (Ferguson plots). Below is given a rapid method for the dissociation of virus to release nucleic acid and protein for gel electrophoresis.

4.5.1 *Rapid Dissociation of Virus*

(i) Make up dissociation buffer:

10 x gel buffer	5 ml
10% SDS	10 ml
10 M urea	7.5 ml
2-mercaptoethanol	0.5 ml
sucrose	4.5 g

Aliquot and store frozen.

(ii) Add an equal volume of dissociation buffer to virus. For nucleic acid, heat to 50–60°C for 10 min; for protein, heat to 100°C for 2 min. Add 1/5 volume 0.06% bromophenol blue solution and load for electrophoresis.

(iii) This will give non-denatured nucleic acid. If denaturation is required follow the methods given above.

4.6 Gel Electrophoresis of Virus Particles

Gel electrophoresis can be used on many viruses with small isometric particles and some with rod-shaped particles to determine the isoelectric point of the particle or to separate particles of different sizes or net-surface charges. Separations are best in acrylamide gels, although agarose gels can be used for certain viruses. As a guide for isometric particles of about 30 nm diameter, gels of 3% acrylamide (5% cross-linked) are suitable. Tris/calcium lactate buffers (17) can give a range of pHs for these gels. Two stock solutions are needed.

(i) 0.6 M Tris with the pH adjusted with lactic acid for pHs 4.0–6.0 or with cacodylic acid for pHs 6.0–7.5. For the pH range 7.5–8.5, 0.6 M lactic acid should have its pH adjusted with Tris solution.

(ii) 60 mM calcium lactate which is $Ca(OH)_2$ with its pH adjusted with lactic acid (or cacodylic acid) to the required pH.

Mixing these solutions 1:1 will give a 20 times concentrated gel and running buffer.

5. REFERENCES

1. Murant,A.F. and Harrison,B.D., eds. (1970 to date) *CMI/AAB Descriptions of Plant Viruses* Nos. 1-275.
2. Kurstak,E., ed. (1982) *Handbook of Plant Virus Infections,* published by Elsevier/North Holland Biomedical Press, Amsterdam.
3. Bar-Joseph,M. and Hull,R. (1974) *Virology,* **62**, 522.
4. Takanami,Y. and Kabo,S. (1979) *J. Gen. Virol.,* **44**, 153.
5. Brunt,A.A. and Kenton,R.H. (1963) *Virology,* **19**, 388.
6. Dunn,D.B. and Hitchborn,J.H. (1965) *Virology,* **25**, 171.
7. Rickwood,D., ed. (1984) *Centrifugation: A Practical Approach,* 2nd Edition, published by IRL Press Ltd., Oxford and Washington, D.C.
8. Markham,R. (1962) *Adv. Virus Res.,* **9**, 241.
9. Wrigley,N.G. (1968) *J. Ultrastruct. Res.,* **24**, 454.
10. Maniatis,R., Fritsch,E.F. and Sambrook,J. (1982) *Molecular Cloning: A Laboratory Manual,* published by Cold Spring Harbor Laboratory Press, NY.
11. Rickwood,D. and Hames,B.D., eds. (1982) *Gel Electrophoresis of Nucleic Acids: A Practical Approach,* published by IRL Press Ltd., Oxford and Washington, D.C.
12. Wilcockson,J. and Hull,R. (1974) *J. Gen. Virol.,* **23**, 107.
13. Burton,K. (1956) *Biochem. J.,* **62**, 315.
14. Markham,R. (1955) in *Modern Methods of Plant Analysis,* Vol. **4**, Paech,K. and Tracey,M.V. (eds.), p. 246.
15. Lin,R.I.-S. and Schjeide,O.A. (1969) *Anal. Biochem.,* **27**, 473.
16. Hames,B.D. and Rickwood,D., eds. (1981) *Gel Electrophoresis of Proteins: A Practical Approach,* published by IRL Press Ltd., Oxford and Washington, D.C.
17. Lane,L.C. and Kaesberg,P. (1971) *Nature New Biol.,* **232**, 40.

CHAPTER 2

Growth, Assay and Purification of Picornaviruses

P.D. MINOR

1. INTRODUCTION

Picornaviruses are small lipid-free viruses containing a single strand of messenger sense RNA (mol. wt. $2.6-3 \times 10^6$ daltons) enclosed in a protein capsid which is typically 25 – 30 nm in diameter. They are widespread in man and other mammals, and viruses with many similar features have been found in insects and plants. This chapter will be confined to picornaviruses of mammals, which may be classified into the acid-stable enteroviruses, growing principally in the gut and the acid-labile rhinoviruses, found principally in the nasopharynx (1). Enteroviruses are further subdivided into polioviruses, Coxsackie viruses, echoviruses (or enterocytopathic human orphan viruses), numbered enteroviruses and the cardioviruses of mice. The aphthoviruses, which are the causative agents of foot-and-mouth disease of cattle, are included in the rhinoviruses. The classification is summarised with examples in *Table 1*.

The virus preparation required varies with the study to be undertaken. Unpurified pools of infectious virus are required in serological or host-range studies and as seeds for virus production (2). Highly purified virus free of host cell or other contaminants may be required as an antigen or immunogen (3), for studies of the viral RNA by cloning or sequencing (4) or as a source of messenger RNA (mRNA) in *in vitro* systems. Radiolabelled purified virus has been used to study the fate of the infecting virus particles (5), for characterising poliovirus by oligonucleotide mapping of its RNA (6,7), for examination of the virion proteins at various levels of detail (8) and in certain types of antigenic studies (9).

This chapter is divided into four sections dealing with safety, growth, assay and purification of picornaviruses. Poliovirus type 3 will be used as the principal example of a picornavirus throughout as it has been most studied in this laboratory, with other viruses discussed where they present significantly different problems.

2. SAFETY CONSIDERATIONS

Man is the natural host of many of the picornaviruses of particular interest. While infection with enteroviruses is usually inapparent, severe diseases can ensue, examples of which are given in *Table 1*. Infants are particularly at risk and pregnant women or parents of young children should exercise extreme care; Coxsackie B4, for example, has been implicated in congenital heart disease following

Table 1. Classification of Picornaviruses.

Group	*Subgroup*	*Exemplar*	*Disease caused in man*
Enteroviruses	Poliovirus	P3/Leon/USA (1937)	Paralytic poliomyelitis
	Coxsackievirus	Coxsackievirus A9	Fever, sore throat, aseptic meningitis, rhinitis
		Coxsackievirus B4	Aseptic meningitis, myocarditis (especially in infants), pericarditis, orchitis, possible adult onset diabetes
	Echovirus	Echo 11	Aseptic meningitis
	Enterovirus	Enterovirus 70	Haemorrhagic conjunctivitis, benign paralysis
	Cardiovirus	EMC	Possible fever with CNS involvement
	Enterovirus	Mengo	
	?	Hepatitis A	Hepatitis
Rhinovirus	Rhinovirus	HRV 1a	Common cold
		HRV 14	
	Aphthovirus	FMDV SAT 1	–

in utero infection, and laboratory infections with Coxsackie A viruses are known to occur, although the incidence is not well documented. It is good practice to change clothing completely and wash thoroughly before leaving the laboratory building, and to check from time to time that serum antibody levels are high.

Vaccines for human use against the Coxsackie and enteroviruses listed in *Table 1* are not available at the time of writing. However, extremely safe and effective poliovirus vaccines are available and the importance of checking and boosting antibody levels before any work commences cannnot be overstressed, as a high level of immunity will prevent an individual developing either the disease or an inapparent infection which could pass the virus on to others. It is probable that Salk inactivated vaccine is the most satisfactory for boosting serum antibody titres against poliovirus in adults. Work with foot-and-mouth disease virus is subject to strict legal control because of its importance as a disease of animals.

3. GROWTH OF PICORNAVIRUSES

Three specific protocols for the growth of different types of poliovirus preparation are given after some general considerations.

3.1 Choice of Cell

While egg-adapted strains have been reported (1) most picornaviruses will grow only in a homologous cell culture system, i.e., a human picornavirus will require a human or primate cell of some kind. One of the major difficulties in working with picornaviruses is the selection of a suitable cell and this factor makes polioviruses among the easiest to study as they will yield titres of $10^7 - 10^9$ infectious units per ml ($\sim 10 - 1000$ infectious units produced per cell) in a variety of easily cultured cell types. Other enteroviruses and rhinoviruses, on the other hand, often yield titres of 10^4 infectious units per ml (i.e., 0.03 units per cell) from certain common cell lines (e.g., Echo 7 in RD cells, Rhinovirus 1b in Hep 2c cells),

Table 2. Suitable Host Cells for Virus Growth

Virus	*Cell*	
Poliovirus	Hep 2c (probably now HeLa)	Human epithelial carcinoma
	HeLa	Human carcinoma
	HeLa (Mandel)	Human carcinoma
	Secondary monkey kidney	Monkey kidney
Coxsackie A	RD	Human rhabdosarcoma
Coxsackie B	HeLa (Mandel)	Human carcinoma
	LLCMK2	Rhesus monkey kidney embryonal
Echo 11	RD	Human rhabdosarcoma
Enterovirus 70	RD	Human rhabdosarcoma
EMC	Krebs II mouse ascites	Mouse tumour
	Secondary mouse embryo	Mouse embryo
Mengovirus	L cells	Mouse tumour
	Novikoff rat heparoma cells NISI-67	Rat liver carcinoma
	Secondary mouse embryo	Mouse embryo
Hepatitis A	Primary marmoset liver	Marmoset liver
	FRMK-6	Foetal rhesus monkey kidney
	MRC-5	Human embryo lung
Rhinovirus	HeLa 0	Human carcinoma
	MRC-5	Human embryo lung
FMDV	KB	Human carcinoma
	BHK	Syrian hamster kidney

and it is difficult to obtain acceptable quantities of purified virus from such low titre material. Studies of these preparations are therefore restricted to examination of neutralisation by antibodies or immunofluorescent studies of replication (10). A strain of HeLa cells (HeLa Ohio) has been recommended for rhinovirus growth, but it is commonly found that cell strains vary between laboratories and different types of rhinoviruses may grow to different titres in the same cell type. *Table 2* summarises some of the cell lines used for the growth of a selection of some of the commonly studied picornaviruses. The list is not intended to be comprehensive but to suggest readily available cell lines. Information on the most suitable cell for growth of a particular picornavirus will probably be obtained from the original supplier of the virus seed.

3.2 Growth Conditions

Although the infectivity of most picornaviruses is reasonably stable the virus seed is best stored frozen at −70°C. However, the stock should be allowed to thaw at room temperature because picornaviruses tend to be heat labile and repeated thawing at 37°C appears to reduce the titre. The multiplicity of infection used can be high if necessary as, in contrast to other viruses, defective interfering particles are not readily generated by picornaviruses (however, see 11).

The host cell line of choice may be grown on the surface of a bottle or culture flask, in some kind of rolled vessel (e.g., bottle or test tube) or in suspension with gentle agitation. Surface grown cells may either be grown to confluence on the vessel or seeded at confluent levels from trypsinised cell cultures the day before use [despite the protease sensitivity of picornavirus-receptor sites (12)]. In our hands rolled test tubes have been the method of choice for the production of up to 100 μg of poliovirus, as a relatively large area of cells (66 cm^2, 10^7 cells) can be covered by a small volume of medium (1 ml), leading to an economical use of isotope and a virus preparation which does not need concentration to a small volume before purification. The yield and specific activity of poliovirus grown in this way is generally the same as that of virus grown on flasks at similar concentrations of cells, isotope and virus. Rolled cultures are, however, definitely advisable for rhinoviruses, which grow best in well oxygenated systems at slightly acid pH, although the virus itself is acid labile.

Rhinoviruses appear to grow best at 33°C, although they can be grown satisfactorily at 35°C (13), and the Sabin vaccine strains of poliovirus are all temperature sensitive, growing optimally at 33–35°C. Calf and horse sera frequently contain inhibitors of poliovirus replication, and cell cultures for their growth should be established using foetal calf serum or maintained on serum-free medium for 24 h before inoculation.

3.3 Harvesting Virus

Virus growth and release is generally accompanied by cell degeneration, and a cytopathic effect (cpe) is most readily seen in cells grown as a monolayer. Initially it is advisable to maintain an uninfected culture for comparison, although degeneration is usually unambiguous, resulting in the loss of the cells and total destruction of the cell sheet. *Figure 1* illustrates the cpe produced by poliovirus. However, visible effects are not always produced by picornaviruses. Hepatitis A virus produces no detectable cpe in FRhK-6 cells and low amounts of virus may be repeatedly harvested from the culture medium.

Generally speaking, virus is completely released from cells infected with enteroviruses when maximum cpe is observed, while rhinoviruses tend to remain more cell-associated. In some instances the timing of the harvest is crucial; foot-and-mouth disease virus particles are reported to have a virion-associated ribonuclease activity which degrades the genomic RNA and thus reduces viral infectivity if the harvest is delayed from 8 to 12 h (14). Similar findings are reported for other rhinoviruses.

It is customary to disrupt the cells by one to three cycles of freeze/thawing followed, if required, by aspiration of the cell suspension through a syringe needle. Cell debris is then usually pelleted by low speed centrifugation.

3.4 Growth of Seed Stocks of Poliovirus Type 3

Observe sterile technique throughout.

(i) Seed a 16 oz glass bottle or 150 cm^2 plastic tissue culture flask with Hep 2c cells ($\sim 5 \times 10^6$ cells) in 100 ml of Eagle's minimal essential medium (MEM) supplemented with 4% foetal calf serum, 2.2 g/litre sodium bicarbonate 120 mg/litre penicillin and 100 mg/litre streptomycin. Grow to confluence at 35°C or 37°C.

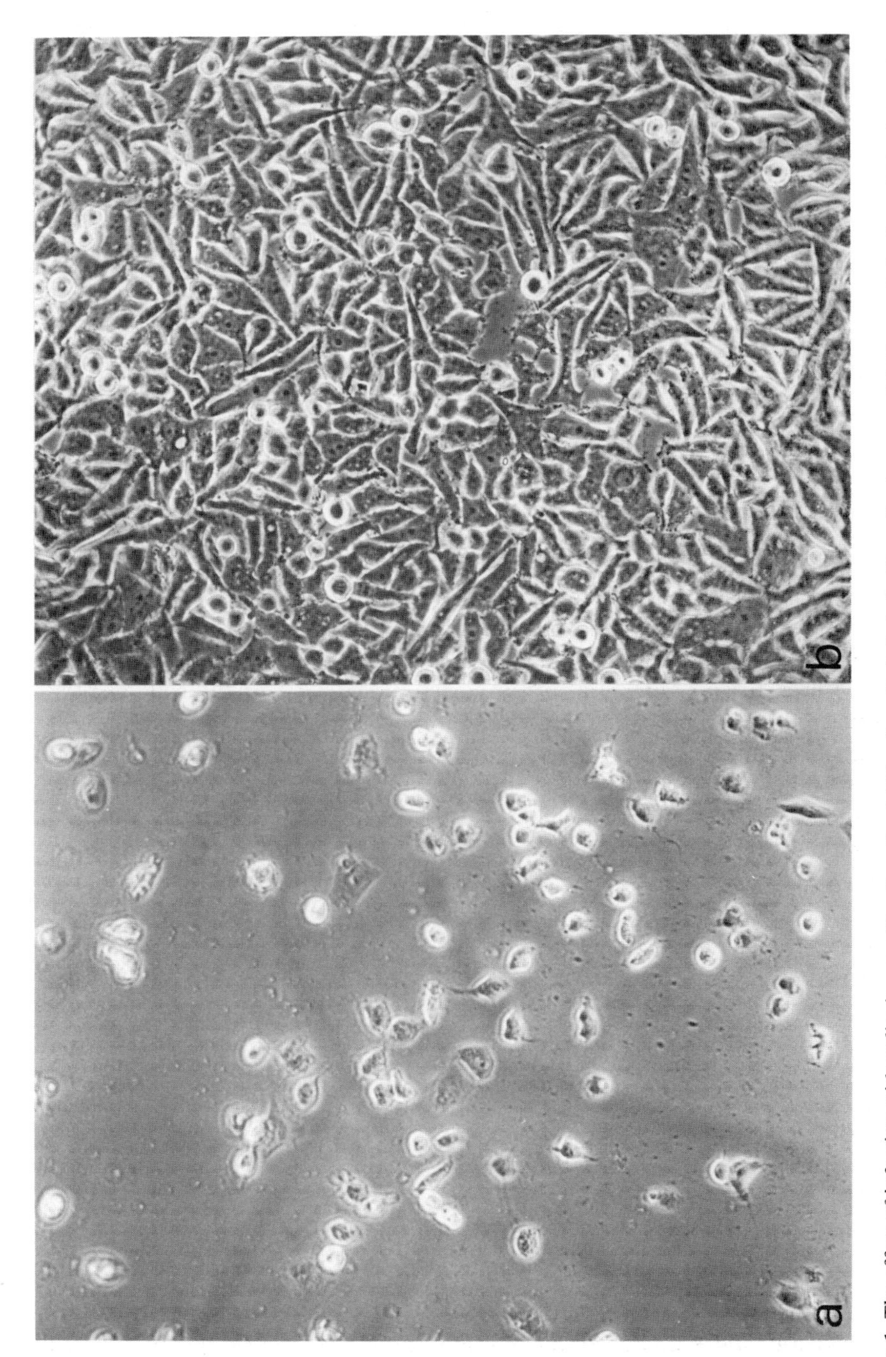

Figure 1. The effect of infection with poliovirus on Hep 2c cells. **(a)** Hep 2c cells infected with poliovirus type 3 24 h post-infection. **(b)** Uninfected control culture. Magnification x 200.

(ii) Decant the medium. Incubate the cell sheet with 0.1 ml or less of type 3 poliovirus-infected tissue culture fluid. Adsorb at room temperature for 30 min with occasional agitation.

(iii) Add 50 ml of MEM with bicarbonate and antibiotics as in (i) but without serum. Incubate at 35°C until the cell layer is completely destroyed (1 – 3 days).

(iv) Freeze at – 70°C, thaw at room temperature. Remove the cell debris from the medium by low-speed centrifugation (2000 g, 5 min). Decant and keep the supernatant.

(v) Store aliquots of supernatants at – 70°C.

This procedure yields pools at $10^8 - 10^9$ p.f.u./ml of type 3 poliovirus.

3.5 Production of [^{35}S]Methionine-labelled Type 3 Poliovirus from Cell Monolayers in Test Tubes

Observe sterile technique throughout.

(i) Harvest Hep 2c cells by trypsinisation into MEM with serum, antibiotics and bicarbonate as in Section 3.4(i). Dilute the suspension to 2 x 10^6 cells per ml. Seed acid-washed sterile test tubes (13.5 cm x 1.5 cm) with 5 ml of cell suspension and stopper with silicone rubber bungs. Place in a roller drum on a roller machine adjusted to hold the tubes horizontally and turn at about 5 revolutions per hour in a hot room or incubator at 35°C. Allow the cells to attach overnight.

(ii) Decant the medium and infect each tube with 0.1 – 0.3 ml of undiluted type 3 poliovirus-infected tissue culture fluid (~ 10^8 infectious units per tube). Roll the tubes for 1 h at 35°C, then add 1 ml of MEM lacking methionine, but supplemented with 1% foetal calf serum, 2.2 g/litre bicarbonate, 120 µg/litre penicillin and 100 µg/litre streptomycin. Roll the tubes for a further 5 h at 35°C to allow the infection to become established and host cell protein synthesis to be suppressed.

(iii) Add 25 µCi of [^{35}S]methionine (~ 2 µl) to each tube. Incubate on the roller overnight at 35°C.

(iv) At 24 h post-infection, the cytopathic effect should be complete with no cells attached to the glass. Freeze the tubes at – 70°C, thaw in a water bath at room temperature. Pool the medium and remove the cell debris by low-speed centrifugation (2000 *g*, 5 min).

(v) Purify the virus in the supernatant further (see Section 5). Typically three tubes (3 x 10^7 cells) yield approximately 10^{10} p.f.u. of virus with a specific activity of 2 x 10^{-4} c.p.m. per p.f.u.

3.6 Production of Virus in HeLa Cell Suspension Cultures

Observe sterile technique throughout.

(i) Specific HeLa cell strains (e.g., HeLa S3) may be grown in medium low in calcium ions (e.g., CaS2MEM). It is essential for efficient infection to have the cells growing well as a monodisperse suspension. Pellet the cells by low-speed centrifugation (900 *g*, 15 min). Resuspend at approximately 2 x 10^7 cells per ml (~1/20 starting volume) in CaS2MEM.

(ii) Add the virus to 10–50 p.f.u./cell; stir for 1 h at 35°C on a magnetic stirrer.

(iii) Dilute to 2 x 10^6 cells per ml in CaS2MEM. Add 100 μCi of [^{3}H]uridine.

(iv) Incubate for 8 h: pellet the cells by low-speed centrifugation (900 *g*, 15 min) and wash once in phosphate buffered saline (PBS) deficient in magnesium and calcium ions.

(v) Suspend at 5 x 10^7 cells per ml in PBS. Freeze and thaw three times.

(vi) Remove the debris by centrifugation (900 *g*, 15 min).

(vii) Retain the supernatant for further purification.

4. ASSAY OF VIRUS

Many picornaviruses will agglutinate erythrocytes of appropriate species under suitable conditions, and this could be used for the detection of virus. The more commonly used alternatives are to assay for infectivity, antigenicity or the presence of virus material (such as RNA or protein) after fractionation.

4.1 Infectivity Assays

All infectivity assays require a susceptible cell line and good sterile technique.

4.1.1 *$TCID_{50}$ Assay for Poliovirus Type 3* (15)

(i) Set up sterile tubes containing 0.9 ml of dilution medium (Eagle's MEM supplemented with 1.3 g/litre of sodium bicarbonate 4% foetal calf serum, 120 μg/litre penicillin, 100 μg/litre streptomycin).

(ii) Make 10-fold serial dilutions of the virus preparation by inoculating 0.1 ml into the first tube, mixing and transferring 0.1 ml to the next tube, etc. Use a clean pipette for each transfer.

(iii) Sterile disposable microtitre plates suitable for tissue culture are available from, for example, Linbro. Each plate contains 96 flat-bottomed wells, each with a capacity of 0.3–0.4 ml. Inoculate four wells with 0.025 ml for each dilution of virus, using a dropping pipette or micropipette. The plates may be covered with sterile blotting paper to protect them between additions.

(iv) Harvest Hep 2c cells by trypsinisation and dilute to 5000–10 000 cells per ml in dilution medium. Inoculate 0.1 ml cell suspension per well. Include a row of wells which have no virus as a negative control.

(v) Seal the plate with pressure-sensitive adhesive film and incubate at 35°C.

(vi) Examine the wells for degenerative changes in the cell sheet on days 4 and 7 using an inverted microscope.

(vii) Calculate the virus titre expressed as the dilution of the virus required to infect 50% of the cultures ($TCID_{50}$). For example, if half (2/4) of the wells show degeneration at a dilution of 10^{-7}, the titre is 10^7 $TCID_{50}$ units in 0.025 ml. If all wells show degeneration at 10^{-7} and none at 10^{-8}, the titre is $10^{7.5}$ $TCID_{50}$ units in 0.025 ml. One well corresponds to approximately $10^{0.25}$ $TCID_{50}$ units. Titres are usually expressed as $TCID_{50}$ per ml.

4.1.2 *Plaque Assay of Poliovirus Type 3*

(i) Seed tissue culture trays containing six 3.5 cm diameter wells (e.g., Linbro FB6) with 3 x 10^6 Hep 2c cells per well in 3 ml medium [Section 4.1.1 (i)] for use the next day. Alternatively, seed at 5 x 10^5 cells per well and allow to grow to confluence (3 days) at 35°C in an incubator gassed with 5% CO_2 in air.

(ii) Make up sufficient agar (Noble's Agar; Difco) at 2% in distilled water to allow 1.5 ml per well. Autoclave (15 p.s.i., 15 min). Keep molten at 56°C.

(iii) Make up sufficient 2 x concentrated Liebowitz L15 medium supplemented with 2% foetal calf serum, 240 μg/litre penicillin, and 200 μg/litre streptomycin to allow 1.5 ml per well. Keep warm at 37°C.

(iv) Make serial 10-fold dilutions of virus as described in Section 4.1.1(ii), but using a dilution medium of Liebowitz L15 medium supplemented with 1% foetal calf serum, 120 μg/litre penicillin and 100 μg/ml streptomycin.

(v) Remove the medium from the plates. Inoculate 0.1 – 0.25 ml samples of virus dilutions onto duplicate cell sheets. Distribute evenly and allow to adsorb at room temperature for 15 – 60 min with shaking at intervals. Do not allow the cell sheet to dry out.

(vi) Mix equal volumes of the 2% agar and the 2 x concentrated L15 medium. Allow to cool slightly (to ~40°C) and overlay each well with 3 ml. Agitate the dish to mix the inoculum of virus and the agar. Allow to solidify at room temperature.

(vii) Incubate the plates upside down to prevent condensation seeping between the agar and the cell sheet. Incubation (at 35°C) may be in a humidified container in a hot room or in a humidified incubator gassed with 5% CO_2 in air.

(viii) Plaques should be of suitable size by 3 – 4 days. Stain with Naphthalene black, Leishmann's stain, Giemsa crystal violet or another suitable stain. Naphthalene black stain is made up from 1 g of Naphthalene black (British Drug Houses Ltd., Poole, Dorset, UK), 13.6 g of sodium acetate, 60 ml of glacial acetic acid and distilled water to 1 litre.

(ix) Calculate the virus titre. If a dilution of 1 in 10^6 yields 19 plaques when 0.1 ml is inoculated, the titre is $\frac{19 \times 10^6}{0.1} = 1.9 \times 10^8$ p.f.u./ml.

4.1.3 *Comparisons of $TCID_{50}$ and Plaque Assay of Infectivity*

The $TCID_{50}$ assay has the advantage that it is easily performed with a minimum of equipment, is economical on cells and, possibly most significantly for picornaviruses, requires a cell that will show a degeneration whereas the plaque assay requires a host cell which will allow the formation of well-defined plaques in a thick cell sheet. If a permanent record is required, it can be obtained by staining the cells although degenerative changes can be obvious to the experienced eye without total destruction of the cell sheet. The $TCID_{50}$ assay as described here is relatively imprecise compared with a plaque assay, giving a figure which is probably accurate to within $10^{0.5}$ at best. However, the accuracy can be improved by using a larger number of wells per dilution. The $TCID_{50}$ assay also requires more dilutions of virus to establish the titre.

4.2 Assays of Antigenicity

The antigenic structure of picornaviruses is unusual in that empty capsids and certain other non-infectious particles are generally held to express an antigen on their surfaces which is not found on infectious particles. For poliovirus this antigen is 'C' or 'H' antigen. Likewise, infectious particles express an antigen not found on empty capsids, known as 'D' or 'N' antigen for poliovirus. It is possible that there are common antigenic determinants present on both types of particle. The infectious 'D' or 'N' antigen is readily destroyed. Heating poliovirus at 56°C for 30 min converts essentially all 'D' or 'N' antigen to 'C' or 'H' antigen, and the same conversion can be achieved to a variable degree by adsorption to the surfaces used in solid phase radioimmunoassay (RIA) or enzyme-linked immunosorbent assay (ELISA). ELISA and other immunoadsorbent assays have been used in the detection of hepatitis A (16), poliovirus (17) and foot-and-mouth disease (18) virus antigens. We have experienced considerable difficulty with this type of assay, partly, we believe, because of the variable adsorption and alteration of antigen which occurs, and partly because of non-specific effects attributable to the fairly small proportion of material derived from the virus rather than the host even in high titre virus preparations.

Immunofluorescent techniques have also been applied to follow virus replication (10), but the assay for purified antigen routinely employed by this laboratory has involved single radial immune diffusion (3).

4.2.1 *Single Radial Immunodiffusion Assay for Poliovirus Type 3*

(i) Dissolve 10 g of Seakem Agarose ME (Marine Colloids Inc., USA) in 1 litre of Dulbecco's phosphate buffered saline A (PBSA) over a boiling water bath while stirring.

(ii) Filter through Whatman Grade 113r filter paper, and add sodium azide to 0.05%. Aliquot into 20–25 ml amounts.

(iii) Clean 12 cm x 12 cm glass plates using detergent; rinse and dry.

(iv) Melt enough agarose for the test allowing 14 ml per plate. Each plate will accommodate 16 samples. Dispense 12 ml amounts into pre-warmed bottles at 60°C; wipe the surface of each glass plate with tissue paper dipped in the remainder and allow to dry.

(v) Apply silicone grease lightly to the inside edge of the mould (*Figure 2*) and place the mould (internal diameter 9 cm) centrally on the coated glass plate (as in *Figure 2*) on a level surface.

(vi) Introduce hot agarose around the inside edge of the mould to seal the mould to the glass plate. Allow to set (5 min).

(vii) Add a suitable amount of hyperimmune anti-polio serum (usually 10–15 μl) to each 12 ml of warm agarose. Mix well avoiding bubbles and pour into the mould, removing any bubbles with a Pasteur pipette.

(viii) Allow to set for about 10 min, remove the mould by a springing action using slight pressure from above. Punch out 16 symmetrically distributed

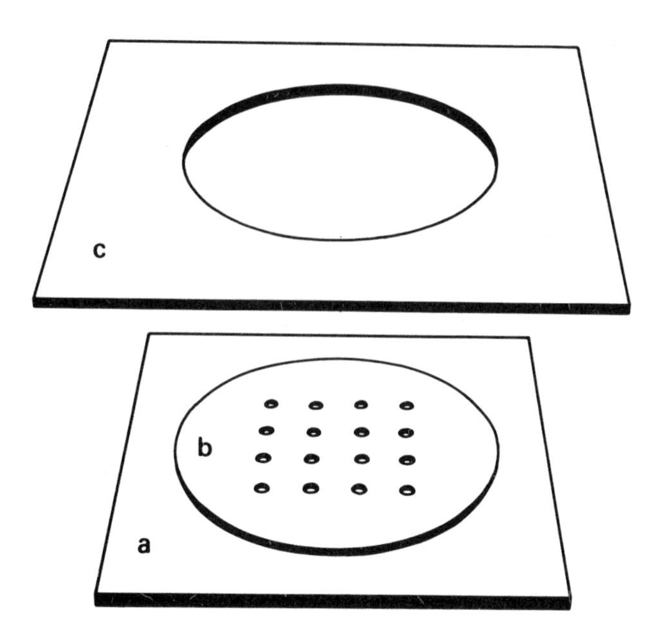

Figure 2. Apparatus for casting gels for single radial diffusion. **a:** glass plate (12 cm x 12 cm). **b:** agarose slab after casting. **c:** Perspex mould (16 cm x 16 cm).

holes (b in *Figure 2*) using a former and a cork borer or suitable punch 4 mm in diameter.

(ix) Add two volumes of antigen to one volume of 20% (v/v) Mulgofen detergent [Mulgofen BCV-720, GAF (Great Britain) Ltd., Manchester, UK]. A sample of 20 μl is sufficient to fill a well once. It is possible to refill the well after the liquid has been adsorbed.

(x) Cover the agarose sheet with a 9 cm Petri dish lid and stack the plates in a humidified container. Allow to diffuse for 1–3 days.

(xi) Wash by immersion in PBSA overnight to remove excess serum.

(xii) Drain and place a circle of filter paper on the agarose. Dry in a hot room (37°C) under a fan. Stain with a suitable protein stain (e.g., 3% Coomassie brilliant blue in 20% acetic acid, 50% methanol) for 5 min at room temperature. Destain in 20% acetic acid, 50% methanol. Allow to dry. This method is extremely adaptable and can be used to detect radiolabelled virus autoradiographically or adapted to the sensitive assay of antibodies including monoclonal antibodies. The antigens used must, however, be purified to avoid non-specific effects.

4.3 Other Methods for Detection of Picornaviruses

After infected tissue culture harvest has been subjected to purification by differential centrifugation or other fractionation process it is necessary to identify the fractions containing the desired material. The methods given in Sections 4.1 and 4.2 can be used, if necessary, but are time consuming and alternative methods are to screen the gradient fractions for peaks of RNA or protein by u.v. scanning (13) or by incorporating radiolabelled virus in the material to be purified as a marker. Aliquots of each fraction may then be assayed for radioactivity by any of the standard methods (e.g., liquid scintillation counting). This is the method of choice where radiolabelled material can be obtained in that it is rapid, accurate and simple.

5. PURIFICATION OF VIRUS

5.1 **General Considerations**

Picornavirus tissue culture preparations are associated with host components including membranes, and if the virus is required for antigenic or biochemical analysis it must be purified, usually by some form of density gradient centrifugation. If substantial amounts of virus are involved, purification is likely to be preceded by concentration, generally best done by precipitation with ammonium sulphate or with polyethylene glycol in the presence of sodium chloride followed by a low-speed centrifugation. Pelleting by ultracentrifugation is also used, especially where small volumes are involved.

5.2 **Concentration of Picornaviruses**

5.2.1 *Concentration by Ammonium Sulphate*

(i) Weigh out 0.4 g of ammonium sulphate (Analytical Reagent grade) per ml of tissue culture fluid.

(ii) Add tissue culture fluid, which should have 1% serum to aid precipitation. Shake thoroughly to dissolve the ammonium sulphate crystals. The solution will become slightly acid and cloudy.

(iii) Spin at 2000 *g* for 1 – 2 h at 4°C in, for example, an MSE 4L centrifuge.

(iv) Decant the supernatant and take up the pellet in at least 10% of the original volume to ensure dilution of the ammonium sulphate. This method has been used for the acid-labile rhinoviruses as well as the acid-stable enteroviruses. The concentration routinely achieved is at most 10-fold, but the method is rapid. Recoveries are 99% or better.

5.2.2 *Concentration by Polyethylene Glycol and Sodium Chloride*

(i) Add 2.22 g of sodium chloride per 100 ml to tissue culture fluid at 4°C on a magnetic stirrer.

(ii) When the sodium chloride is dissolved, add 7 g of polyethylene glycol 6000 (British Drug Houses Ltd., UK) per 100 ml of tissue culture fluid.

(iii) Stir for at least 4 h, preferably overnight.

(iv) Centrifuge at 2000 *g* for 2 h at 4°C in, for example, an MSE 4L centrifuge.

(v) Decant the supernatant and resuspend the pellet in PBS, supplemented with 2% bovine serum albumin, using 1/50 to 1/100 of the starting volume. Sonicate or homogenise the suspension to break up large fragments.

(vi) Centrifuge at 3000 *g* for 10 min to remove solid material. All the virus may be recovered by this method and a greater concentration can be achieved than with ammonium sulphate without the fluctuations in pH associated with that method. It is, however, slightly more time-consuming.

5.2.3 *Concentration of Picornaviruses by Ultracentrifugation*

Virtually all poliovirus can be pelleted from a tube with a path length of 5 cm by a 1 h spin at 100 000 *g*, or from a path length of 12 cm by a 3 h spin at 100 000 *g*. This method has the advantage that no chemicals are added, and there is less chance of precipitating unwanted proteins than by the methods given above. However, only small volumes can be conveniently handled and there may be difficulties in resuspending the virus.

5.3 **Purification of Picornaviruses**

5.3.1 *General*

While they are themselves lipid free, picornaviruses are frequently associated with membranes when released from the host cell, and detergent is usually added to a concentration of 1% to free the virus particle. Resolution of the virus in subsequent purification steps is poor or non-existent without removal of host membranes. The detergents which have been used vary from the mild non-ionic Nonidet P-40 (NP-40) type to sodium lauryl sulphate or sarcosyl. It is probably good policy to use a mild detergent, as empty capsids and some viruses such as certain strains of type 3 poliovirus tend to be destroyed by the more vigorous compounds. We routinely employ 1% NP-40.

Picornaviruses are usually purified by some form of density gradient centrifugation (although gel filtration has been used commercially). The small size of the particles gives a sedimentation coefficient of 150 – 160S, which is lower than is found for many other viruses, but the absence of lipid and the tightly integrated structure gives them a high buoyant density (~1.34 g/cm^3 in caesium chloride gradients; but see Section 5.3.3 below). Such a density is unattainable on sucrose gradients. Purification therefore involves separation by sedimentation rate on sucrose gradients or by buoyant density on caesium chloride or caesium sulphate gradients.

Gradients may be poured using commercially available equipment or some variation on the apparatus illustrated in Chapter 1, *Figure 3*.

The RNA-containing poliovirus particles which may be identified include the infectious virus particle, sedimenting at 150 – 160S and expressing 'D' antigen, and a non-infectious particle probably produced by failure of virus to penetrate the cell after adsorption (3) which sediments at 130S, expressing 'C' antigen. Empty capsids expressing 'C' antigen sediment at approximately 80S, but another 'D' antigen-expressing component may also be encountered, sedimenting at 70S; smaller components (14S) may also be produced. If the particles are required for antigenic analysis it is clearly important to separate these components.

5.3.2 *Purification of Picornaviruses on Sucrose Gradients*

(i) Prepare a solution to give final concentrations of 15 g of sucrose per 100 ml, 10 mM Tris-HCl pH 7.4, 50 mM NaCl. Prepare a similar solution at final concentrations of 45 g of sucrose per 100 ml, 10 mM Tris-HCl pH 7.4, 50 mM NaCl.

(ii) Prepare 30 ml linear gradients of sucrose from 15 to 45% in an ultracentrifuge tube of 35 – 40 ml capacity (e.g., Beckman SW28) using 15 ml of each of these solutions per gradient and a suitable gradient maker.

(iii) Prepare a solution of 10% NP-40 (BDH) in PBS. Add one-tenth of a volume to the sample, whose volume should not exceed 6 ml. The sample should clarify visibly as it is shaken with the detergent.

(iv) Layer the sample carefully onto the pre-formed gradient. Balance the tubes with liquid paraffin.

(v) Centrifuge the tubes at 4°C at 80 000 *g* for 4 h (for example at 25 000 r.p.m. in a Beckman L8 ultracentrifuge, SW28 rotor).

(vi) Harvest the gradients as described in Section 5.4 or by other methods. Assay for virus. The infectious virus peaks should be about 2/3 down the gradient.

Sucrose purification of viruses in this way is rapid and inexpensive and has little effect on the virus (see below). However, some impurities are likely to remain and if extreme purity is required it may be necessary to re-purify the virus on caesium chloride gradients. This is usually unnecessary, although the virus may be pelleted to remove the sucrose (Section 5.2.3).

5.3.3 *Purification of Picornaviruses on Caesium Chloride Gradients*

Pre-formed gradients may be used as follows.

(i) Prepare a solution of 40% (w/w) caesium chloride by dissolving 4 g of solid caesium chloride in 6 ml of 0.01 M Tris-HCl pH 7.4.

(ii) Prepare a solution of 5% (w/w) caesium chloride by dissolving 0.5 g of solid caesium chloride in 9.5 ml of 0.01 M Tris-HCl pH 7.4.

(iii) Pour a 10 ml 5 – 4% linear caesium chloride gradient in an ultracentrifuge tube of 12 ml capacity (e.g., Beckman SW41 tube) using 5 ml of each solution and a suitable gradient maker. Layer the sample containing 1% NP-40 on the gradient in a volume of 1 ml. Alternatively prepare the gradient by substituting 6 ml of sample containing 1% NP-40 for the 5% caesium chloride solution and using 6 ml of 40% caesium chloride solution. Clean and disinfect the gradient maker after use. Balance the tubes with liquid paraffin.

(iv) Centrifuge at 120 000 *g* for at least 4 h and preferably overnight at 4°C in, for example, a Beckman L8 ultracentrifuge, SW41 head at 30 000 r.p.m.

(v) Harvest the gradients as described in Section 5.4, or by other methods. Assay for virus.

Samples may be fractionated on gradients formed in the course of centrifugation (field formed gradients) as follows.

(i) Add solid caesium chloride to 0.46 g/ml of final sample volume. Thus, if the capacity of the centrifuge tube is 4.5 ml, add 2.07 g of solid caesium chloride to the sample and make it up to 4.5 ml with PBS or 0.01 M Tris. Add NP-40 to a final concentration of 1% v/v.

(ii) Centrifuge for 24 h at 4°C at 140 000 *g* (for example, in the Beckman L8 ultracentrifuge, SW50 head at 38 000 r.p.m.).

(iii) Harvest the gradient as described below, or by other methods. Assay for virus.

Methods involving caesium chloride are expensive, but can yield an extremely pure product. Moreover it is possible to use large sample volumes in either pre-formed or field formed gradients as the virus migrates to its equilibrium density. The use of SDS as a detergent is inadvisable, as caesium lauryl sulphate is extremely insoluble. The principal disadvantage of caesium chloride gradients apart from their expense, however, is that they may modify the virus. There is evidence that the rhinoviruses in particular can adsorb caesium ions to give a particle with an artefactually high buoyant density of 1.4 g/ml instead of the normal picornavirus density of 1.34 g/ml (19). A similar phenomenon is reported for poliovirus.

5.4 Harvesting Gradients

Commercial apparatus is available for harvesting centrifugation gradients, but simple home-made apparatus has been found to give equally satisfactory results, and is both less expensive and easier to maintain. Three are illustrated in *Figure 3*.

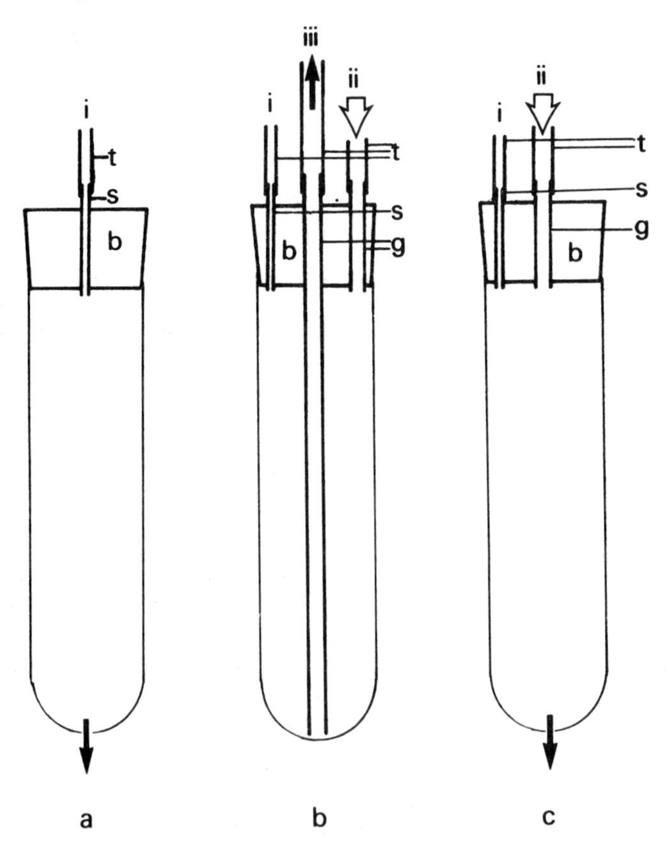

Figure 3. Apparatus for collecting fractions from density gradients after centrifugation. In each case the filled arrow indicates the path of the harvested fractions, and the open arrow the path of the displacing liquid paraffin where applicable. b:bung, g:glass tube, s:stainless steel tube, t:silicone rubber tubing. For details of operation see text.

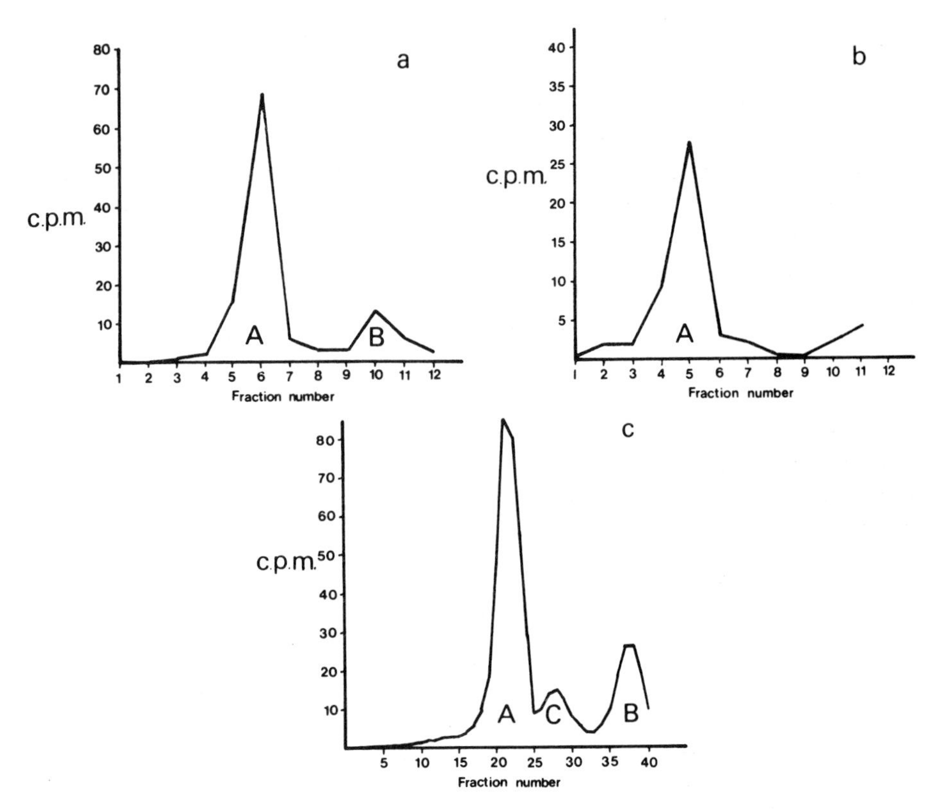

Figure 4. Specimen gradients of radiolabelled virus collected by the apparatus shown in *Figure 3*. **(a)** [^{35}S]methionine-labelled virus collected by apparatus (a) (2 ml fractions). **(b)** ^{32}P-labelled virus collected by apparatus (b) (2 ml fractions). **(c)** [^{35}S]methionine-labelled virus collected by apparatus (c) (0.5 ml fractions). Abscissa, radioactivity in 50 μl aliquot in c.p.m. x 10^{-3}. Ordinate, fraction number. The direction of centrifugation was from right to left in all cases and fractions from the top of the gradient are not shown. A:infectious virus peak (160S). B:empty capsid peak (80S). C:'eluted' virus peak (130S).

The apparatus (*Figure 3a*) consists of a piece of narrow-bore steel tubing pushed through a rubber bung of a size to fit the centrifuge tube. The steel tubing is attached to a short piece of silicone rubber tubing.

(i) Insert the bung into the centrifuge tube held vertically in a clamp.
(ii) Pinch off the rubber tubing and pierce the bottom of the tube with a heated needle.
(iii) Position a vessel (tube or vial) beneath the hole and control the outflow of the gradient by pressure on the silicone rubber tubing.

It is easy to estimate fractions of about 0.5–2 ml by eye, and this method has been used where peak fractions are required without any need for an accurate sedimentation value. A sample of [^{35}S]methionine-labelled virus fractionated on a 36 ml gradient in this way into approximately 2 ml fractions is shown in *Figure 4a*. The infectious virus peak (155S) is labelled A, and the empty capsid peak (80S) is labelled B.

The gradient may be fractionated without piercing the tube by the apparatus shown in *Figure 3b*. It consists of a rubber bung through which are inserted a bleed tube of narrow bore steel tubing attached to (i) a short piece of silicone rubber tubing, (ii) a glass inlet tube attached by silicone rubber tubing to a reservoir holding liquid paraffin and (iii) a glass tube which reaches to the bottom of the centrifuge tube at one end, and is attached to a piece of silicone rubber tubing at the other.

(a) Clamp the centrifuge tube vertically.
(b) Position clips on tubes (ii) and (iii). Carefully insert the rubber bung into the centrifuge tube.
(c) Remove the clip on tube (ii) and allow the liquid paraffin to flow into the centrifuge tube and overflow through the bleed tube (i).
(d) Position the clip on bleed tube (i).
(e) Remove the clip on tube (iii) and collect 0.5 – 2 ml fractions as they are displaced from the bottom of the tube.

This method enables the tube to be re-used, but might be thought likely to disrupt the gradient and contaminate the virus band. In fact these problems have not arisen in practice. A ^{32}P-labelled poliovirus preparation fractionated on a 36 ml gradient fractionated into 2 ml fractions is shown in *Figure 4b*. The infectious virus peak is labelled A.

A more accurate fractionation of a gradient into known volumes may be achieved by the apparatus shown in *Figure 3c*, which consists of a rubber bung as above, a bleed tube of steel tubing attached to a short piece of silicone rubber tubing (i), and a glass inlet tube attached to a repeating syringe adjustable to deliver 0.5 – 2 ml of liquid paraffin (ii).

(a) Clamp the centrifuge tube vertically.
(b) Insert the rubber bung.
(c) Pump liquid paraffin into the tube until it emerges from the bleed tube. Clamp the bleed tube.
(d) Pierce the bottom of the centrifuge tube.
(e) Collect fractions of known volume by displacement with a known volume of liquid paraffin delivered by the repeating syringe.

Virus labelled with [^{35}S]methionine and resolved in this way into 0.5 ml fractions from a 36 ml SW28 gradient of [^{35}S]methionine-labelled type 3 poliovirus is shown in *Figure 4c*. The infectious virus peak (155S) is labelled A, the empty capsid peak (80S) is labelled B, and the non-infectious RNA containing particle peak (130S) is labelled C.

6. REFERENCES

1. Andrewes,C., Pereira,H.G. and Wildy,P. (1978) *Viruses of Vertebrates,* 4th edition, published by Bailliere Tindall, London.
2. Crainic,R., Coullin,P., Blondel,B., Cabau,N., Boué,A. and Horodniceanu,F. (1983) *Infect. Immun.,* **41**, 1217.
3. Minor,P.D., Schild,G.C., Wood,J.M. and Dandawate,C.N. (1980) *J. Gen. Virol.,* **51**, 147.
4. Cann,A.J., Stanway,G., Hauptmann,R., Minor,P.D., Schild,G.C., Clarke,L.D., Mountford,R.C. and Almond,J.W. (1983) *Nucleic Acids Res.,* **11**, 1267.
5. Crowell,R.L. and Philipson,L. (1971) *J. Virol.,* **8**, 509.

6. Nottay,B.K., Kew,O.M., Hatch,M.H., Heyward,J.T. and Obijeski,J.F. (1981) *Virology,* **108**, 405.
7. Minor,P.D. (1982) *J. Gen. Virol.,* **59**, 307.
8. Semler,B.L., Anderson,C.W., Kitamura,N., Rothberg,P.G., Wictor,W.L. and Wimmer,E. (1981) *Proc. Natl. Acad. Sci. USA,* **78**, 3464.
9. Ferguson,M. Qi Yi-Hua, Minor,P.D., Magrath,D.I., Spitz,M. and Schild,G.C. (1982) *Lancet,* **II**, 122.
10. Provost,P.J. and Hilleman,P.R. (1979) *Proc. Soc. Exp. Biol. Med.,* **60**, 213.
11. McClaren,L.C. and Holland,J.J. (1974) *Virology,* **60**, 579.
12. Zagac,I. and Crowell,R.L. (1965) *J. Bacteriol.,* **89**, 574.
13. Erickson,J.W., Frankenbergen,E.A., Rossmann,M.G., Font,G.S., Medappa,K.C. and Rueckert,R.R. (1983) *Proc. Natl. Acad. Sci. USA,* **80**, 931.
14. Denoya,C.D., Scodeller,E.A., Vasquez,C. and La Torre,J.L. (1978) *Arch. Virol.,* **57**, 153.
15. Domok,I. and Magrath,D.I. (1979) *WHO Offset Publication,* No. 46.
16. Mathieson,L.R., Freestone,S.M., Wong,D.C., Skinboej,P. and Purcell,R.H. (1978) *J. Clin. Microbiol.,* **1**, 184.
17. Hagenaars,A.M., van Delft,R.W., Nagel,J., van Steenis,G. and van Wezel,A.L. (1983) *Intervirology,* **6**, 233.
18. Doel,T.R. and Collen,T. (1983) *J. Biol. Stand.,* **10**, 69.
19. Medappa,K.C. and Rueckert,R.R. (1974) *Am. Soc. Microbiol. Abstr.,* 207.

CHAPTER 3

Growth, Titration and Purification of Alphaviruses and Flaviviruses

E.A. GOULD and J.C.S. CLEGG

1. INTRODUCTION

The viruses to be described herein fall into two different viral genera, Alphaviruses (family Togaviridae) and Flaviviruses (family Flaviviridae), formerly known as the group A and B viruses, respectively. Viruses within these genera are antigenically related, have similar replication pathways and morphologic properties, but are antigenically distinct from each other (1). They are all loosely referred to as arboviruses (arthropod-borne viruses) in view of their ability to grow in arthropods as well as mammalian hosts. Other genera of these families do not come within the scope of this chapter.

For many years the most common choice of host for isolating wild strains of arboviruses has been the newborn mouse, and in general this is still true, although nowadays mosquitoes, mosquito larvae and cell cultures derived from invertebrates are becoming more widely used. This is particularly true for some of the most fastidious arboviruses such as the dengue viruses.

The methods employed for titration of virus infectivity depend partly upon the facilities available, partly upon the virus and partly upon the type of titration needed. For example, in some countries the newborn mouse provides a cheaper and more convenient laboratory titration system than mosquito cell lines, whilst in others exactly the opposite situation exists. Furthermore, where maximum sensitivity is required, mosquito cell lines can yield the highest infectivities. However, it has not yet been established that all arboviruses grow in these cell lines.

Where possible, several methods will be described to cover each of the possibilities that may be encountered and an indication of the particular application will be provided, together with the advantages and disadvantages.

Preparation of purified virus particles may be required for many different purposes, either for study of the composition or architecture of the viruses themselves, or as the starting material for preparation of individual structural components. Depending on the aims of the experiments, preparations of virus may be labelled by growth in the presence of radioactive isotopes, e.g., [^{3}H]uridine or [^{32}P]orthophosphate to label the RNA, [^{35}S]methionine to label proteins or [^{3}H]-glucosamine or mannose to label glycoproteins. While purification of labelled virus generally involves only small-scale culture and correspondingly small volumes of liquid, unlabelled virus required in large quantities may involve the processing of several litres of cell culture medium, and extremely high concentrations of infectious virus may be obtained in the later stages of purification.

It is important to note that many of these viruses must be considered dangerous. Therefore appropriate measures should be taken to ensure that a physical barrier always exists between the operator and the virus. In most countries very strict regulations apply to ensure that only some strains can be used for general laboratory purposes such as teaching and fundamental research. The names of the viruses, their known invertebrate host, the relative level of safety with which they can be used and the most common laboratory hosts or cell lines in which they can be grown are presented in the Catalogue of Arthropod-Borne Viruses (2).

2. GROWTH OF ARBOVIRUSES

2.1 Growth in Newborn Mice

Intra-cerebral inoculation of newborn mice with togavirus-infected material can yield high titres of virus. Many of the viruses kill the mice within 2–8 days but some specimens of field material may require one or two 'blind passes' (see below) before they produce high yields. Collect brains from infected mice either a few hours before, or close to, the time of death. Usually during the last 24 h the newborn mice become disarranged in the nest, unable to balance properly, the hind legs often becoming paralysed, and within a few hours they die. Observe the mice daily and record those that die within the first 24 h as non-specific deaths.

If there are no obvious symptoms after the first 6 days, process one half of the litter and re-inoculate into more newborn mice. This is known as blind passage. If no symptoms are observed after two serial blind passes, record the sample as negative. Field material, unlike laboratory-adapted virus, may not produce a uniform pattern of illness in the entire litter; it is therefore a good policy to collect, process and blind pass, individually, any sick mice that are seen from 2 days post-inoculation.

2.1.1 *Preparation of Material for Inoculation into Newborn Mice*

Place pregnant mice, at or close to full term, in separate clean cages and check daily. Record the date of birth and on either day 2 or 3 post-partum, reduce to a maximum of 10 per mother. Wear disposable gloves whilst handling the newborn mice to reduce the risk of the mother killing them. If several litters of the same age are available, pool them and distribute 8–10 mice per cage. As a general rule do not use mice older than 4 days for isolating and growing the viruses unless this is a specific requirement of the experiment. For isolation of virus from serum samples, inoculate the serum without dilution. Prepare tissue and brain material as a 10% suspension in diluent containing 25% heat-inactivated foetal calf serum (FCS) (see Appendix). Isolation of virus from mosquitoes or mosquito larvae is carried out after a suspension of the mosquitoes or larvae has been prepared by grinding them in a mortar. Add 2 ml of diluent containing 25% FCS. After grinding, collect the suspension into tubes. Centrifuge at 1500 *g* for 15 min at 4°C and use the supernatant as the inoculum. Store for long periods at −70°C. If samples are known to contain high virus titres dilute them in serial 10-fold steps.

The highest dilution that produces infection of the whole litter will usually yield the highest titre of virus. Perform all mouse inoculations in a total exhaust safety cabinet that complies with the regulations of the country.

The following additional equipment is required: disposable gloves, 1 ml sterile disposable syringes, 26-gauge sterile disposable needles, paper towels, cork inoculation board and methylated spirits.

2.1.2 *Method of Inoculation*

(i) Place a paper towel on the cork board inside the safety cabinet.

(ii) Wearing disposable gloves and with hands well inside the safety cabinet, load syringe with the appropriate inoculum.

(iii) Place a newborn mouse on the paper towel.

(iv) Using *Figure 1* as a guide, grip the mouse firmly but carefully and inoculate 0.01 ml of sample into one of the cerebral hemispheres. Ensure that the needle only just penetrates the brain and leave it in the brain for 2 sec before removing it slowly. Repeat the procedure until the complete litter has been inoculated. Place your arms on the outside lip of the cabinet to steady them whilst carrying out the inoculations.

(v) For virus isolations from field material use at least two litters. When performing virus titrations, commence with the highest dilution so that the same needle and syringe can be used with each virus.

(vi) Return each litter to its mother. Ensure that the cage is properly labelled and return the cage to the rack.

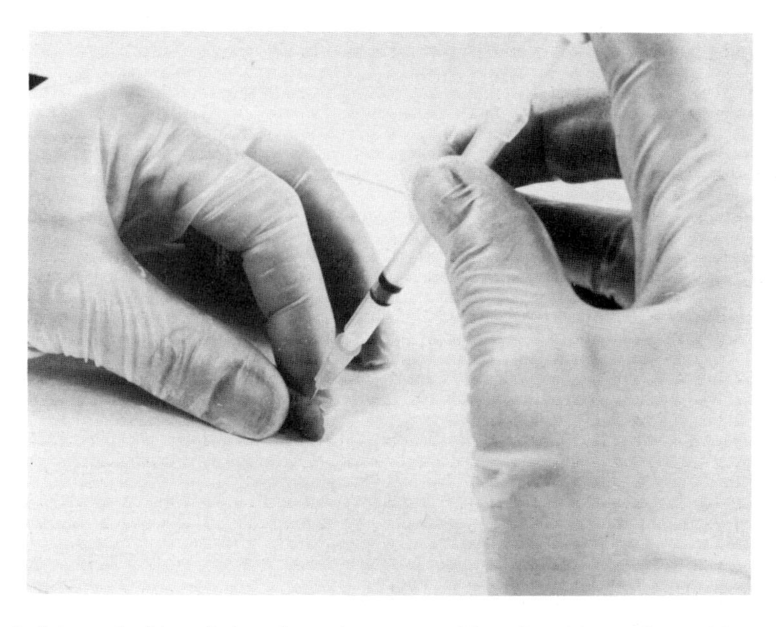

Figure 1. Intracerebral inoculation of a newborn mouse. The point of the needle should enter the left or right cerebral hemisphere and penetrate to a depth of no more than 1.5 mm.

2.1.3 *Method of Inspection and Virus Collection*

(i) Inspect the mice daily, preferably early in the morning and record the conditon of each one according to the following procedure. D, dead; S, sick; N, normal; M, missing. Remove the dead mice with forceps, using separate forceps for each cage.

(ii) Mice showing severe symptoms, or those that have died within 1 or 2 h of observation, should contain large amounts of virus in the brain. Kill the mice by immersion in a closed vessel containing either chloroform or ether fumes.

(iii) Cover the cork board with two paper towels, one on top of the other.

(iv) Pin the dead mouse to a cork board by placing one pin through the nose and the other through the base of the tail as shown in *Figure 2*.

(v) Soak the mouse with plenty of methylated spirits, then remove the scalp using one pair of sterile scissors and forceps. Then, with a separate pair of sterile instruments for each mouse, remove the skull cap as follows. Cut across the back of the scalp, using scissors that have a pointed and a blunt blade. Insert the pointed blade of the scissors into the soft rear centre point of the skull and cut down the outside at the centre line towards the nose (*Figure 3a*). Reverse the position of the scissor blades by turning the hand over so that the other side of the skull can be cut. The skull cap can now be elevated (*Figure 3b*) exposing the brain which can be lifted out on the end of the closed scissors.

(vi) Immediately place the brain in a sterile 25 ml glass bottle containing glass beads (3 – 5 mm diameter). Up to 15 brains may be placed in one bottle.

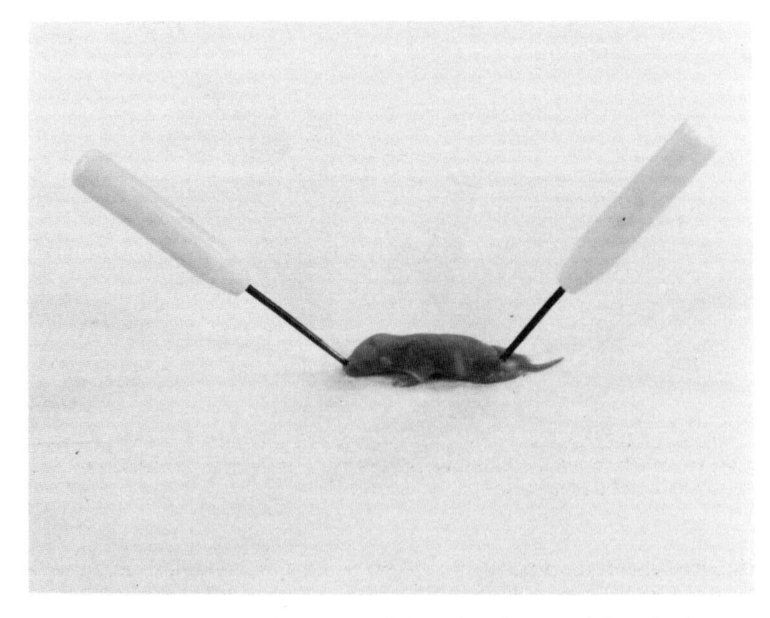

Figure 2. Newborn mouse, killed using ether and then pinned to a cork board prior to removal of brain.

a

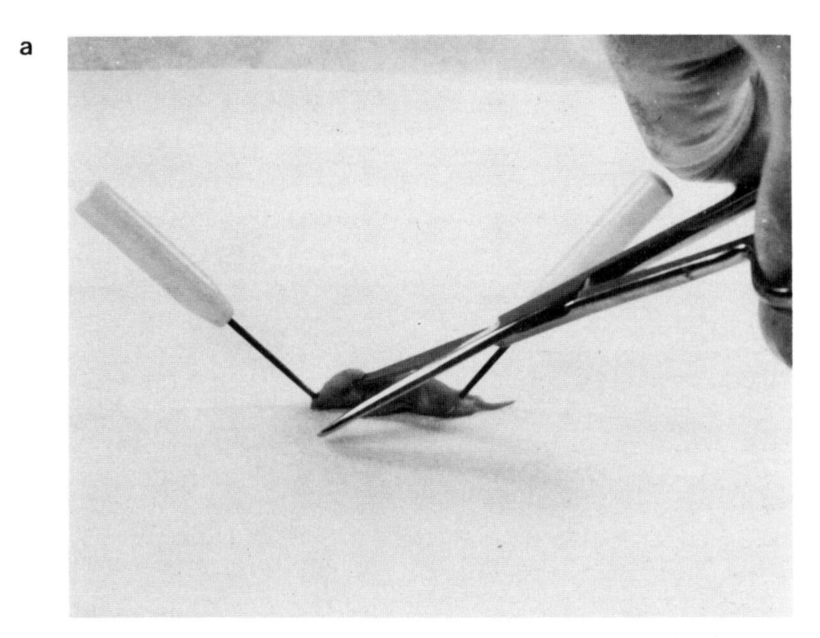

b

Figure 3. Method of dissecting skull to allow its elevation. The pointed blade of the scissors is used to penetrate the rear of the skull and then the cutting action is down one side before reversing the position of the scissors and cutting down the other side.

(vii) Ensure that the top of the bottle is screwed down tightly and mix vigorously either by shaking the bottle or placing it on a vortex mixer for 30 sec.

(viii) With the bottle in the safety cabinet and the operator wearing disposable

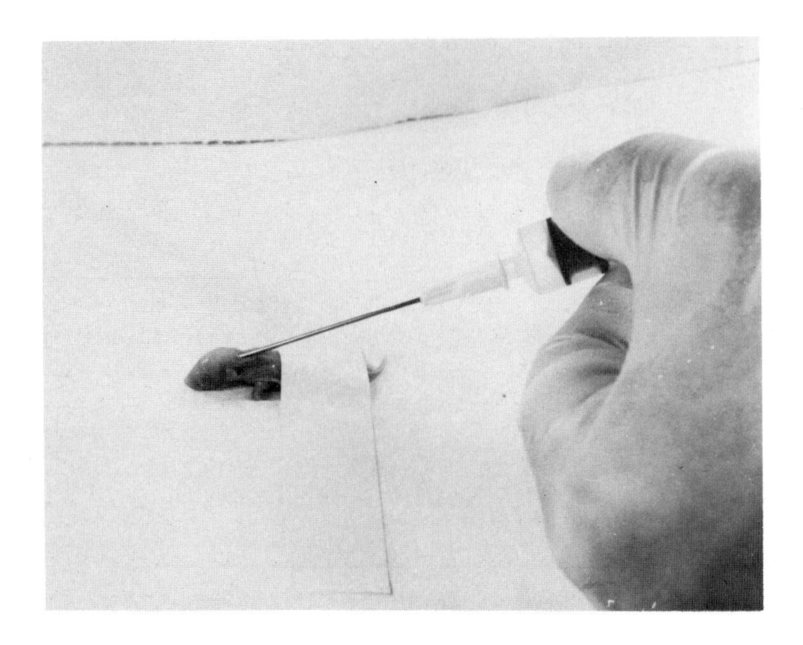

Figure 4. Method of removing brains from newborn mice after they have been frozen at −70°C.

gloves, add diluent at the rate of 1 ml per newborn mouse brain, or 1.5 ml per weanling mouse brain.

(ix) Replace the top on the bottle and screw it down tightly, then mix vigorously as before for 30 sec.

(x) Centrifuge the suspension at 1500 *g* for 15 min at 4°C.

(xi) Working in a safety cabinet, collect the semi-clear supernatant medium and dispense for storage at −70°C. This is a 20% suspension of suckling mouse brain virus antigen.

Step (v) can be modified provided that newborn mice were inoculated and that they were not left to develop beyond 5 or 6 days old; when the mice become sick freeze them at −70°C.

(i) Place the frozen mouse in the safety cabinet until completely thawed.

(ii) Pin or tape the mouse down as previously shown then, using a 1 ml disposable syringe with an 18-gauge needle, pierce the back of the skull (*Figure 4*) and withdraw the semi-solid brain material into the syringe.

(iii) Without emptying this syringe and keeping it in the safety cabinet, use it to collect 0.5 ml of diluent (2% FCS) from a tube containing 2 ml of diluent. Gently eject the mixture of brain and diluent back into the same tube making sure that the tip of the needle remains below the surface of the fluid. In this procedure it is important not to create aerosols with the syringe and needle.

(iv) Centrifuge the suspension as before. The clarified supernatant corresponds approximately to a 10% suckling mouse brain antigen preparation.

After either procedure for preparing mouse brain antigen, incinerate all mice, paper towels, gloves and disposable instruments. Sterilise the cages and instruments and thoroughly clean the safety cabinet with disinfectant.

2.2 Isolation and Growth using Mosquito Cell Lines

Invertebrate cell cultures have been found in many cases to be more sensitive to alpha and flaviviruses, particularly for primary virus isolation, than either the newborn mouse or vertebrate cell cultures. Assuming that standard tissue culture facilities are available, invertebrate cell cultures are easily maintained and can be recommended for routine work with togaviruses. Some mosquito cell lines show high sensitivity to mosquito-borne alpha and flaviviruses (3, 4) and many of these viruses have been shown to produce plaques although the results seem, to some extent, to depend upon the exact conditions chosen (5). If one set of experimental conditions proves unsuitable it is worth attempting one of the others. For example, mosquito cell lines are usually incubated at 27 – 28°C for routine subculture and virus isolation, but when infected with dengue virus it may be advantageous to increase the temperature to 32°C (6). Several cell lines are available and it is worth testing more than one before assuming that virus growth cannot be achieved. For details of the appropriate media see Appendix.

2.2.1 *Growth of Mosquito Cell Lines*

(i) Remove the supernatant medium from confluent monolayers of cells and add 5 – 10 ml of fresh growth medium to each flask.

(ii) Forcibly remove the cells from the surface of the flask either by blowing the medium directly at the cell surface or, using a sterile silicone rubber policeman, rub the cells off the surface.

(iii) Ensure that a uniform suspension has been produced then count the cells using a haemocytometer. Count the four large corner squares (each consisting of 16 small squares). Calculate the average number of cells in one large corner square and multiply by 1×10^4. This gives the number of cells per ml. Normally, the counts are most accurate if there are between 100 and 200 cells in one large corner square, therefore make an appropriate dilution of cells.

(iv) After counting the cells use growth medium to adjust the cell concentration to 2×10^5 cells per ml and dispense into sterile culture vessels. The volume required differs for each type of vessel; aim to produce a depth of 2 – 4 mm.

(v) Incubate the cells at 28°C. If the medium is buffered with sodium bicarbonate, incubate the cells in a humidified 5% CO_2 incubator with the tops of the tubes or flasks loosened.

(vi) Confluent monolayers will normally develop within 3 – 6 days.

2.2.2 *Growth of Arboviruses in Mosquito Cell Lines*

(i) Remove the growth medium from confluent monolayers of cells and inoculate a small volume of the virus directly onto the monolayer. Ensure

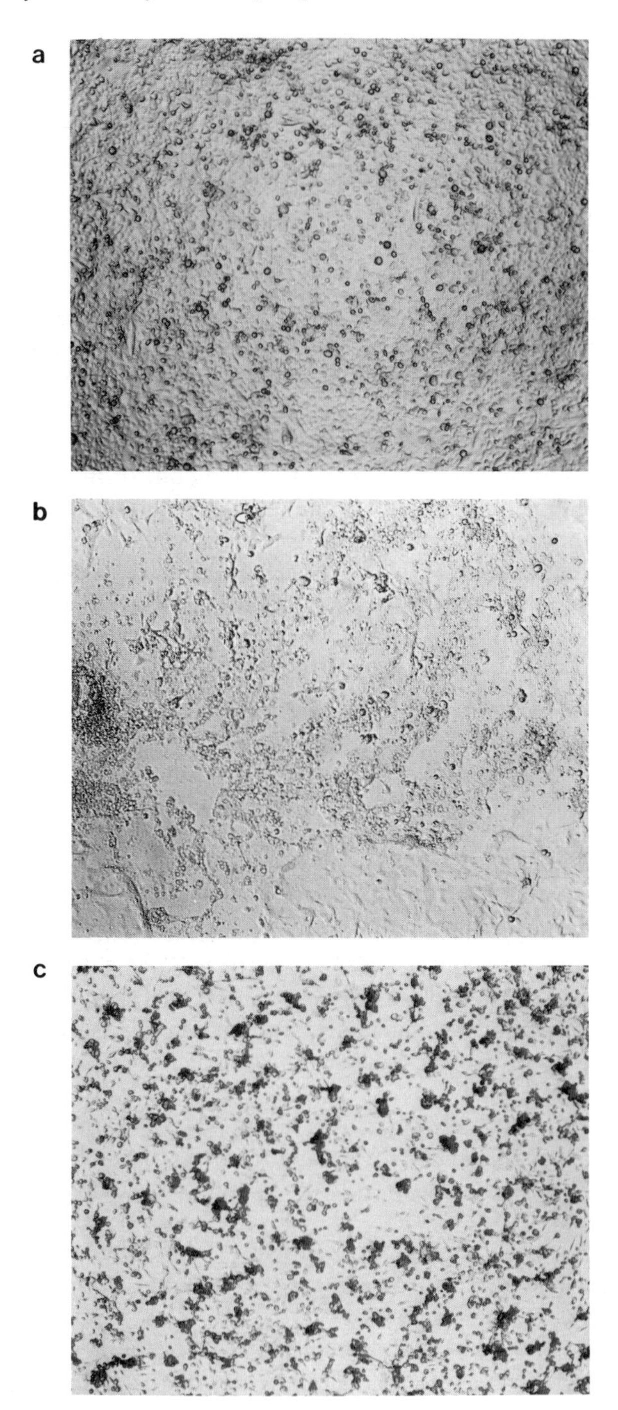

Figure 5. Typical types of mosquito (AP61) cell monolayers after infection with arboviruses. (**a**) healthy monolayer, (**b**) large syncytium produced as the result of fusion between individual cell membranes, (**c**) cells have rounded up and may have become detached from the glass surface.

that the inoculum completely covers the cells and place at room temperature for 1 h, checking every 10 min that the entire monolayer is completely covered. Field material, usually in diluent containing 25% FCS and antibiotics, should not be diluted. However, if the virus has already been adapted to growth either in mice or tissue culture cells dilute it at least 1:100 in the appropriate maintenance medium.

(ii) After adsorption, add more maintenance medium to produce a depth of approximately 2–4 mm and incubate at 28°C, using a CO_2 incubator if necessary. If there is the possibility that the virus may grow better at higher temperatures prepare duplicate cultures and incubate accordingly.

(iii) Inspect monolayers daily using a low-power inverted microscope.

(iv) At the appropriate time, collect the supernatant and centrifuge at 1500 *g* for 15 min at 4°C.

(v) Store the clarified medium at −70°C to retain infectivity.

Control non-infected cultures should remain healthy for up to 14 days. Many of the alpha and flaviviruses do not produce a cytopathic effect (cpe) in mosquito cells at 28°C. In these cases the virus has to be identified by other techniques such as immunofluorescence, haemagglutination or ability to kill newborn mice. The type of cpe that might be expected depends upon the virus and the passage history of the cells. However, the two most common types are either the production of syncytia, or rounding up and detachment of the cells from the surface (*Figures 5a, b* and *c*). Infectious virus is produced into the supernatant medium. With alphaviruses, relatively high titres of virus can usually be obtained from the second day onwards. With flaviviruses, however, it may take 5 days or more before there is a significant yield. If the supernatant medium is collected and replaced daily then virus will continue to be produced. This is particularly useful when large quantities are required for concentration and analysis.

2.2.3 *The Indirect Immunofluorescence Test to Identify Virus when no cpe is Produced*

This test can only be performed if the following are available: a specific antiserum to the virus, a fluorescein isothiocyanate (FITC)-conjugated antiserum specific for the virus antiserum and a fluorescence microscope. The FITC antiserum can be obtained commercially.

(i) Remove some cells from a small area of the infected culture vessel.

(ii) Sediment them by centrifugation at 1500 *g* for 10 min.

(iii) Re-suspend the cells in approximately 0.5 ml of PBS (see Appendix).

(iv) Prepare a clean microscope slide by scoring several small circles using a diamond pen.

(v) Place drops of the cell suspension into the small circles on the slides.

(vi) Allow the spots to dry at room temperature keeping them inside the safety cabinet. Then immerse them in cold (4°C) acetone for 10 min. The slides can either be used immediately or stored at −70°C.

(vii) Add an appropriate dilution of virus-specific antiserum to the spots on the slide and incubate at 37°C for 40 min in a humidified container.

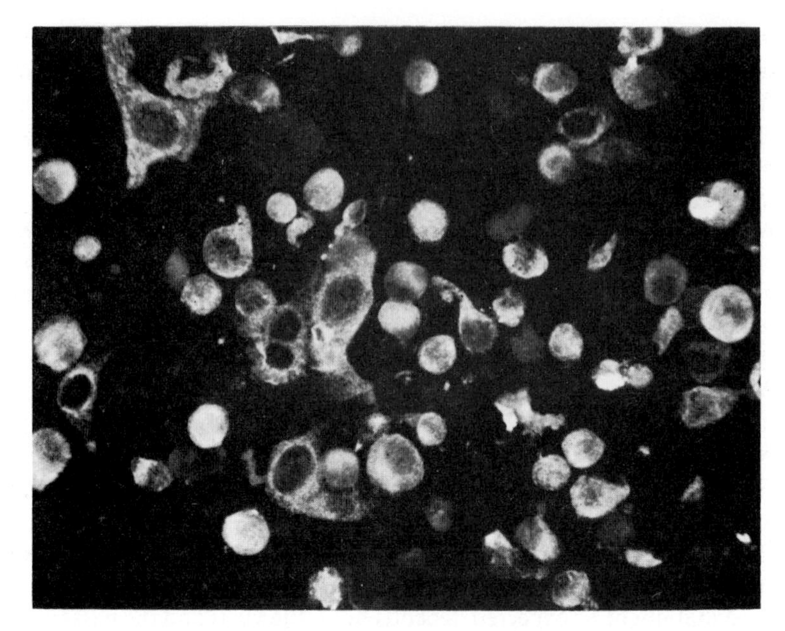

Figure 6. Immunofluorescence of dengue virus infected cells after testing by the 'spot-test' method. Using this method the cells do not flatten out as in a normal monolayer. The infected cells are those that produce bright fluorescence (white in photograph).

(viii) Place the slide in PBS for 5 min. Replace the PBS after 5 and 10 min.

(ix) Blot dry and fill each spot with the fluorescein-conjugated antiserum. Incubate and wash and dry as above.

(x) Add glycerol/saline (90:10) mountant and cover with glass coverslips.

(xi) Examine the slide using a fluorescence microscope with a u.v. light source and the appropriate filters for fluorescein.

(xii) The infected cells will appear fluorescent green and the non-infected cells will be either dark or invisible. This can be checked with control non-infected cells. If there is a generalised fluorescence, dilute the antiserum further until only virus-specific fluorescence remains and that of the cellular background disappears.

(xiii) The typical appearance of spot slides, tested in this way and photographed in black and white, is shown in *Figure 6*.

2.3 Growth in Mammalian Cell Lines

Several mammalian cell lines are susceptible to arboviruses, particularly after the viruses have been isolated and their titres enhanced under laboratory conditions. Since the procedures for culturing these cell lines and growing them are similar, only one method will be presented. In most circumstances any of the commonly used lines of cells (e.g., Vero, LLCMK2, BHK-21 or PK15) can be used. There are many media formulations (see Appendix) and it is usually wisest to use that in which the cells have been routinely subcultured by the supplier. However, Leibovitz-L15 medium has the advantage of not requiring a CO_2 incubator or

Hepes buffer and when supplemented with 10% FCS supports the growth of all the above cell lines. It is also worth noting that Vero cells are very stable and can be held at 37°C for 10–14 days without needing attention.

2.3.1 *Sub-culturing Mammalian Cells*

(i) Remove the growth medium from a confluent monolayer of cells. Add to the monolayer 5 ml of warm (37°C) trypsin-versene (see Appendix), rinse rapidly and repeat twice more.
(ii) Place the flask in the 37°C incubator for 1–2 min, then tap it firmly against the palm of the hand and look for evidence of the cell monolayer slipping away from the surface. As soon as this happens re-suspend the cells in a convenient volume (5–10 ml) of growth medium.
(iii) Count the cells in a haemocytometer, adjust the concentration to 2 x 10^5 cells/ml with growth medium.
(iv) Dispense the cell suspension into culture vessels using a volume that produces a depth of 2–4 mm of medium.
(v) If necessary incubate at 37°C, in a humidified CO_2 incubator. The cells will produce a confluent monolayer within 2–4 days.

2.3.2 *Growth of Arboviruses in Mammalian Cells*

In most routine laboratory situations the infectivity of the virus will already be known and will probably be in the range of 10^6 p.f.u./ml or higher. Unless there is a specific requirement to add very large or very small amounts of virus it is best to add sufficient to infect between 1% and 10% of the cells in the monolayer. If the virus titre is not known, then two or three different input concentrations should be used in the first instance. Overloading the cells with virus will not increase the yield.

(i) To infect cell monolayers, remove the supernatant growth medium from confluent monolayers of cells.
(ii) Add a small volume of the appropriate dilution of virus (diluted in maintenance medium) to the monolayer so that there is sufficient just to cover the cells.
(iii) Leave at room temperature for 1 h, checking every 10 min that the inoculum has completely covered the cells.
(iv) If the cells are used for virus isolation from field samples then add maintenance medium after adsorption for 1 h. If, on the other hand, the virus to be grown is already adapted to cell cultures or newborn mice, remove the inoculum, gently wash the monolayer twice with 10–20 ml of maintenance medium and add sufficient maintenance medium to produce a depth of 2–4 mm.
(v) Collect the supernatant medium when cpe is visible. Centrifuge at 1500 *g* for 15 min. Store at –70°C or use immediately.

Most of the alphaviruses produce a marked cytopathic effect within 2 days. Initially the cells become elongated with very long processes (*Figure 7a* and *b*); they then round up and float free from the surface of the flask. The flaviviruses

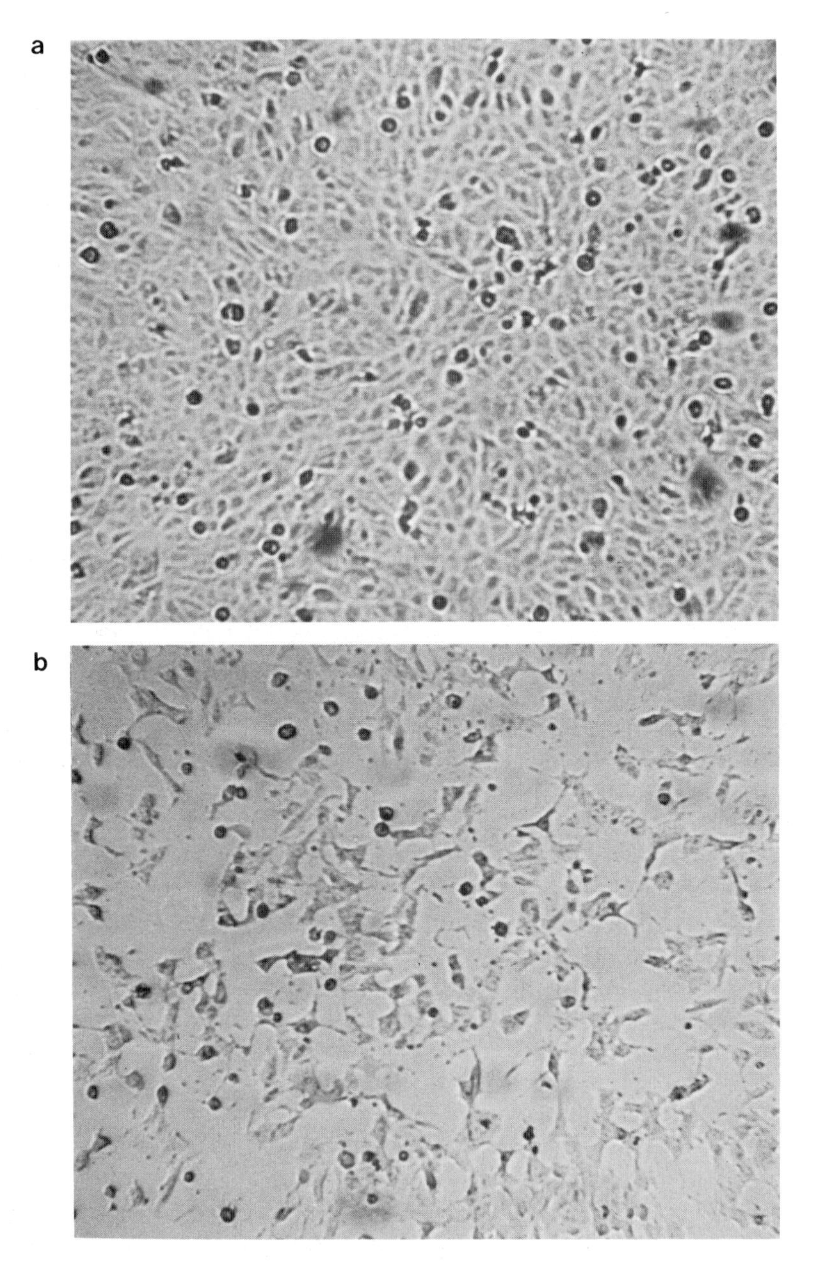

Figure 7. Typical appearance of Vero cell monolayers either as healthy non-infected cells (**a**) or when they are in the relatively early stage of degeneration after infection with an alphavirus such as Sindbis or Semliki Forest virus (**b**). Note the long processes on the infected cells; each of these processes will have many budding virus particles.

tend to produce cpe more slowly, starting from the third or fourth day post-inoculation. In many cases the monolayer remains intact for several days although rounded cells develop and float free. If the cells of the monolayer are subcultured in growth medium they usually produce new monolayers and continue to produce infectious virus. Very large quantities of virus can be obtained if the maintenance medium is collected and replaced daily whilst the cell monolayer remains intact.

As an alternative procedure, it is possible to infect cells whilst they are in suspension and then to seed them into culture vessels. This procedure has several advantages, for example, when infected cells are required for immunofluorescence and when large numbers of samples are to be handled at one time.

(i) Remove the cells from the culture flask using trypsin-versene as described previously.
(ii) Prepare a suspension of cells, count and adjust to 2 x 10^7 per ml using growth medium.
(iii) Add the virus to aliquots of the cells at a dilution that will infect 1 – 10% of the cells. Mix well for 1 h at room temperature by rotating the bottle on an appropriate mixer.
(iv) If the infected cells are required for immunofluorescence, dilute an aliquot to 2 x 10^5 per ml in growth medium and seed into Petri dishes containing glass coverslips.
(v) Incubate at 37°C until 30 – 50% of the cells contain virus-specific antigen when tested by immunofluorescence.

The same principle can be applied to produce infected cells on Teflon-coated multispot slides or Terasaki plates. In addition, persistently infected cell cultures can also be seeded as above. In this case incubate the plates or slides in a humidified incubator at 37°C for 4 – 6 h and then wash and fix in PBS and in acetone in the standard way.

2.4 Growth in Mosquitoes and Mosquito Larvae

These techniques require a suitable insectary and well trained personnel. Considerable dexterity is also needed. Dengue viruses, in particular, but also some of the other flaviviruses cannot be easily isolated and identified from field material. The mosquito and mosquito larvae inoculation methods have been found to be sensitive and in the latter case relatively rapid (7). The procedures are described briefly but it is advisable to use them only if facilities are adequate and some time can be spent in perfecting the methods.

2.4.1 *Intra-thoracic Inoculation of Mosquitoes*

The equipment required for mosquito inoculation is described in detail by Rosen and Gubler (8). Place the mosquitoes in a cooled box or tube on ice. This will make them inactive without killing them. Ensure that the syringe to be used is filled with air (using the 3-way tap to isolate the syringe from the capillary tube). Place an inactive mosquito on the stage of the microscope and, looking through the eyepieces, place a finely drawn pipette tip just through the surface of the thorax. Switch the inoculation tap so that when the plunger of the syringe is depressed it causes the inoculum to be expelled from the needle. Inject approximately 0.2 μl of virus sample into the thorax, stopping the injection by switching

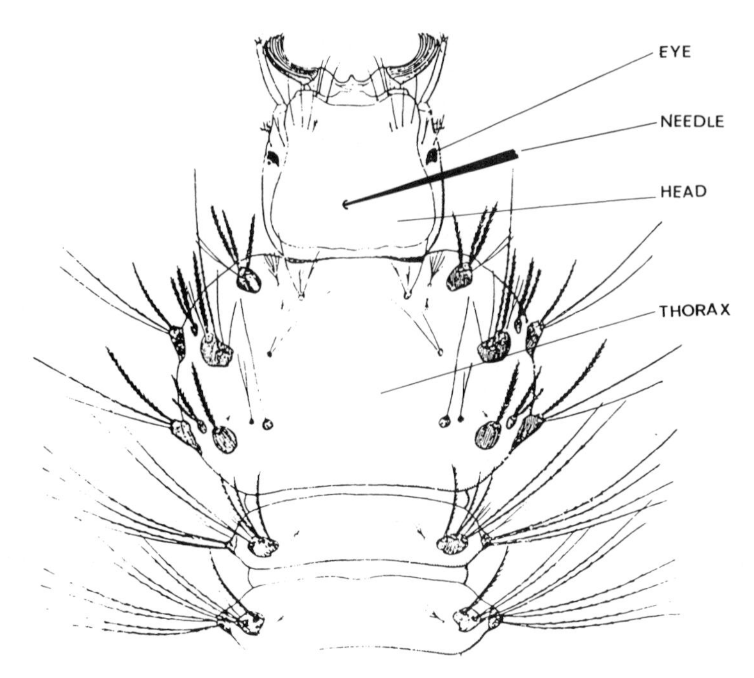

Figure 8. Intra-cerebral inoculation of larvae. The diagram shows the point at which the needle should enter the larva.

the tap to the open position. Remove the tip of the pipette and place the mosquito into a holding chamber. The entire operation must be carried out as quickly as possible to reduce the risk of the mosquito escaping. When the inoculations have been carried out return the mosquitoes to the appropriate incubators.

2.4.2 *Intra-cerebral Inoculation of Mosquito Larvae*

The equipment for inoculation is similar to that described for mosquito inoculations. Place a piece of moist gauze on the microscope stage. Place a larva on the gauze and hold it gently with forceps. Insert the needle, to a depth of approximately 0.5 mm, in the dorsal epitome of the head capsule in the area behind the eyes and between the frontal ecdysial lines (*Figure 8*). Inject approximately $0.1-0.2\ \mu l$ then remove the needle. Return the inoculated larva to the appropriate incubator. At daily intervals after inoculation prepare smears of either mosquitoes or larvae and test for the presence of viral antigen by immunofluorescence as described below.

2.4.3 *Testing for the Presence of Viral Antigen by Immunofluorescence*

(i) Prepare microscope slides by scoring four circles on each slide, making each circle approximately 1 cm in diameter and ensuring that the circles are well separated.

(ii) Place a mosquito on one circle and, using a scalpel, cut the head away from the thorax and the thorax from the abdomen.

(iii) Remove the thorax but leave the head in the upper half of the circle and the abdomen in the lower half.
(iv) Repeat the procedure for each mosquito.
(v) Place a microscope slide on top of the slide containing the four mosquitoes and press them together firmly.
(vi) Remove the top slide by sliding it over the surface of the lower one for approximately 3 mm, then lift it away. This produces duplicate smears of each head and abdomen.
(vii) Use fine forceps to remove any obvious scaly parts of the mosquitoes, such as wings, etc. Allow the slides to dry at room temperature then place them in cold acetone (4°C) for 10 min.
(viii) Either use the slides immediately for immunofluorescence tests or store at −70°C.

The procedure for mosquito larvae is almost identical, the head being separated and squeezed onto duplicate slides and fixed in acetone. Carry out immunofluorescence tests as previously described. A more rapid fluorescence test can be performed if a fluorescein-conjugated antiserum that is specific for the virus is available.

It is worth pointing out that considerable patience and experience are required in interpreting the results, therefore negative controls must be prepared with every test run.

3. TITRATION OF ARBOVIRUSES

The methods for titrating arboviruses depend upon the objectives to be achieved. Traditionally, newborn mice were most commonly used; however for many purposes they have been superseded by cell cultures. In some circumstances mice are still required, for example, when estimating protective antibody levels.

3.1 Titration by Intra-cerebral Inoculation of Newborn Mice

3.1.1 *Titration of Virus*

(i) Pool newborn mice (1−3 days old) and place with the mothers in litters of not more than 10. Label the cages as appropriate.
(ii) Make serial 10-fold dilutions of the virus to be titrated using maintenance medium.
(iii) Place a cork inoculation board, covered with disposable paper towels, in the safety cabinet.
(iv) Wearing disposable gloves place the first litter, without the mother, onto the covered inoculation board inside the safety cabinet. Inoculate, intracerebrally, each mouse with 0.02 ml of the highest virus dilutions using a 26-gauge needle and a 1 ml disposable syringe. Immediately return the inoculated mouse to its mother in the cage. The procedure for intracerebral inoculation has already been demonstrated (*Figure 1*) and described in Section 2.1.2 on virus growth. The same syringe and needle can be used for all dilutions of each virus sample provided that the operator always commences at the highest dilution.

Table 1. Method for Calculating the Virus Infectivity Titre.

Virus dilution (10-fold log_{10})	*Dead*	*Live*	*Cumulated scores*			
			Total[a] dead	*Total[a] live*	*Mortality*	
					Ratio	*%*
−2	10	0	28	0	28/28	100
−3	10	0	18	0	18/18	100
−4	6	4	8	4	8/12	66.66
−5	2	8	2	12	2/14	14.29
−6	0	10	0	22	0/22	0

[a]Cumulated scores for the number of animals that died or survived are obtained from the previous columns by adding upwards and downwards, respectively.

(v) If more than one virus is to be titrated, change the gloves, syringe, needle and inoculation board. Swab the cabinet carefully with disinfectant before commencing with the next virus.

(vi) Incinerate or sterilise all equipment at the end of each experiment.

(vii) Observe the mice daily and record the number of deaths. Most titrations can be regarded as completed within two weeks.

(viii) The dilution of virus that kills 50% of the mice is its titre. The result is expressed as lethal doses (LD) at the 50% end-point (LD_{50}). An example of the method (9) for calculating the virus infectivity titre is given in *Table 1*.

The 50% end-point is between dilutions 1×10^{-4} and 1×10^{-5}. It is calculated as follows:

$$\frac{\text{\% mice dead at dilution next above 50\%} - 50}{\text{\% mice dead at dilution next above 50\%} - \text{\% mice dead at dilution next below 50\%}}$$

or in the example:

$$\frac{66.66 - 50}{66.66 - 14.2} = \frac{16.66}{52.46} = 0.32$$

The LD_{50} is therefore $1 \times 10^{4.32}$ LD_{50}

If the volume inoculated was 0.02 ml per mouse

Then the LD_{50} per ml $= 1 \times 10^{4.32} \times 50$ LD_{50}

$= 1 \times 10^{6.02}$ LD_{50} per ml

3.1.2 *Titration of Virus-specific Antibody Titre by Mouse Protection Experiments*

The ability of an antiserum to protect newborn mice against lethal infection with an arbovirus can be assessed using a modification of the method described above.

(i) From the titre that was obtained in the above titration calculate the dilution of virus necessary to produce an inoculum of 100 LD_{50} per 0.02 ml. For example, if the virus had a titre of 5×10^6 LD_{50} per ml then a dilution of 1 in 1000 would be required.

(ii) Make the appropriate dilution using maintenance medium.
(iii) Make serial 10-fold dilutions of the antiserum to be tested.
(iv) Mix 0.1 ml aliquots of the diluted inoculum with 0.1 ml of each dilution of the antiserum to be tested.
(v) Incubate the mixtures at 37°C for 60 min. Include a control mixture consisting of virus mixed with pre-immune antiserum at the highest antibody concentration.
(vi) Inoculate the mixtures intracerebrally as 0.02 ml doses into groups of pooled newborn mice using the procedures described earlier. Start at the lowest dilution of antibody and change the syringe and needle for each mixture.
(vii) At daily intervals examine and record deaths in the litters. All controls should die. The antibody titre is the dilution of antibody that protects 50% of the mice. This 50% protective dose (PD_{50}) can be calculated using the method described previously for LD_{50} (9).

3.2 Titration using Cell Cultures

Although methods are described for Vero cells, they can also be used with BHK-21, LLCMK2, PK15 and several other mammalian cell lines. The first method is ideal for assaying large numbers of samples and can easily be adapted from microtitre plates, as described here, to plates with larger cups or dishes if, for example, plaque morphology is under investigation. The titration system utilises carboxymethylcellulose (CMC). Subsequently, a different procedure is described in which agarose is used and the cells are stained with neutral red. This is particularly useful for studying plaque populations when it is important to maintain virus infectivity.

3.2.1 *Plaque Assay using Carboxymethylcellulose Overlay*

(i) Prepare eight serial 10-fold dilutions of virus in maintenance medium.
(ii) Starting at the eighth virus dilution and using a Pasteur pipette, place one drop of virus suspension into a cup of the eighth row of a flat-based microtitre plate.
(iii) Discard any remaining virus suspension from the pipette and repeat the procedure with the seventh dilution into the equivalent cup of the seventh row.
(iv) Repeat this procedure with each dilution.
(v) Immediately after completion of the plate add one drop of Vero cells using a suspension (4 x 10^6/ml) prepared in maintenance medium.
(vi) Incubate the plates at 37°C for 3 h. Then to each cup add two drops of maintenance medium containing 1.5% CMC (see Appendix).
(vii) Incubate the plates in a sealed box at 37°C.
(viii) Most alphaviruses produce plaques within 2–3 days whereas the flaviviruses take longer. Examine the plates daily until the characteristics of the viruses are known.
(ix) When cpe is clearly visible at low-power magnification (using an inverted microscope) remove the supernatant medium from each cup of each plate.

(x) Add 2–3 drops of formal-saline, leave for 10 min and then replace this with naphthalene black stain (see Appendix). After 30 min at room temperature wash the plates with tap water.

(xi) Estimate the infectivity (titre; p.f.u.) of the virus samples by counting the number of plaques produced at each dilution. For example:

Number of plaques produced	9
Dilution of virus	1×10^7
Volume of inoculum	0.04 ml

$$\text{Virus titre} = 9 \times 1 \times 10^7 \times 1/0.04 \text{ p.f.u. per ml}$$
$$= 1.75 \times 10^9 \text{ p.f.u. per ml}$$

It must be emphasised that titrations should be carried out at least in triplicate to avoid problems arising from small errors made in the titration procedure. However, when large numbers of samples are being compared, the above technique as described will usually produce satisfactory results.

3.2.2 *Plaque Assay using Agarose Overlay*

(i) Prepare confluent monolayers of cells in either small plastic disposable 25 cm² flasks or dishes.

(ii) Prepare serial 10-fold dilutions of virus in maintenance medium.

(iii) Remove culture medium and add 0.2 ml of virus inoculum, starting from the highest dilution. Ensure that a film of medium completely covers the cell sheet.

(iv) Either place the flasks on a rocker or every few minutes rock the flasks by hand. Leave at room temperature for 1 h.

(v) Remove the inoculum, preferably with a pipette or an aspirator, then add 5 ml of agarose overlay medium (see Appendix).

(vi) Ensure that the overlay medium has spread evenly over the monolayer, leave at room temperature for 10 min then incubate at 37°C.

(vii) Different viruses will require different lengths of time before plaques develop. Determine the optimum incubation time by daily examination of the monolayers, starting from the second day for alphaviruses and the fourth day for flaviviruses.

(viii) When plaques begin to appear add 2 ml of agarose overlay medium containing 0.02% neutral red. Allow 10 min for the agarose to set and then incubate the flask at 37°C in the dark.

(ix) Within a few hours living cells will be stained by the neutral red and the plaques will appear clear. Do not expose the plates to light for more than a few seconds if it is intended to collect infectious virus from the plaques.

(x) For permanent staining, once the plaques have developed and have been stained with neutral red, add 2–4 ml of 10% formal-saline to the dish. After 30 min at room temperature, remove the formalin and agarose and wash the monolayers with tap water.

(xi) Estimate the virus titre as plaque forming units per ml (p.f.u./ml) as described above by counting the number of plaques at an appropriate dilution.

3.2.3 *Plaque Reduction Neutralisation Test for Arbovirus Antibodies*

As before, the method for Vero cells will be described, however many other mammalian and mosquito cell lines are suitable. The microtitre method described provides the titre of the antibody as either a neutralisation index or as a dilution end-point titre, the latter being more accurate. However, it also requires more time and equipment.

(i) Place the serum to be tested at 56°C for 30 min to inactivate some of the complement components and also to destroy non-specific inhibitors.
(ii) Prepare serial 10-fold dilutions of virus and antibody in maintenance medium. If the antibody titre is thought to be relatively low, make serial 2-fold dilutions of antibody instead of 10-fold dilutions.
(iii) Place one drop per cup of the highest dilution into row 'H' of the microtitre plate, then the next dilution into row 'G', repeating this procedure with successive dilutions to row 'A'.
(iv) Put one drop per cup of maintenance medium into rows 11 and 12 then add one drop per cup of the highest antibody dilution to row 10. Repeat with the next dilution in row 9 and continue through to row 1 with successive antibody dilutions.
(v) Incubate the plates at 37°C for 60 min then add 1 drop per cup of Vero cells from a suspension (4×10^6 cells/ml) prepared in maintenance medium.
(vi) Incubate at 37°C for 3 h then add two drops per cup of maintenance medium containing 1.5% CMC and incubate in a sealed box at 37°C or in a CO_2 incubator if the medium is bicarbonate-buffered.
(vii) When plaques have formed in the control non-antibody treated rows, remove the supernatant medium and replace with 10% formal-saline for 10 min followed by naphthalene black for 30 min at room temperature. Then wash the plates with tap water.

The antibody neutralisation titre can be calculated by either of two methods if this type of checkerboard titration has been performed.

The first method estimates the dilution of antibody that reduces the plaque numbers by 50%. For example, if a 1/1000 dilution reduced the number of plaques from 20 to 10 then the plaque reduction neutralisation titre would be 1000. This 50% reduction end-point can be determined by plotting antibody dilution against number of plaques.

In the second method, which is less precise, the difference between the virus titre at the highest antibody concentration and the non-antibody treated control indicates the serum neutralisation index (SNI). For example:

If the titre at 1/20 antibody dilution	=	1×10^2 p.f.u./ml
and the titre in the control	=	1×10^5 p.f.u./ml
The SNI (expressed as $\log_{10}$ units)	=	3.0

3.2.4 *Virus Infectivity Enhancement*

This is a specialised method that utilises the Fc-receptors on the surface of either mouse or human macrophage cell lines (10). Viruses that fail to adsorb to the cell receptors may gain access to the cytoplasm as the result of forming complexes with virus-specific antibody. The antigen-antibody complexes attach to the Fc-receptors and the infectivity of the virus pool is therefore enhanced. Utilisation of this principle is particularly valuable for work with monoclonal antibodies since surface reactive antibodies are identified. Three sets of data can be obtained using the enhancement technique, i.e., enhancement of virus infectivity by antibody, neutralisation titre of antibody and standard virus infectivity. Many togaviruses can replicate in the mouse macrophage cell line P388 D1 and these cells have Fc-receptors on their surface. They also adhere to the surface of tissue culture flasks and can therefore be treated in the same way as ordinary cell lines.

(i) Make serial 10-fold dilutions of virus in L15 maintenance medium.
(ii) Using a sterile Pasteur pipette dispense 1 drop of each virus dilution into 12 replicate cups of a microtitre plate starting at the highest dilution of virus.
(iii) Add 1 drop of L15 maintenance medium to rows 11 and 12 of the virus dilutions. These will be the control, non-antibody treated, rows.
(iv) Prepare serial 2-fold dilutions of the antiserum in maintenance medium.
(v) Add 1 drop of each antibody dilution to eight replicate cups of the virus dilutions to produce a checkerboard titration, i.e., row 10 will receive one drop per cup of the highest antibody dilution, row 9 receives one drop per cup of the next antibody dilution and so on.
(vi) Incubate the covered plate in a sealed lunch box at 37°C for 30 min.
(vii) Meanwhile prepare a suspension of P388 D1 cells by decanting the growth medium from confluent monolayers. Forcibly remove the cells from the surface of the flask either by blowing medium at them through a pipette or using a silicone policeman.
(viii) Count the cells and adjust the concentration to 1×10^6 cells/ml then dispense one drop into each cup of the microtitre plate.
(ix) Incubate the plates at 37°C in a sealed box for 3 h to enable them to settle out then add 2 drops of L15-CMC overlay medium (see Appendix) to each cup and incubate at 37°C. Most alphaviruses will produce cpe within 3 days whereas flaviviruses usually take between 4 and 7 days.
(x) Determine the optimum time of incubation by daily examination of the monolayers. When there are distinct cytopathic changes, remove the supernatant medium and replace it with 10% formal-saline (see Appendix).
(xi) After 10 min, replace this with napthalene black stain for 30 min at room temperature.
(xii) Wash the plates with tap water, allow to dry and count the plaques in each microtitre cup. The virus titre can then be calculated as follows:

Number of plaques per cup x dilution factor x 1/ input volume per cup.

If there were 15 plaques per cup at a dilution of 1×10^6 and if the input volume per cup was 0.04 ml,

$$\text{Titre of virus} = 15 \times 1 \times 10^6 \times 1/0.04 = 3.75 \times 10^8 \text{ p.f.u./ml}$$

(xiii) Calculate the virus enhancement of infectivity by estimating the titre of virus in each row. The highest titre obtained is compared with the control titre. The enhancement titre is the difference between these. For example:

Control titre = 6×10^6 p.f.u.

Maximum enhanced titre = 3.6×10^7 p.f.u.

Therefore the infectivity was enhanced 6-fold.

(xiv) A neutralisation index for the antibody can also be obtained from the same set of results. Compare the lowest titre of the antibody treated rows with the control, for example:

Control titre = 6×10^6 p.f.u.

Lowest antibody treated titre = 6×10^4 p.f.u.

Neutralisation index = 2.0

3.3 Titration using the Haemagglutination Test

All arboviruses agglutinate a broad range of species of erythrocytes although the exact experimental conditions required for each arbovirus differ slightly. The two most commonly used species are ganders and newborn chicks since these two types seem to be the most sensitive. Hameagglutination (HA) titrations with arboviruses that have not previously been characterised in the laboratory are usually performed at a range of pHs. Once the optimum pH has been deduced, further titrations can then be carried out at the appropriate optimum pH.

As is the case with all viruses that agglutinate erythrocytes, the HA test measures particle numbers, not infectious titre. Nevertheless for many purposes this is a good guide as to the quality of the virus preparation.

3.3.1 *Preparation and Concentration of Haemagglutinin from Cell Monolayers*

The method that will be described utilises Vero cell monolayers for the production of HA. In our hands this has been found to be satisfactory for a wide variety of alpha and flaviviruses provided that the serum level is kept very low. However, high yields can also be obtained from other mammalian and mosquito cell lines. The most common methods employed in the past used suckling mouse brain antigen which usually had to be treated with organic solvents before use. These latter procedures are all described very clearly by Clarke and Casals (11) and they will not be described here. The HA that is produced from cell monolayers does not require treatment with organic solvents. Yields with many of the viruses can be greater than 10 000 haemagglutinating units (HAU).

(i) Remove the supernatant medium from confluent monolayers of cells.
(ii) Add a small volume of virus, diluted in maintenance medium, to produce an input multiplicity of approximately 0.1 (10% of cells infected). Ensure that the inoculum covers the monolayer and leave at room temperature for 1 h.
(iii) Every 10 min agitate the flasks so that the inoculum is well distributed.
(iv) Add sufficient maintenance medium to cover the cells to a depth of 2–3 mm and incubate at 37°C (or 28°C if mosquito cells are used).

(v) Most alphaviruses will produce marked cpe within 2–3 days. Collect the supernatant medium at this stage.
(vi) In monolayers infected with flaviviruses remove the supernatant medium and replace with medium containing only 0.5% FCS as soon as cpe starts to appear. This will usually be between days 2 and 6 post-infection.
(vii) At daily intervals, collect all the supernant medium from these flasks and replenish it immediately with fresh medium containing 0.5% FCS.
(viii) Centrifuge the pooled supernatant medium at 2000 *g* for 15 min at 4°C. Collect the clarified fluids and add 23.36 g/l of NaCl and 60 g/l of polyethylene glycol 6000.
(ix) Stir until dissolved then leave at 4°C overnight.
(x) Sediment the precipitated virus at 5000 *g* for 1 h. Pour off the supernatant medium leaving the centrifuge tubes inverted until all the medium has been removed.
(xi) Re-dissolve the precipitate in borate-buffered saline (BBS; see Appendix) using 1% of the original volume that was recovered from the monolayers, thus producing a concentration factor of 100.
(xii) This haemagglutinin can then be stored almost indefinitely at −70°C or for several months at 4°C.

3.3.2 *Titration of HA*

The volumes used in the method described below can be altered as long as the relative volumes remain the same. Furthermore, after red blood cells (rbc) have been washed in dextrose-gelatin buffer (DGV, see Appendix) they can be kept at 4°C for several days. However, the rbc suspensions prepared at pHs 6.0–7.0 should be prepared immediately before they are dispensed into the microtitre plates.

(i) Wash the gander or chick erythrocytes (rbc) in DGV buffer by centrifugation at 1500 *g* for 5 min. Repeat this procedure three more times, centrifuging the cells for 10 min on the last occasion. Re-suspend the cell pack to a concentration of 10% in DGV.
(ii) Dispense 25 μl aliquots of BBS at pH 9.1 into each cup of V-based microtitre plates.
(iii) Add 25 μl of the HA sample to be tested into the first cup of each row.
(iv) Make serial 2-fold dilutions of the HA (in BBS) using a multichannel micropipette with disposable tips. There is no need to change the tips until the dilutions on each plate are completed.
(v) Using the buffers prepared to give pHs ranging from 6.0 to 7.0 and the 10% suspension of rbc make a 1/20 dilution of rbc in each buffer.
(vi) Immediately dispense the 0.5% suspensions as 25 μl aliquots into a row of virus dilutions and run a control containing BBS without virus.
(vii) Cover the plates and incubate at 37°C for 40 min. Cups in which no haemagglutination takes place will have small buttons of rbc with a clear supernatant fluid whilst those in which agglutination has taken place will appear uniformly coloured with no cell button.

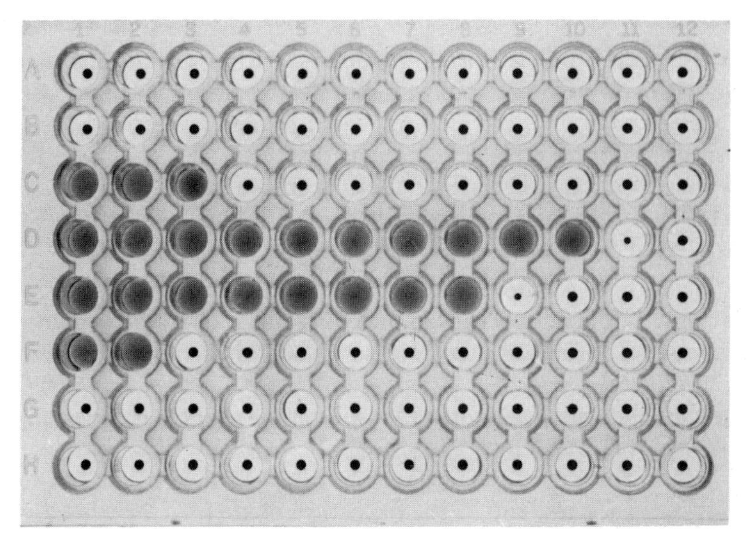

Figure 9. Haemagglutination titration performed at different pHs to determine the optimum conditions.

The HA titre is the highest dilution that produces very slight, but not complete, button formation. The optimum pH is that producing the highest HA titre. The typical appearance of a titration, performed as described here, is shown in *Figure 9*. Some arboviruses agglutinate optimally at either 4°C or 22°C; therefore these temperatures should also be tried in initial tests.

3.3.3 *Titration of Arbovirus Haemolysin*

Agglutination of rbc is often followed by lysis of the cells. This is due to the envelope glycoprotein of the virus and is distinct from the agglutinating component (12). Haemolysis is particularly noticeable with alphaviruses. Although titrations of haemolysin (HL) are not performed routinely in most laboratories, they can be useful as an additional test for studies of the envelope glycoprotein.

(i) Perform the HA titration as described above but only use the optimal pH conditions.
(ii) As soon as the HA titre has been estimated, re-suspend the rbc in the microtitre plate by gently tapping the sides of the plate. Ensure that all of the cells are re-suspended.
(iii) Incubate the plates at 37°C for 4–6 h then centrifuge them at 1000 *g* for 1–2 min. With practice, the HL titre can be estimated by examination of the plates at this stage. Cups in which lysis has taken place will be clear-red with no or little button formation. Those in which no lysis has occurred will have a colourless supernatant and a distinct button of cells.
(iv) A more precise estimation can be carried out using larger volumes of each reagent in glass Kahn tubes. After lysis has occurred, centrifuge the tubes, collect the supernatant fluids and estimate the percentage lysis spectrometrically at 410 nm (or 540 nm). Prepare a range of tubes containing lysis from 0 to 100% by sonication of the corresponding rbc suspensions.

3.3.4 *Titration of Antibody by Inhibition of Haemagglutination and Haemolysis*

Prior to use, sera for haemagglutination-inhibition (HAI) and haemolysis-inhibition (HLI) titrations normally need to be treated to remove non-specific inhibitors of haemagglutination (see Appendix). In the presence of HAI antibodies haemolysis does not take place; consequently, the HLI test as described below can only be performed either with antisera prepared against individual antigenic components that have no HAI antibody or with monoclonal antibodies that do not show HAI activity.

(i) Estimate the HA titre of the virus as described above.
(ii) Dispense 25 μl aliquots of BBS into each cup of a V-based microtitre plate.
(iii) Make serial 2-fold dilutions of the treated antisera using a multichannel micropipette.
(iv) Estimate the dilution of virus that will produce 4 HAU; for example, if the HA titre was 1/256 then a dilution of 1/64 of the virus will give 4 HAU.
(v) Make the appropriate dilution using BBS and then dispense 25 μl of this virus into each test cup of the microtitre plate, i.e., the cups that contain the antibody dilutions.
(vi) Incubate the covered plates at 37°C for 1 h.
(vii) Prepare a 0.25% suspension of rbc in buffer at the optimal pH for the virus.
(viii) Immediately add 50 μl of the rbc suspension to each test cup and incubate the covered plate at 37°C (or 22 or 4°C as appropriate) for 40–60 min. The highest dilution that just inhibits haemagglutination is the HAI titre.
(xi) Run a control titration of the 4 HAU each time HAI titrations are carried out.
(x) If the HLI titre is to be determined then the above test must produce a negative result, i.e., no HAI activity. Re-suspend the rbc by gently tapping the plates, incubate at 37°C for 2–4 h then centrifuge at 1000 *g* for 1–2 min. The highest dilution that just inhibits lysis of the rbc is the HLI titre.

As for the HL titration, this test can be performed in tubes using increased volumes of reagents. Also with some of the viruses it may be necessary to increase the standard dose of virus from 4 HAU to 8 or 16. This should be done if there is difficulty in detecting lysis when 4 HAU is used.

4. PURIFICATION OF ARBOVIRUSES

4.1 Growth of Virus

4.1.1 *Culture Systems*

The growth of arboviruses has been discussed in detail elsewhere in this chapter. The best system to use will have to be determined in each particular case. In general, the system giving maximum titres of infectious virus is appropriate although it must be borne in mind that purification of virus from tissues of infected animals, e.g., brain extracts, is often difficult due to the large amount of host-specific material present and radioactive labelling of virus in animals is usually not feasible. In cases where virus grows well in tissue culture, it is worth while spending some time optimising the yield of infectious virus. Parameters such as host cell line, culture medium, multiplicity of infection, incubation temperature and time of harvest are important variables affecting the virus yield. It is important to use an assay which actually measures the quantity of virus present in the starting material for purification while optimising the conditions. It can be misleading to rely, for example, on measuring the proportion of infected cells by immunofluorescence.

The scale of culture that is required is obviously determined by the use to which the purified virus material is to be put and by the yield of virus per cell. Adequate quantities of radioactively-labelled virus may often be obtained from stationary cultures in 75 cm^2 flasks but generally culture of cells in roller bottles provides more flexible facilities for small- or large-scale experiments. Plastic disposable roller bottles are currently available with growth areas between 490 and 1750 cm^2 and glass bottles of similar sizes can also be used. The cell growth area of roller bottles can be considerably increased by the addition of microcarriers such as Cytodex beads or, if facilities are available, growth of host cells in suspension or microcarriers may result in significant simplification of liquid handling and reduction in the volume of culture medium to be processed. Useful information on the use of microcarriers in cell culture, the details of which are outside the scope of this chapter, may be found in the technical literature of Pharmacia (13).

4.1.2 *Radioactive Labelling*

(i) RNA. The genomes of arboviruses can be labelled by growth in the presence of [^{3}H]uridine or [^{32}P]orthophosphate. [^{3}H]Uridine can be simply added to normal culture medium and is taken up by the cells *via* a scavenging pathway. Radioactive concentrations between 5 and 20 μCi/ml are usually used to label arbovirus RNA. Add the labelled uridine at the time of infection and leave until the virus is harvested. Two approaches are used for labelling with ^{32}P. Normal growth medium can be used, in which case rather large quantities of isotope (100 – 200 μCi/ml) have to be added to achieve a useful specific activity because of the high phosphate content of normal media formulations. Alternatively, media containing no unlabelled phosphate can be used together with more moderate concentrations of isotope, e.g., 10 – 25 μCi/ml. While the second approach considerably reduces the radioactive hazard, the virus may not grow as well in phosphate-free medium as it does when the cells are maintained in normal medium, and unacceptably low levels of incorporation of isotope may result. Pilot experiments with both systems may be necessary to determine the best strategy in each case. Phosphate-free media formulations are commercially available, e.g., from Flow Laboratories. If this approach is adopted, dialyse any serum in the medium to remove phosphate and sterilise by filtration.

(ii) Proteins. The most common label for proteins is [^{35}S]methionine, due to the high specific activity and relatively low cost of commercially available preparations. Use concentrations of 5 – 50 μCi/ml to label arbovirus proteins. Label proteins deficient in methionine with [^{35}S]cysteine, which is available at similar

specific activities, or with a mixture of ^{3}H-labelled amino acids. Labelled amino acids are usually used in conjunction with media deficient in the amino acid carrying the label. Such formulations are commercially available, or alternatively labelling for short periods (a few hours or less) can be more conveniently carried out in buffered saline, e.g., Hanks' salt solution containing 20 mM Hepes buffer, 2% dialysed bovine serum or 0.1% bovine serum albumin (BSA).

(iii) Glycoproteins. The sugar residues of glycoproteins are generally labelled by the addition of ^{3}H-labelled sugars in the medium. [^{3}H]Mannose and [^{3}H]glucosamine are the most common choices, used at concentrations of 50 – 100 μCi/ml. Incorporation can sometimes be improved by using media containing low concentrations of glucose, or another sugar such as galactose or fructose as an energy source.

4.2 Concentration of Virus

The first step required in most virus purification schemes, certainly those starting from tissue culture fluid, is the concentration of the virus. Since most, if not all, of the methods used take advantage of the large size of the virus particles relative to most other components present in the spent medium, there will be a concomitant partial purification of the virus away from smaller molecules, e.g., any compounds used for radioactive labelling.

4.2.1 *Removal of Cell Debris*

Spent tissue culture medium may contain, in addition to virus, considerable quantities of debris consisting of dead or dying cells and particles derived from them. These are best removed by centrifugation in a high-speed refrigerated centrifuge. Spin at 5000 – 10 000 *g* for 10 – 15 min. A high capacity rotor such as the 6 x 500 ml GS-3 for the Sorvall RC-5B will be required to process large volumes of medium. Use tubes fitted with sealing cap assemblies and, when working with a virus known to be pathogenic, do not fill the tubes above the level which would result in leakage in the event of cap failure. This volume may usually be found in the User's Manual for the centrifuge or rotor involved. After centrifugation decant the virus-containing supernatant from the tightly packed small pellet into a container packed in ice.

4.2.2 *Pelleting Virus by Ultracentrifugation*

When relatively small quantities of tissue culture fluid are involved, virus particles can be concentrated by centrifuging them to form a pellet. With arboviruses 2 h at 50 000 *g* (average) are sufficient. Either swing-out rotors or angle rotors may be used, the choice depending on the volume to be processed and the equipment available. Tubes for angle rotors should be carefully sealed. The best method is to heat seal them using tubes and equipment available from Beckman.

Some degree of purification can be obtained at this stage by layering the virus-containing fluid on top of a small volume of buffer containing 20% sucrose. The virus then sediments through the sucrose layer and forms a pellet while low molecular weight contaminants, including soluble proteins, are stopped. A suitable buffer is GTNE (200 mM glycine, 50 mM Tris, 100 mM NaCl, 1 mM EDTA, pH adjusted to 7.5 with HCl). Place an appropriate volume of buffered sucrose solution in each tube to be used (10–20% of the total volume of the tube). When using an angle rotor ensure that there is enough sucrose solution to cover the pellet even after the rotor has stopped. Then carefully layer the virus-containing fluid onto the sucrose by running it slowly down the side of the tube. To help to avoid mixing when running the rotor up to speed, use a slow acceleration facility if one is available, although this is not strictly necessary.

Under these conditions the virus will form a firm pellet. Remove the supernatant by decanting and allow residual fluid to drain from the tube by inversion. Remember the supernatant will still contain residual infectious virus so it should be disposed of in a suitable manner. Gently suspend the virus pellet in a small volume of GTNE by submerging it in buffer and either allowing it to stand overnight at 4°C or putting the tube in a sonicating water bath for a few minutes.

4.2.3 *Polyethylene Glycol Precipitation*

Instead of directly ultracentrifuging the virus it can be aggregated by the addition of PEG and sedimented in a high-speed centrifuge. Precipitation is often enhanced by addition of NaCl with the PEG. This method is usually more convenient for the treatment of large volumes of material since most laboratories are more likely to be equipped with high capacity rotors capable of the relatively low centrifugal forces required. This method is also more gentle and usually results in a lower loss of infectivity compared with direct ultracentrifugation. The procedure described below works for several arboviruses (and other enveloped viruses) and the concentrations of PEG and NaCl do not appear to be critical, but in individual cases some variations may be required for optimum yield.

Measure the volume of the virus-containing fluid and place it in an ice bath on a magnetic stirrer. With constant stirring slowly add solid NaCl to a concentration of 0.5 M and solid PEG 6000 to a concentration of 7%. Stir gently for 20–30 min until all the PEG has dissolved and allow to stand a further 2 h at 4°C to allow aggregation of the virus. Centrifuge at 10 000 *g* for 30 min. The precipitate is generally distributed on the side of the centrifuge pot or tube if an angle rotor has been used, so carefully remove the supernatant by decanting it with the precipitate uppermost. Remove as much residual supernatant as possible with a pipette. Resuspend the precipitate in GTNE buffer using a volume about 1% of the initial virus-containing fluid. The precipitate can usually be redissolved quite easily by rinsing it down the sides of the tubes with buffer, followed by rapid passage up and down a pipette several times or through a wide-bore syringe needle. Remove material that remains insoluble by centrifugation for 5 min at 10 000 *g*.

4.3 **Purification of Virus by Gradient Centrifugation**

Material prepared by either of the methods described above will usually be contaminated by non-viral material. However, it may be useful at this stage in applications where purity is not crucial, for instance, as antigen in a radioimmunoassay or enzyme-linked absorbance assay for antibody. For most purposes though, the virus must be purified further by sedimentation on rate-zonal and/or isopycnic gradients. In a rate-zonal gradient the virus moves to a position in the gradient governed by its sedimentation coefficient, while in an isopycnic gradient it moves to a position governed by its buoyant density. For maximum resolution of virus particles from impurities, the concentration of the gradient material should vary continuously down the gradient, but as an intermediate step in a purification scheme discontinuous gradients are sometimes employed, where the concentration varies in a step-like manner. In some systems rate-zonal and isopycnic gradients are combined, e.g., in glycerol-potassium tartrate gradients (14). In these gradients the potassium tartrate concentration rises towards the bottom of the tube eventually reaching densities at which virus will band isopycnically. The glycerol concentration is highest at the top of the tube where the increased viscosity impedes the movement of slowly sedimenting material down the tube. All types of gradients are usually run in swing-out rotors, but similar or superior results with isopycnic gradients can be achieved using angle or vertical rotors when facilities are available for slow acceleration and deceleration to allow reorientation of the gradients without mixing. In a purification scheme for a particular virus it will frequently be necessary to put the virus through a number of successive gradient purification steps to achieve the required degree of purity.

4.3.1 *Discontinuous Sucrose Gradients*

The lower sucrose concentration is arranged to be denser than the virus so that during centrifugation it moves through the top layer, leaving more slowly sedimenting particles behind, and forms a sharp band at the interface between the sucrose concentrations. Concentrations of 20% sucrose and 50% sucrose (w/w) for the upper and lower parts of the gradient respectively, dissolved in GTNE, are useful for arbovirus work. Place some 50% sucrose solution in a suitable centrifuge tube so that it occupies about 1/10th of the total volume of the tube, and carefully layer a 3/10th volume of 20% sucrose solution above it. Tilting the tube and running the solution slowly down the side will help to avoid mixing of the layers. Carefully overlay the step gradient with the crude virus preparation. Centrifuge the gradients at 4°C at 80 000 – 100 000 *g* for 2 – 3 h. Recover the virus from the interface between the two sucrose layers (see Section 4.3.4).

4.3.2 *Rate-zonal Sucrose Gradients*

These are best performed in swing-out or vertical rotors to avoid interfering effects from the walls of the tube. Make linear gradients from 5% to 20% sucrose (w/w) in GTNE using any of the standard methods (see Chapter 1, Section 2.5.1) and carefully layer the virus solution on the top. Since there will be no concentrating effect due to banding of the virus at its buoyant density, the volume of the sample should not exceed 1/10th the volume of the gradient. Centrifugation should be carried out at 4°C. Since the time and speed of centrifugation will depend critically on the geometry of the rotor used and on the sedimentation rate of the viruses, only a rough guide can be given. Values used initially will probably need to be modified in the light of the initial results in a particular system. In a swing-out rotor such as the SW27, 2 h at 27 000 r.p.m. should provide adequate separation of most arboviruses.

4.3.3 *Isopycnic and Combined Rate-zonal/Isopycnic Gradients*

Continuous gradients suitable for the isopycnic banding of arboviruses are linear from 20 to 50% w/w sucrose in GTNE. Combined glycerol-potassium tartrate gradients (14) are made from 30% glycerol (w/w) in 50 mM Tris-HCl, 1 mM EDTA (pH 7.5), forming the less dense end of the gradient, and 50% (w/w) potassium tartrate, forming the denser end of the gradient. Do not try to reduce the pH of the potassium tartrate solution below pH 8.5 since a precipitate will form. Form linear gradients from these solutions by the normal means. Since the virus will be concentrated by banding at its buoyant density in these gradients, the volume of virus suspension that can be layered on top is not critical, and can reach a quarter of the volume of the gradient. Centrifuge at 4°C at 80 000 – 100 000 *g* overnight.

4.3.4 *Detection and Recovery of Virus in Gradients*

(i) Fractionation of gradients. When reasonably large quantities of material are being purified, virus banding in isopycnic gradients can usually be seen as a bluish-white opalescent layer. To help visualise the virus band, shine a narrow beam of light up the tube (a small torch is suitable) while holding it against a dark background. If the virus can be definitely identified by this procedure, it can be recovered from the tube by side-puncture. If the tube has been heat-sealed make a small hole at the top to allow entry of air as liquid is removed. Insert a hypodermic needle, mounted on a 5 ml or 10 ml disposable syringe, bevel upward, immediately below the position of the virus band. Slowly withdraw fluid until most of the virus has been recovered. If several bands can be seen on the gradient, collect each one separately starting at the top and identify the virus band by a specific assay (see below). With pathogenic viruses, side puncture of the tube is too hazardous an operation, and bands should be collected with a long blunted needle inserted at the top of the tube (see Chapter 2, Section 5.4).

In cases where the position of the virus is not apparent visually, it will be necessary to fractionate the entire gradient. Either (a) smear the bottom of the tube with silicone grease, insert a wide-bore needle, being careful not to introduce air bubbles which may disturb the gradient, and collect fractions of equal size by counting drops, or (b) carefully introduce a thin probe attached to a peristaltic pump from the top of the gradient without disturbing it unduly. With the probe at the bottom of the gradient tube, pump the contents slowly (2 – 3 ml/min) to a fraction collector.

(ii) Location of virus in gradient fraction. The methods used to determine which fractions of a gradient contain virus can be divided into those that specifically assay viral functions, e.g., infectivity or haemagglutination, and those that do not, e.g., u.v. absorbance or radioactivity (in the case of labelled virus). While setting up a purification scheme for a virus it will probably be necessary to use a specific method to locate the fractions of interest, but it may be sufficient for

later routine use to rely on a non-specific method. Methods for assay of infectivity and haemagglutination have already been described (Section 3) and are applicable, without modification, to purified virus preparations. When needed, measure the absorbance of undiluted fractions at 260 or 280 nm. To locate radiolabelled virus, spot small aliquots (up to 25 μl) on numbered 1 cm squares of Whatman 3MM paper. Wash the squares from a gradient together in three changes of 5% trichloroacetic acid (TCA) (~5 ml/filter), two changes of ethanol, and allow to dry. Immerse in scintillation fluid and count. This procedure removes any residual unincorporated label and avoids difficulties caused by quenching by components of the gradient.

(iii) Recovery of virus from gradient fractions. Removal of gradient-forming components is necessary at the end of the purification process. At intermediate stages it is usually sufficient to dilute the virus-containing pooled fractions of a gradient with an equal volume of buffer so that the material can be directly layered on to a further equilibrium gradient. Since a rate-zonal gradient requires the sample to be in a relatively small volume, dilution of material from previous gradients is usually not possible, so it is preferable to run rate-zonal gradients before isopycnic gradients where both are required in a purification scheme. Complete removal of gradient-forming components may be achieved either by dialysis, passage over a column of Sephadex G-50 or dilution followed by ultracentrifugation. Dialyse the material against several changes of suitable buffer, e.g., GTNE, although the subsequent use of purified virus may require the use of some other buffer. Equilibrate Sephadex columns, using a volume about 5 times that of the sample, with the desired final buffer. Monitor the appearance of virus in the excluded volume of the column by u.v. absorbance or radioactivity as appropriate, and pool virus-containing fractions. Ultracentrifugal pelleting of virus has the additional merit of concentrating the virus. Dilute the pooled gradient fractions with at least an equal volume of buffer to reduce the density, and centrifuge at 4°C at 80 000 – 100 000 *g* for 2.5 h. Discard the supernatant, taking care to drain as much as possible from the virus pellet, which should be colourless and translucent. Resuspend the virus in buffer using a sonicating water bath to give a bluish-white opalescent suspension and store at −70°C.

4.3.5 *Assessment of Purity*

The most useful criterion by which purity of arbovirus preparations can be assessed is electrophoresis of the virus structural proteins on polyacrylamide gels containing sodium dodecyl sulphate (SDS). The alphaviruses and flaviviruses that have been examined so far have a small number of structural proteins of characteristic size that form a recognisable pattern on the gel. The alphaviruses have a nucleocapsid protein with a molecular weight of 30 000 – 35 000 and two larger envelope glycoproteins (which may not be resolved from each other on the gel) with molecular weights of 45 000 – 55 000. The flaviviruses have an envelope protein, that is usually glycosylated, with a molecular weight of 50 000 – 55 000, and two smaller internal structural proteins with molecular weights of about 15 000 and 7000.

The discontinuous gel system of Laemmli (15) is suitable. Gel slabs are better than gels cast in tubes, since several samples may be run side by side so that progress of the purification can be accurately monitored. A separating gel concentration of 10% acrylamide is effective for the resolution of alphavirus proteins, but a gradient gel of 8–20% acrylamide is necessary for adequate resolution in the wide molecular weight range occupied by flavivirus proteins. Disrupt the sample to be analysed (which should contain 10–50 μg protein in a volume of 30 μl or less) by adding 1/5 volume of a solution containing 10% SDS, 5% 2-mercaptoethanol, 50% glycerol, 0.05% bromophenol blue. Heat in a boiling water bath for 2 min, cool and apply to the gel, together with a set of molecular weight standards. When the tracking dye has moved to the end of the gel, stain in a solution of 1.0% Coomassie blue in 25% isopropanol, 10% acetic acid for 2 h with constant agitation, and de-stain overnight in 10% isoproppranol, 10% acetic acid. In the case of radioactive viral proteins, the gel should be dried and autoradiographed (16).

5. GENERAL CONCLUSIONS

Isolation and growth of the arboviruses presents no major difficulties since there are so many possible choices of host. For molecular studies very high titres of virus are usually required and these can often be produced in cell cultures if alphaviruses are used. Many of the flaviviruses, however, produce lower titres and it may be necessary to design the experiment accordingly. For example, relatively high virus input multiplicities can be achieved if the cell monolayers are produced in small culture vessels.

There is considerable scope for research on the flaviviruses partly because not all of them are safe to work with and partly because there are so many viruses in the genus. Purification procedures are very similar to those used for viruses of other families. Here again, however, there is scope for development of improved methods to enable the identification of non-structural proteins and for the detailed investigation of replicative pathways.

Many of the titration methods described have been developed with diagnosis in mind; nevertheless, they are equally applicable to investigations concerned with antigenic relationships and virus replication. The infectivity titre of any one virus will differ if titrated by two or more methods. Generally, virus produced and titrated in newborn mice yields the highest titre but mouse brain material may not be suitable for use without further purification steps.

Another problem that has not yet been resolved is that of virus variation, arising as the result either of serial passage in one host or of changing the host. It is known that many arboviruses establish persistent infections in mosquito cell lines even though they may not do so in mammalian cells. The virus that results from these persistent infections often shows reduced virulence for mice and humans. This type of change should be borne in mind if, for example, HA is produced by a long term culture as described above.

Finally there is no doubt that some of the mosquito cell lines are very sensitive to wild-type viruses and this can be used to advantage, but in many cases mammalian cells do perform equally well and are probably more easily managed in most laboratories.

6. ACKNOWLEDGEMENT

The authors are deeply indebted to Mavis Edmonds for arranging and typing the manuscript and to Alan Buckley for assistance with the photographs.

7. REFERENCES

1. Porterfield,J.S. (1980) in *The Togaviruses*, Schlesinger,R.W. (ed.), Academic Press Inc., London and New York, p.13.
2. Berge,T.O., ed. (1975) *International Catalogue of Arboviruses* 2nd ed., DHEW Publ. No. (CDC) 75-8301, published by US Department of Health, Education and Welfare, Public Health Service, Washington, DC.
3. Varma,M.G.R., Pudney,M. and Leake,C.J. (1974) *Trans. R. Soc. Trop. Med. Hyg.*, **68**, 374.
4. Singh,K.R.P. (1972) *Adv. Virus Res.*, **17**, 187.
5. Pudney,M., Leake,C.J. and Varma,M.G.R. (1979) in *Arctic and Tropical Arboviruses*, Yunker,E. (ed.), Academic Press Inc., London and New York, p. 245.
6. Tesh,R.B. (1979) *Ann. J. Trop. Med. Hyg.*, **28**, 1053.
7. Pang,T., Lam,S.K., Chew,C.B., Poon,G.K., Ramalingam,S. (1983) *Lancet*, **(i)**, No. 8336, 1271.
8. Rosen,L. and Gubler,D. (1974) *Am. J. Trop. Med. Hyg.*, **23**, 1153.
9. Reed,L.J. and Muench,H. (1938) *Am. J. Hyg.*, **27**, 493.
10. Peiris,J.S.M. and Porterfield,J.S. (1979) *Nature*, **282**, 509.
11. Clarke,D.H. and Casals,J. (1958) *Am. J. Trop. Med. Hyg.*, **7**, 561.
12. Chanas,A.C., Gould,E.A., Clegg,J.C.S. and Varma,M.G.R. (1982) *J. Gen. Virol.*, **58**, 37.
13. *Microcarrier Cell Culture*, Principles and Methods, Pharmacia Fine Chemicals.
14. Obijeski,J.F., Marchenko,A.T., Bishop,D.H.L., Cann,B.W. and Murphy,F.A. (1974) *J. Gen. Virol.*, **22**, 21.
15. Laemmli,U.K. (1970) *Nature*, **227**, 680.
16. Laskey,R.A. (1980) *Methods in Enzymology*, Vol. **65**, Grossman,L. and Moldave,K. (eds.), Academic Press Inc., London and New York, p. 363.
17. Varma,M.G.R. and Pudney,M. (1969) *J. Med. Ent.*, **6**, 432.

APPENDIX

1. TISSUE CULTURE GROWTH MEDIA

All media are shown with FCS; however, newborn calf serum is usually satisfactory. It is also possible to reduce the concentration of serum to 5% without harmful effects on cell growth, furthermore tryptose phosphate broth (TPB) is not used in all laboratories. Therefore it is worth testing the cells to see which conditions best suit them.

(i)	Eagle's medium (Glasgow modification)	80 ml
	FCS	10 ml
	TPB	10 ml
	Penicillin	100 units/ml
	Streptomycin	100 μg/ml
	Sodium bicarbonate (if not added to Eagle's)	0.275%

This medium will support the growth of all mammalian cells listed in the text but a 5% CO_2 incubator is required.

(ii)	Leibovitz L-15 medium (pH 7.2)	80 ml
	FCS	10 ml
	TPB	10 ml
	Penicillin	100 units/ml
	Streptomycin	100 μg/ml

This medium supports the growth of all cells listed in the text although some may require adaptation. A CO_2 incubator is not required. The pH of the medium should be adjusted using HCl before sterilisation by filtration.

(iii)	RPMI-1640 (Dutch modification)	80 ml
	FCS	10 ml
	TPB	10 ml
	Sodium bicarbonate (if not in the RPMI)	0.1%
	Penicillin	100 units/ml
	Streptomycin	100 μg/ml

This medium contains Hepes buffer and therefore a CO_2 incubator is not required.

Suitable for the growth of C6/36 cells at 28–33°C.

(iv)	MM/VP12 medium	85 ml
	FCS	15 ml
	Penicillin	100 units/ml
	Streptomycin	100 μg/ml

Cells grown in this medium do not require CO_2.

The AP61 cell line derived from *Aedes pseudoscutellaris* grows particularly well in this medium at 28°C. The formulation of this medium is described by Varma and Pudney (see ref. 17).

2. MAINTENANCE MEDIA

(i) With the exception of MM/VP12 medium, it is normal to reduce the concentration of FCS to 1% or 2% for cell maintenance. In the case of AP61 cells, Leibovitz L-15 medium containing 2% FCS is substituted for the MM/VP12.

(ii) Carboxymethylcellulose (CMC) overlay medium (L15-CMC)

Leibovitz L-15 medium (double strength)	44 ml
CMC (3%) in distilled water	44 ml
TPB	10 ml
FCS	2 ml
Penicillin	100 units/ml
Streptomycin	100 μg/ml

The 3% CMC is prepared by adding 3 g CMC per 100 ml of distilled water. The mixture is stirred at 56°C overnight until dissolved and then dispensed into 44 ml aliquots, autoclaved at 10 lb for 10 min and stored at 4°C.

(iii) Agarose overlay medium

Agarose*	10 g
Distilled water	100 ml

*Several special preparations of agarose are available and most are suitable; however, it is a good idea to test them before ordering large quantities.

To prepare the agarose, place the mixture into a boiling water bath until it has melted. Autoclave for 10 min at 10 lb then adjust the temperature to 42–44°C and add 10 ml of this to 90 ml of maintenance medium at 44°C.

3. DILUENT FOR FIELD MATERIAL

Hanks balanced salt solution	75 ml
FCS	25 ml
Penicillin	500 units/ml
Streptomycin	500 μg/ml

4. PREPARATION OF SUSPENSIONS OF FIELD TISSUE

Add the tissue sample to diluent to give 1 g/10 ml, using a screw-capped bottle containing 8–10 glass beads (5–6 mm). With the cap screwed down tightly, shake the bottle vigorously or preferably use a vortex mixer for about 30 sec. Centrifuge the sample, collect the supernatant and either use it immediately or store at –70°C.

5. TREATMENT OF SERA FOR USE IN HAI TESTS

Heat the serum at 56°C for 30 min. Add 0.1 ml of serum to 1.9 ml of acid-washed kaolin (25% in BBS buffer). Shake the mixture vigorously for 15 min at room temperature then centrifuge at 200 *g* for 10 min. Collect the clear supernatant fluid and add washed and packed gander or chick erythrocytes to the supernatant fluid, to produce a 10% suspension of cells. Leave at room temperature for 1 h resuspending the erythrocytes every 10 min then centrifuge at 1500 *g* for 5 min. Collect the supernatant and either use immediately for HAI test or store at –20°C (or below).

6. REAGENTS AND BUFFERS

Naphthalene black stain:

Naphthalene black	1 g
Glacial acetic acid	60 ml
Sodium acetate	13.6 g
Distilled water to	1000 ml

Borate-buffered saline (BBS):

Sodium chloride	7.02 g
Boric acid	3.09 g

Dissolve both chemicals in a small volume of warm distilled water and make up to 100 ml. Adjust pH to 9.0 with concentrated sodium hydroxide, sterilise at 10 lb for 10 min then add 4 g of bovine albumin powder per 100 ml.

Dextrose-gelatin veronal buffer (DGV):

Veronal (barbital)	0.58 g
Gelatin	0.60 g
Sodium veronal	0.38 g
Calcium chloride (anhydrous)	0.02 g
Magnesium sulphate (hydrated)	0.12 g
Sodium chloride	8.5 g
Dextrose	10.0 g
Add distilled water to 1 litre	

Dissolve the veronal and gelatin in about 250 ml of water at 56°C then add the other reagents. Adjust the pH after sterilisation at 10 lb for 10 min.

pH Adjusting diluents for erythrocyte suspensions:

Solution A	
Sodium chloride	8.76 g
Di-sodium hydrogen ortho-phosphate	28.39 g
Add distilled water to 1 litre	

Solution B	
Sodium chloride	8.76 g
Sodium di-hydrogen ortho-phosphate	31.20 g
Add distilled water to 1 litre	

To obtain the appropriate pH for the HA test mix the two solutions A and B as shown below.

Final pH* required	Solution A (ml)
6.0	12.5
6.2	22.0
6.4	32.0
6.6	45.0
6.8	55.0
7.0	64.0

Make each solution up to 100 ml by adding the appropriate amount of solution B.

*The pH is that obtained when one volume of the mixture is added to one volume of BBS (pH 9.0).

Phosphate buffered saline (PBS):

This can be obtained commercially as tablets, add each tablet to 100 ml of distilled water.

Versene solution:

PBS	100 ml
Sodium versenate	0.2 g
Glucose	0.2 g

Sterilise by filtration:

Trypsin solution (2.5%) (T/V): obtain commercially	2 ml
Versene solution	18 ml

Store the trypsin in 2 ml aliquots at −20°C. Thaw it, add to the versene, warm it to 37°C and use the mixture immediately. The versene can be stored at 4°C for long periods. Do not store the T/V at 4°C.

Glycerol-saline mountant:

Glycerol	90 ml
PBS	10 ml

Adjust pH to 8.3.

CHAPTER 4

Growth, Purification and Titration of Rhabdoviruses

W.H. WUNNER

1. INTRODUCTION

Currently, more than 80 viruses belonging to the family rhabdoviridae are known to cause infections in vertebrates, arthropods or plants. Many of those viruses also grow with remarkable ease outside their natural host in a variety of tissue culture systems. Rhabdoviruses which appear to be more difficult to propagate to high titre *in vivo* usually require considerable adaptation for growth *in vitro*. Experimental infections with rhabdoviruses, particularly those which grow rapidly and to high titre in cell culture, have played an important role in studies aimed at elucidating viral replication strategies and virion morphogenesis. Chemical studies in which the structure and function of rhabdovirus components have been investigated have also been heavily dependent on tissue culture systems that produce virus particles in large numbers. Since rhabdoviruses are one of the best sources of easily purified proteins and nucleic acid (RNA), some of the most significant advances in the understanding of virus structure and replication have come from studies of these distinctively bullet-shaped particles (for reviews, see 1–6). The methods described in this chapter for the growth, purification and titration of rhabdoviruses have been selected to emphasise certain practical aspects of virus production rather than give a detailed account of a very long history and widespread experience in methodology associated with growing and assaying the broad spectrum of rhabdoviruses. Two of the rhabdoviruses of vertebrates, vesicular stomatitis virus (VSV) and rabies virus, are discussed in some detail as models. VSV, however, is by far the most convenient of rhabdoviruses and the more suitable of the two model viruses for study in the laboratory. Investigators should be aware that both viruses are serious animal pathogens, and that rabies virus infection in animals, and humans especially, is usually fatal. Permission may be required from Public Health authorities to possess and work with such viruses and, moreover, precautions must be taken for the safe handling and containment of these viruses in the laboratory. Rabies virus should be handled *only* by personnel who are vaccinated against rabies and who maintain an antibody titre above the World Health Organization accepted level for protection (7). Additionally in this chapter, special consideration is given to the purification of rhabdoviruses from plant tissues, since considerable difficulties have been encountered with the purification of plant rhabdoviruses.

2. GROWTH OF RHABDOVIRUSES IN TISSUE CULTURE

The standard procedure for the propagation of rhabdoviruses in the laboratory is to inoculate cells in culture either with tissue culture-adapted virus or unadapted virus from original infected tissue or secretions. Many primary cell culture systems and established cell lines which have been used for this purpose are identified in a recent review of systems for growth and assay of rhabdoviruses (8). VSV replicates efficiently in almost any known type of mammalian and avian cell culture. Particularly high yields of virus are consistently obtained whether from primary cultures of chick fibroblast cells (9) or from the more commonly used BHK/21 cell line (10), and equally high titres of virus can be obtained in cells whether maintained in suspension cultures or in monolayer cultures (11). Single-step growth curves for VSV in suspension cultures and in monolayer cultures are virtually identical. Released progeny VSV is detectable within 2 h after the initiation of infection; thereafter virus yield increases at a logarithmic rate for 6 – 8 h. Maximum yields of up to 1000 p.f.u. per cell are obtained by 10 – 12 h post-infection.

2.1 One-step Growth Curve

To measure the growth cycle of rhabdoviruses in cell culture, the inoculum should be adjusted to infect all cells. After an initial period for viral adsorption to cell monolayers, the cells are washed to remove unadsorbed virus. Warmed growth medium is added and incubation is continued at 37°C. At intervals, a cell culture is sampled for virus either by scraping cells into the medium and then disrupting the cells by three freeze-thaw cycles (alternately freezing the cells in a methanol/dry-ice bath at – 60°C and rapidly thawing at 37°C) or by sonication for 1 – 2 min to release cell-associated virus, or by simply harvesting the culture fluid containing the released progeny virus. The number of cells infected with virus (infectious centres) is also measured so that the yield of virus per cell can be calculated.

The growth cycles of some rhabdoviruses such as VSV are short, being measured in hours, whereas others such as rabies virus may have much longer growth cycles spanning several days. It is not known why some viruses replicate more efficiently, although it has been suggested that the growth efficiency is related to the activity of the virion-associated transcriptase.

Set up the experiment by preparing a series of plates (60-mm Petri dishes are convenient) each containing 10^6 cells in monolayer culture, or a culture large enough to yield a conveniently countable number of infectious virus particles, and mark them IC (for infectious centre assay), 0, 2, 3, 4, 6, 8, 10, 12, and 18. BHK/21 cells (12) grown either in Dulbecco's modified Eagle's medium (MEM), or Glasgow-MEM (Gibco Laboratories), supplemented with 10% foetal calf serum (MEM-10) are frequently used for this purpose with VSV. The time course may be altered to suit the growth rate of other rhabdoviruses.

The cells should be subconfluent at the time of infection. With some rhabdoviruses it is difficult to obtain strict one-step conditions and achieve a synchronous infection even with very high input multiplicities of virus. Approximate one-step conditions are, however, realised by infecting the monolayer cultures with a large inoculum, which leaves no uninfected cells. When all cells are infected at once, multiplication is necessarily confined to a single cycle. The size of the inoculum also has its limitations. High multiplicities of infection may result in abnormal multiplication, caused by generation of defective-interfering particles (see Section 3.1.1), and extensive re-adsorption of progeny virions to cell debris. It may be best to choose an input multiplicity of 10 p.f.u. per cell initially, and then increase or decrease the multiplicity by 10-fold if required to ensure infection of all cells in subsequent growth curve experiments. Alternatively, certain polyions added to virus or tissue culture medium enhance the binding of virus to cells and increase the infectivity of the virus in cells (13). Diethylaminoethyl (DEAE) dextran and protamine sulphate at concentrations of 50 μg/ml and 100 μg/ml, respectively, have an enhancing effect as long as they are added before (up to 4 h) the time of adsorption of the virus. The polyions have no enhancing effect when added after initial exposure to virus. It is important to realise with some polyions that, while they may increase attachment of virus to cells, they may not necessarily lead to increased infectivity of the virus (14).

(i) To begin the experiment, remove the medium from the cell monolayer with a Pasteur pipette, preferably in a laminar flow hood to maintain sterile tissue culture conditions and to ensure safe handling of virus.
(ii) To each plate add 0.2 ml of virus inoculum which has been diluted to give the desired input multiplicity of infection (e.g., 10 p.f.u./cell).
(iii) Rock the plates gently to spread the virus inoculum evenly and incubate at 37°C in a 5% CO_2-incubator for 1 h.
(iv) After virus adsorption, suck off excess virus with a Pasteur pipette.
(v) Wash the plates gently three times with 1.5 ml of PBS/serum diluent transferring wash fluid to a waste container which will be treated later to inactivate the virus.
(vi) Add 5 ml of growth medium (MEM) supplemented with either 2% foetal calf serum or 0.2% bovine serum albumin to all plates except the one marked IC and return those marked 2 – 18 to the incubator at 37°C.
(vii) Add 0.5 ml trypsin to plate IC and place it in the incubator for 5 min or leave it at room temperature until cells detach.
(viii) Add 4.5 ml of growth medium supplemented with 10% foetal calf serum (MEM-10) to the detached cells, transfer the cells to a fresh flask and make two dilutions, 10^{-2} and 10^{-3}.
(ix) Immediately plate the freshly diluted cells by adding the dilutions to BHK/21 cells in suspension and measure infectious centers by the virus plaque assay technique as described below in Section 2.2.
(x) Take the plate marked '0' and scrape the cells into the medium with a rubber policeman.
(xi) Transfer the cells and medium to a correspondingly marked container and freeze-thaw three times or sonicate for 1 – 2 min.
(xii) Store the sonicated or freeze-thawed cells at 4°C (or store frozen) with the other samples as collected.
(xiii) At 2, 3, 4, 6, 8, 10, 12 and 18 h, take the appropriately marked plate and

harvest the virus in medium and treat the cells in fresh medium as described for the '0' plate.

(xiv) When the growth curve is completed, thaw out the samples for plating and prepare serial dilutions as shown below.

PBS/serum (ml)	1.98	1.8	1.8	1.8	1.8	1.8
Virus (ml)	0.02	0.2	0.2	0.2	0.2	0.2
Dilution	10^{-2}	10^{-3}	10^{-4}	10^{-5}	10^{-6}	10^{-7}

The appropriate dilutions for the different sample times might be as follows:

Sample Time (h)	Dilution
0	10^{-2}, 10^{-3}
2	10^{-2}, 10^{-3}
3	10^{-2}, 10^{-3}, 10^{-4}
4	10^{-3}, 10^{-4}
6	10^{-3}, 10^{-4}, 10^{-5}
8	10^{-4}, 10^{-5}
10	10^{-4}, 10^{-5}, 10^{-6}
12	10^{-5}, 10^{-6}, 10^{-7}
18	10^{-5}, 10^{-6}, 10^{-7}

2.2 **Virus Titration**

2.2.1 *Plaque Assay*

Rhabdoviruses which produce cytopathic effects in cell culture systems form plaques the number of which is proportional to the amount of virus present. Cell degeneration, evident after virus infection giving rise to plaques, provides the most accurate titration method. The number of plaques, x, produced in a monolayer of cells by plating a volume, v, of a dilution, y, of virus can be used to determine the titre, n, in plaque-forming units (p.f.u.) per ml according to the equation, $n = x/vy$.

VSV is titratable using the standard plaquing technique. However, this practical and precise assay of infectious virus may not be as applicable for some rhabdoviruses. Alternative methods including modified plaquing techniques have been successfully applied instead to facilitate accurate and reproducible assays for these viruses.

The classical plaque assay method for animal viruses, as described by Dulbecco (15), involves adsorption of the virus on a monolayer culture of susceptible cells, followed by the addition of a nutrient agar medium, which becomes solid when cooled. Plaques are visible when the cells are stained with neutral red after an appropriate incubation time.

In the plaque assay for VSV, cultures of BHK/21 cells are inoculated either in monolayer or in suspension with samples of virus. Suspension cultures are often preferred for inoculation since virus adsorbs more uniformly to cells in suspension. Chick embryo fibroblasts or mouse L cells, rather than BHK/21 cells, may be used to determine p.f.u.s of VSV (10, 16).

(i) To set up the assay label a vial for each dilution to be plated, e.g., 0 h 10^{-2}, 0 h 10^{-3}, etc.

(ii) To each vial add 2 ml of the cell suspension (containing 4 x 10^6 cells/ml).
(iii) Add 0.4 ml of the appropriate virus dilution.
(iv) Shake the virus-cell suspension at 37°C for 20 min.
(v) After the period of shaking at 37°C, 1 ml of infected cells is added to a labeled 60-mm plastic Petri dish containing 4 ml of growth medium with antiserum. The antiserum contains antibodies that neutralise unadsorbed virus and prevent secondary plaque formation.
(vi) Incubate the cultures for 10–24 h at 37°C. Longer times may occasionally be required to visualise plaques.
(vii) After incubation, remove the medium and wash the cell monolayer carefully with PBS. Stain the cells with neutral red solution or Giemsa stain, and count plaques.

Certain fixed rabies viruses also form clear plaques by the standard procedure on monolayers of hamster CER cells, a cell line now widely used to propagate and plaque non-rhabdoviruses as well as rabies virus (17). The plaque system employing CER cells is particularly sensitive for the CVS (challenge virus standard) strain of rabies virus. This plaque-forming system uses the monolayer culture and an agarose overlayer and has been conveniently set up in 35-mm wells of 6-well plastic plates.

(i) The day before virus inoculation, add 1–1.25 x 10^6 cells in 3 ml of MEM-10 to each well; the cell monolayer is usually confluent by the next morning.
(ii) To infect the cells, drain the wells of medium and add 0.1 ml of virus in PBS or MEM supplemented with 0.2% bovine serum albumin.
(iii) After adsorption for 1 h at 37°C in a 5% CO_2-incubator (tilting the plates often to distribute the virus evenly), overlay the infected cells with 2 ml of 0.5% agarose in medium for agarose (see Section 2.2.2) and incubate the cultures at 37°C for 4–5 days in 5% CO_2 atmosphere and then score for virus plaques, which are normally stained with neutral red in liquid or solid (in agarose) overlay or formalin crystal violet following removal of agarose.

2.2.2 *Modified Plaque Assay using Agarose Suspension*

Various modifications of the plaque technique including agar suspensions of cells or agar overlays have been described (16, 18). One of these has been developed to accommodate the slow growth of rabies virus (19) and is performed as follows. A subline of BHK/21 cells, clone 13S, was adapted to be maintained in suspension culture in the presence of agarose for 1.5–2 weeks (20). While the cells will grow in monolayer culture, they can be grown in suspension cultures in MEM-10. Cells stored in liquid nitrogen are first propagated in monolayer culture and subcultivated by regular trypsinisation. About 30 subcultivations can be made from frozen stock; after this, the ability of cells to produce plaques may be affected.

(i) To prepare cells for the plaque assay, trypsinise several T-150 flasks of monolayer cultures of BHK/S13 cells and resuspend the cells in MEM supplemented with 2% foetal calf serum (MEM-2) for suspension of cells, using 10 ml of medium per flask.

(ii) Count the cells and adjust the concentration to $8 \times 10^6 - 10 \times 10^6$ cells per ml using MEM-2. 50 ml of cell suspension are necessary for 100 plates. Keep this cell suspension at 37°C.

(iii) Prepare an agarose underlayer by mixing 250 ml of medium for agarose consisting of Basal Medium (21) in a double concentration, without phenol red, supplemented with 4% inactivated foetal calf serum, penicillin (200 IU/ml), streptomycin (200 μg/ml; 156 IU/ml) and mycostatin (100 IU/ml), from a 37°C water bath with 250 ml of 1% (w/v) agarose (Seakem, Bausch and Lomb, Rochester, NY) which has been first boiled to dissolve and sterilise the agarose and then allowed to cool to 42°C. Immediately pipette 4-ml aliquots of this mixture into Petri dishes. Allow the agarose to harden on a level surface.

(iv) Mix the 50 ml of cell suspension with 50 ml of 0.5% agarose in medium above and pipette 1 ml of this mixture per plate on to a hardened agarose base.

(v) Spread the cell mix evenly over the base of agarose. Allow to harden. The final concentration of agarose is 0.25%, in which the cells are maintained. Use the plates immediately or after incubation for 24 h in a humidified 5% CO_2 incubator.

(vi) Infect the cells in agarose by depositing 0.1 ml of virus inoculum on the hardened agarose. Place the infected cultures in a humidified incubator containing 5% CO_2 in air for 5–7 days depending on the strain of virus assayed.

(vii) The plaques are made visible at this time by adding 2 ml of 0.5% agarose containing 1:10 000 dilution of neutral red per ml to the plates. Count the plaques after 2–4 h.

2.2.3 *Consideration of the Infectivity Titration Based on Plaque Counts*

When plaque counts are made on replicate plates at each dilution of a series of dilutions of the virus, the concentration of p.f.u. in the undiluted suspension may be calculated from the sum of all plaques counted provided the following is true.

(i) The mean plaque number must be inversely proportional to the dilution. This relation is unlikely to hold true where overcrowding effects are present on plates with larger numbers of plaques. Therefore, all dilutions that give rise to overlapping and unrecognisable plaques should be eliminated from the calculation.

(ii) The variation between plaque counts on plates of the same dilution should not exceed that to be expected from random fluctuation alone. Differences due to other sources of variation such as lack of homogeneity in these monolayer preparations or technical errors in pipetting must also be eliminated from the calculation.

2.2.4 *Preparation of Plaque-purified Virus*

Well-separated virus plaques are usually found at the highest virus dilution that is capable of forming visible plaques. To isolate plaque-purified virus, pick up the well-separated plaques with different Pasteur pipettes and disperse them individually in 5 ml of medium containing 1×10^6 freshly trypsinised BHK/21 cells in T-25 Falcon plastic tissue culture flasks. The contents are dispersed by vigorous pipetting. To monitor the infection of these cells, transfer 0.5 ml amounts of the infected cells from the T-25 flasks into wells of 4-chamber Lab-Tek tissue culture slides. Incubate the T-25 flasks and Lab-Tek slides for 3 days and monitor the presence of infected cells in the Lab-Tek cultures by an appropriate fluorescent antibody staining technique.

2.2.5 *Assay by Fluorescent Antibody Technique*

Virus may also be titrated, albeit with less accuracy than by the plaque technique, in a microtitre assay by infecting cells with serial dilutions of virus and counting the number of infected cells by fluorescent antibody (FA) staining. One advantage of this technique is that fewer numbers of cells per virus dilution are required as virus dilutions are performed in 6 mm wells of 96-well flat bottom Microtest II tissue culture plates (Falcon) and there is a significant saving of reagents in each test. Prepare serial 5- or 10-fold dilutions of virus for assay in the 96-well plate as follows.

(i) Set the plate before you with the long side or short side going from left to right, depending on the number of dilutions of virus required, and place 80 μl of diluent (MEM-10) in each well of the top row. Repeat this procedure for each row across the plate; each row being used for a different virus sample.

(ii) With a micropipette and changing the tip at each virus transfer, add 20 μl of undiluted virus suspension to the first well in the row and serially transfer 20 μl of diluted virus to the next well in the row.

(iii) Using freshly trypsinised BHK/21 cells, add 1 drop from a Pasteur pipette (or 25 μl) of cell suspension containing 2 x 10^6 cells per ml. The final virus-cell suspension at each dilution will contain 5 x 10^4 cells in 105 μl.

(iv) For serial 10-fold dilutions, fill the plate's 96 wells with 90 μl of MEM-10 and transfer 10 μl of virus as described above.

(v) After gentle agitation of the plate, remove 10-μl aliquots of each virus-cell suspension (containing 4.7 x 10^3 cells in the 5-fold dilution series) and transfer them in duplicate or triplicate to individual wells of a 60- or 72-well Microwell (Terasaki) plate (Nunc).

(vi) Cover and incubate the infected cells at 37°C in a 5% CO_2 atmosphere for 1 – 2 days. Incubate for 3 – 4 days at 37°C in a CO_2 incubator if the BHK/21 cell suspension is 1 x 10^6 cells per ml. The 3- to 4-day incubation period is preferred for more accurate titres.

Note: for a fast growing cytopathic rhabdovirus such as VSV, it may be possible that by simply observing the cytopathic effect of the virus at different dilutions after 2 days, one can determine the end-point dilution for the cytopathic effect and calculate the virus titre of the original virus suspension from the number of dilutions it takes to reach the end-point as described below. For a number of the slower-growing non-cytopathic viruses requiring several days of incubation, titrations may be determined based on the end-point dilution of virus that produces FA-positive cells.

Virus is detected after fixation and FA staining of infected cells in the wells of the Terasaki plate.

(i) To fix the cells, remove the cover from the plates and add 10 ml of PBS.
(ii) Gently agitate the PBS by aspirating the solution with a Pasteur pipette to mix the medium with PBS.
(iii) Remove the PBS-diluted medium by suction and rinse the plate by dipping into 80% acetone-20% H_2O at 4°C for 1 min. Immediately transfer the plate to a second container of cold 80% acetone for 30 min.
(iv) After fixation remove the plate from acetone, invert and bang the plate dry.
(v) Continue to dry the plate in an incubator at 37°C for 30 min.
(vi) Recycle the acetone into stocks after filtering through Whatman No. 1 filter paper.
(vii) Apply fluorescent antibody to detect viral antigen in infected cells. Place 5 μl of fluorescein-conjugated antibody directed against viral antigen (e.g., nucleocapsid protein) in each well of the Terasaki plate and incubate at 37°C for 30–60 min in a humidified box. Make certain that the wells of the plate are free of trapped air bubbles. Alternatively, viral antigen can be characterised indirectly using mouse monoclonal antibodies followed by anti-mouse fluorescein-conjugated antibody.
(viii) After staining, wash the plate twice with 10 ml PBS and once with tap water, and drain.
(ix) Read the plates while still wet using a fluorescence microscope.
(x) To determine the titre of the undiluted virus, record the highest dilution of virus still capable of infecting less than 100% of cells. At this dilution, it can be assumed that each well is infected by 1 infectious unit.

The infectivity of the undiluted inoculum is calculated using the Spearman-Kärber formula (22) which determines the 50% end-point dilution, by the addition of two values; the first is the lowest dilution of virus at which the cells of all replicate wells are FA-positive; the second is the total number of positive wells ('positives') at this dilution and at all higher dilutions. The starting point of the calculation of 50% end-point dilution is the dilution showing FA positives next below 50% (starting point dilution).
Using the formula,

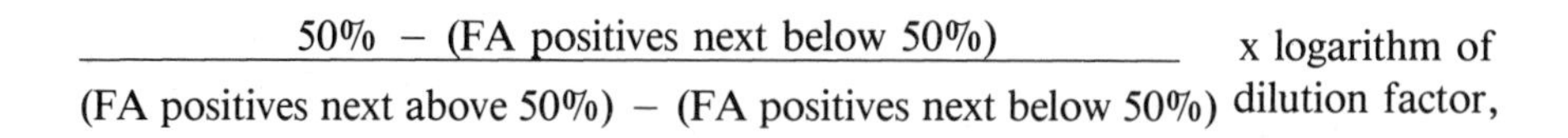

$$\frac{50\% - (\text{FA positives next below } 50\%)}{(\text{FA positives next above } 50\%) - (\text{FA positives next below } 50\%)} \times \text{logarithm of dilution factor,}$$

determine the difference between the logarithm of the starting point dilution and the logarithm of the 50% end-point dilution (difference of logarithms). If the FA positives decrease with increasing dilution, the 50% end-point dilution will be lower than the starting point dilution. The difference of logarithm has therefore to be subtracted from the logarithm of the reciprocal of the starting point dilution. This is known as the method of Reed and Muench (22).

3. VIRUS ISOLATION AND PURIFICATION

3.1 Animal Rhabdoviruses

Mammalian rhabdoviruses of the *Vesiculovirus* and *Lyssavirus* genera which have been known to replicate intracerebrally in the adult mouse or rat brain may also be propagated in the intracerebrally inoculated animal. Field strains of virus may be isolated in the laboratory after one adult mouse brain passage of inoculated specimen. The suckling mouse, however, is a more susceptible animal than an adult mouse for virus detection and recovery (23). Depending on the incubation of the virus, the duration of each mouse brain passage may be from 7 to 18 days and occasionally longer in the case of the slower growing rabies virus. VSV generally requires a much shorter (2-day) period. Virus is identified by FA staining.

(i) To isolate virus, prepare a 10% suspension of infected brain tissue, using MEM-10 as diluent, by homogenisation in a Ten Broeck homogeniser or by grinding with a pestle and mortar. Do not use a motor-driven homogeniser which creates aerosols.

(ii) Clarify the suspension by centrifugation (1600 r.p.m. for 10 min) in a refrigerated centrifuge, collect the supernatant and test for sterility. Store the clarified suspension at low temperature (−70°C).

(iii) Prepare a suspension of freshly trypsinised BHK/21 cells in MEM-10 at a concentration of 1 x 10^6 cells per ml. A T-75 tissue culture flask generally has 20 x 10^6 cells in a confluent monolayer.

(iv) To isolate rabies virus, for example, mix 1 ml of BHK/21 cell suspension with 0.25 ml of brain suspension in a T-25 flask and let this stand for a few minutes.

(v) Add 5 ml of MEM-10, mix and transfer 0.5 ml amounts of virus-cell suspension to individual wells of an 8-chamber Lab-Tek tissue culture slide.

(vi) Incubate the cells in the T-25 flask and Lab-Tek slide at 37°C.

(vii) Change the medium after 24 h to remove brain tissue debris.

(vii) After 3 days of incubation, fix the cells on the Lab-Tek slide and, by a direct FA conjugate staining technique (see Section 2.2.5), assess the level of cell infection as expressed by the percent of fluorescent viral inclusions.

(ix) The monitoring of cell infection can also be done on Terasaki plates. In this case, 10 µl aliquots of cells should be transferred before the addition of 5 ml of medium and the Terasaki plates can be fixed and stained with FA conjugate as described in Section 2.2.5.

(x) If less than 50% of the cells are infected, remove the medium from the culture in the T-25 flask and store at −70°C.

(xi) Trypsinise the cells, suspend them in 2 ml of MEM-10, return 0.5 ml of cells to T-25 flasks (1:4 split) and add 5 ml of MEM-10.

(xii) If necessary, repeat the incubation of cells, monitoring the infection by FA staining as described above until all cells in the cultures are infected and show a positive staining by a direct FA staining procedure.

(xii) The titre of virus from the first culture or after an additional infected cell transfer should be determined. If no virus is present after the second infected cell transfer the culture can be considered negative and discontinued (24).

3.1.1 *Purification of Rhabdoviruses*

Rhabdovirus particles have a highly characteristic structure which exhibits, in negatively stained preparations, a rigid cylindrical symmetry of varying length but invariably with one hemispherical and one planar end. These bullet-shaped particles, as they are commonly called, are predominantly 170–180 nm long, 70 nm in diameter, and represent standard infectious particles. Defective-interfering (DI) particles of rhabdoviruses maintain the bullet-shaped morphology characteristic of the infectious particles but are shorter in the long axis. These truncated particles, are non-infectious particles produced in increasingly larger numbers during serial undiluted passage of most standard infectious rhabdovirus as a consequence of successful competition (interference) with production of standard infectious virus. DI particles have the same chemical composition as standard particles but contain less genomic RNA (25). The difference in size between DI particles and their homologous standard virions results in considerably different sedimentation coefficients for the two types of particles and has been used as a basis for their physical separation by rate-zonal centrifugation.

The purification procedure outlined below is effective when infectious tissue culture fluid is used as a source of virus. Since virus is released from the cell by budding from the plasma membrane of the infected cell, infections with virus which cause only weak cytopathic effects yield extracellular virus with relatively little contamination by cellular components. Early harvests may also reduce cell debris in the tissue culture fluid and, so that binding of serum components to the virions and contamination of purified virus preparations with aggregated components will not occur, the virus should be grown in cell culture medium which is supplemented with 0.1–0.3% bovine serum albumin instead of serum. Considering these possible sources of contamination and the process of differential centrifugation, at best only a partial purification of virus may be achieved.

The first step in the purification procedure must be to free virus from the cells and cellular debris contaminants in the culture fluid. This can be accomplished by low-speed centrifugation at 600 *g*. The second step is to concentrate the virus in clarified supernatants and free it of the bulk of low and high molecular weight substances. This may be done by any of the methods suitable for virus concentration, provided the temperature is kept low (4–10°C), the pH of the suspension is kept within the range 6.8–8.0, and extreme ionic strengths are avoided. Two methods discussed below are sedimentation of the virions by high-speed centrifugation and precipitation of the virus by zinc acetate, ammonium sulphate or polyethylene glycol. The sedimentation and precipitation methods are useful for small and preparative size batches of virus in culture fluid and have the advantage of being able to separate virus particles from the bulk of non-virion proteins.

Sedimentation of virions by high-speed centrifugation is the most widely used method of virus concentration providing the laboratory is equipped with a suitable rotor and high-speed centrifuge. The sedimentation coefficient of standard infectious VSV and rabies virus is about 600S; DI particles being shorter in length by approximately one-third to three-fourths of the standard virion have correspondingly smaller sedimentation coefficients (25). The routine procedure adopted in most laboratories for the sedimentation of virus from infectious tissue culture fluid is to transfer the clarified virus suspension in a laminar flow hood to high-speed centrifuge tubes or large size (250 ml) cups, fitted with caps, of a suitable size angle or swinging bucket rotor of a Beckman (Spinco) or other high-speed centrifuge.

(i) Sediment the virus by centrifugation at 48 000 – 50 000 *g* for 90 min at 4°C.
(ii) After centrifugation, return the closed tubes containing virus pellets to a laminar flow hood and carefully remove the supernatant fluid.
(iii) Drain the tubes well.
(iv) Resuspend the virus in the pellet in a small volume of solution containing 0.13 M sodium chloride, 0.05 M Tris-HCl, 1 mM EDTA (NTE buffer, pH 7.8) to 1% of the original volume of culture fluid.
(v) Set the contents of the tubes at 4°C for 3 – 4 h with occasional swirling or overnight without swirling to soften the pellets. Gentle pipetting with a Pasteur pipette is then sufficient to resuspend the virus.
(vi) Remove any cell debris and aggregated virions which cannot be easily resuspended under these conditions by centrifugation at 1000 *g* for 20 min at 4°C.

Almost all the infectivity present in the original preparation is recovered in the resuspended sediment.

(i) To precipitate virions by Zn^{2+} ions, mix 50 parts of the infectious tissue culture fluid (pH 7.4 – 7.8) with 1 part of 1 M zinc acetate (pH 5.0) (26). After mixing, the pH of the virus suspension drops to about 6.8.
(ii) Allow the mixture to stand at 0 – 4°C for 20 – 60 min and collect the precipitate formed by centrifugation at 1000 *g* for 40 min.
(iii) Discard the clear, essentially virus-free, supernatant fluid and resuspend the sedimented material in a saturated solution of the disodium salt of EDTA made with 11.7 g per 100 ml of water adjusted to pH 7.8 by the addition of solid tris (hydroxymethyl) aminomethane (Tris). The volume of the EDTA solution used for resuspending the virus must be at least 1.25% of the original volume of infectious tissue culture fluid.
(iv) Check that the virus suspension is clear; continue to dissolve any residual precipitates, if present, by adding more of the EDTA solution.
(v) Clarify the concentrated virus by centrifugation at 1000 *g* for 20 min, it may still be contaminated with variable amounts of bovine serum albumin and small amounts of cellular components and compounds of low molecular weight which are often trapped in the precipitate.

Precipitation of virions with ammonium sulphate or PEG-6000 offers an alternative method for the concentration of virus.

(i) To the clarified infectious tissue culture fluid, add $(NH_4)_2SO_4$ (300 g per litre of supernatant fluid) or PEG-6000 (30 g per litre) slowly (with stirring) at 4°C keeping the pH at 7 – 8 by adding 1 M Tris-HCl buffer (pH 8).

(ii) Collect the precipitate by centrifugation at 1500 *g* for 30 min, dissolve the ammonium sulphate precipitate in 0.4 M $(NH_4)_2SO_4$, 0.1 M Tris, 0.01 M EDTA (pH 7.4) and centrifuge for 70 min at 82 000 *g* over a 1 ml pad of 100% glycerol in a swinging bucket rotor (Spinco SW 27).

(iii) After centrifugation, completely remove the supernatant above the virus band and collect the virus on the pad by adding $(NH_4)_2SO_4$-Tris-EDTA buffer and swirling as minimally as possible to suspend the particles.

(iv) Dilute the virus suspension 3- to 5-fold for further purification by sucrose density centrifugation as described below.

The third and final step in the purification of virus is to separate standard infectious virus from DI particles which may be present in the extracellular virus harvest. This is accomplished by banding the virus in a linear density gradient, taking advantage of the differential sedimentation coefficients between standard infectious virions and their DI particles. Banding of virus in a sucrose density gradient is usually accompanied by a 2-fold or 3-fold decrease in infectivity, because many of the virus particles are disrupted during removal of the sucrose by dialysis. To minimise inactivation, purified virus has to be dialysed stepwise against sucrose solutions of decreasing concentrations.

(i) Separation of standard virion and DI particles is achieved by layering the concentrated virus on to a linear gradient of 5 – 30% (w/v) sucrose (shallow) or 10 – 60% sucrose (steep) in NTE buffer and centrifuging the gradients for 45 min at 54 000 *g* or for 90 min at 61 000 *g*, respectively.

(ii) After centrifugation, collect 20 – 30 fractions through a hole pierced in the bottom of the centrifuge tube or, if bands of particles are visible, collect these by side-puncture with a needle connected to a syringe.

(iii) Dilute the harvested particles in 3 – 5 volumes of NTE, and sediment the particles at 189 000 *g* for 3 h in a swinging bucket rotor.

(iv) Resuspend the pellets in NTE for further biochemical analysis.

In the absence of carrier protein, the infectivity of virus purified by the procedure outlined above is rather labile. After the addition of 0.5 – 1% of bovine serum albumin, the virus can be stored at – 70°C for several months without essential loss of infectivity.

3.2 **Plant Rhabdoviruses**

Most purification procedures for plant viruses are not suitable for the purification of rhabdoviruses as they involve clarification of crude plant extracts with heat at 55 – 60°C or by treatment with organic solvents (4). These procedures are impractical for purification of rhabdoviruses for two reasons; the first is the possibility of thermal inactivation, and the second is the use of organic solvents will destroy the viral envelope. To circumvent these problems, methods have been developed for purifying the plant rhabdoviruses along the lines used for animal rhabdoviruses. An excellent review of these methods discusses the use of density-gradient centrifugation and electrophoretic techniques for separating virus from tissue debris and cites additional procedures for extracting plant rhabdovirus from infected material (4).

The method which follows is a scheme currently used in the Department of Virology at Agricultural University, Wageningen, The Netherlands (27). Virus is obtained for purification from leaves of infected *Nicotiana glutinosa* or *Nicotiana christii* 10–12 days after inoculation. Healthy plants can be mechanically inoculated with virus from one or two leaves of infected plants after infected leaves have been ground in 0.01 M Na_2SO_3 which acts as a reducing agent. In cases where virus cannot be mechanically inoculated into plants, insect-infected plants have been used. The growth conditions can have an effect on the purification and yield of virus, low yields resulting when conditions with much sunlight are used.

Virus purification begins with an extraction procedure which yields virus from the infected plant.

(i) Harvest and infiltrate approximately 60–80 g of infected leaf material with extraction buffer under vacuum. The extraction buffer consists of 0.1 M glycine (or 0.01 M Tris), 0.01 M $MgCl_2$ and 0.01 M KCN (or 0.01 M Na_2SO_3), pH 8.2–8.4.

(ii) Incubate the leaf infiltrate for 1–2 h before blending the leaves in a Waring blender with three volumes of extraction buffer for 15 sec or less.

(iii) Adjust the pH of the macerate if necessary so that it remains at or above pH 7.2; otherwise it can be difficult to separate the plant constituents from the virus during the purification. If the amount of leaf tissue happens to be larger (e.g., 300 g), blending can be done in an equal volume (300 ml) of cold glycine extraction buffer for 1 min at the highest setting (28).

(iv) Squeeze the brei through four layers of cheese cloth to remove the pulp. A note about the glycine extraction buffer to keep in mind is the buffer containing 0.1 M glycine, 0.1 M NaH_2PO_4, and 0.01 M $MgCl_2$ (adjusted to pH 7.2 with NaOH) can be stored at 4°C, and just before use, Na_2SO_3 can be added as reducing agent. The presence of reducing agent is absolutely required for recovery of virus (29).

(v) Centrifuge the juice at 8000 *g* for 5 min.

(vi) Collect the supernatants and clarify the plant extracts by adding Celite or Hyflo Supercell to a final concentration of 10% (w/v) and filter the mixture by sucking the liquid through a Celite or Hyflo Supercell pad (5–7 mm thick), moistened with extraction buffer, in a Buchner funnel. Sometimes a vacuum from a strong pump is necessary. Pull the vacuum until the liquid reaches the top of the green surface so that the surface is slightly moist. Do not allow the pad to dry!

(vii) Slowly add 60 ml of the glycine extraction buffer to the surface of the pad and pull the liquid under vacuum until the surface is moist.

(viii) Add a second aliquot of extraction buffer and continue the suction until the pad is dry. The resulting filtrate should be a light tan color without any green tinge.

(ix) Light scattering can be detected in preparations containing reasonable amounts of virus.

Note: preparation of the Celite pad is extremely important and it is somewhat of an art. For 300 g of leaf tissue, the Celite pad is prepared in a 19.5 cm diameter Buchner funnel. Usually, 85 g of Celite analytical filter aid is suspended in a sufficient amount of glycine extraction buffer to make a well-suspended slurry. This slurry is poured over two wet Whatman 3MM filter papers (18.5 cm diameter) and gently sucked under vacuum until the surface of the pad just glistens with liquid. The pad should be about 5 – 7 mm thick and uniform in thickness. This pad should be made within an hour of initiating the purification and stored at 4°C until used. One pad should be sufficient for 300 g of leaf tissue (29).

(i) Concentrate the virus from the filtrate with addition of 0.3 volumes of 30% polyethylene glycol (PEG 8000) in water.
(ii) Stir the mixture at 4°C for 1 h to precipitate the virus.
(iii) Recover the precipitate by centrifugation for 20 min at 9000 r.p.m. in a large capacity (250 ml bottle size) rotor, and dissolve the pellet in maintenance buffer (0.1 M Tris-HCl, 0.01 M $MgAc_2$, pH 7.5) equal to one fifth the volume of the clarified filtrate.
(iv) Filter the virus suspension through a thinner Celite pad, prepared as above except that 30 g of Celite is suspended in maintenance buffer, as described for the thicker filter, but use 35 ml of maintenance buffer for each of the washes.
(v) Transfer the filtrate to centrifuge tubes and sediment the virus by centrifugation at 20 000 *g* for 1 h.
(vi) Resuspend the pellets in 1.5 ml of maintenance buffer (this volume may be altered depending on the size of the pellet) and layer the virus suspension on to a linear (15 – 30%) sucrose density gradient in maintenance buffer and band the virus by rate-zonal centrifugation at 67 000 *g* for 25 min.
(vii) Remove any visible band of virus in the gradient by side puncture and purify the virus suspension further through a second steeper density gradient, e.g., 20 – 60% (w/v) sucrose.
(viii) Recover the virus band, and dilute in approximately five volumes of maintenance buffer and sediment the virus by high speed centrifugation through the dilute sucrose.
(ix) To recover the virus pellet, carefully decant the supernatant, draining and wiping the insides of the tube with rolled up filter paper, and resuspend the pellet in 0.5 ml of maintenance buffer and freeze at low temperature (– 20 or – 70°C).

4. REFERENCES

1. Howatson,A.F. (1970) *Adv. Virus Res.*, **16**, 196.
2. Matsumoto,S. (1970) *Adv. Virus Res.*, **16**, 257.
3. Knudsen,D.L. (1972) *J. Gen. Virol.*, **20**, 105.
4. Francki,R.I.B. (1973) *Adv. Virus Res.*, **18**, 257.
5. Emerson,S.U. (1976) *Curr. Top. Microbiol. Immunol.*, **73**, 1.
6. Schneider,L.G. and Diringer,H. (1976) *Curr. Top. Microbiol. Immunol.*, **75**, 153.
7. Recommendation of the Immunization Practices Advisory Committee (1984) in *Morbidity and Mortality Weekly Report*, Vol. **33**, p. 393, published by *N. Eng. J. Med.*
8. Clark,H.F. (1979) in *Rhabdoviruses*, Vol. **1**, Bishop,D.H.L. (ed.), CRC Press, Boca Raton, Florida, p. 23.

9. McClain,M.E. and Hackett,A.J. (1958) *J. Immunol.*, **80**, 356.
10. Takehara,M. (1975) *Arch. Virol.*, **49**, 297.
11. Bachrach,H.L., Callis,J.J. and Hess,W.R. (1956) *Proc. Soc. Exp. Biol. Med.*, **91**, 177.
12. MacPherson,I. and Stoker,M. (1962) *Virology*, **16**, 147.
13. Kaplan,M.M., Wiktor,T.J., Maes,R.F., Campbell,J.B. and Koprowski,H. (1961) *J. Virol.*, **1**, 145.
14. Wunner,W.H., Reagan,K.J. and Koprowski,H. (1984) *J. Virol.*, **50**, 691.
15. Dulbecco,R. (1952) *Proc. Natl. Acad. Sci. USA*, **38**, 747.
16. Franklin,R.M. (1958) *Virology*, **5**, 408.
17. Smith,A.L., Tignor,G.H., Mifune,K. and Motohashi,T. (1977) *Intervirology*, **8**, 92.
18. Cooper,P.D. (1961) *Virology*, **13**, 153.
19. Sedwick,W.D. and Wiktor,T.J. (1967) *J. Virol.*, **1**, 1224.
20. Vaheri,A., Sedwick,W.D., Plotkin,S.A. and Maes,R. (1965) *Virology*, **27**, 239.
21. Eagle,H. (1955) *Science (Wash.)*, **122**, 501.
22. Lorenz,R.J. and Bögel,K. (1973) in *Laboratory Techniques in Rabies*, Kaplan,M.M. and Koprowski,H. (eds.), World Health Organization, Geneva, p. 321.
23. Koprowski,H. (1966) in *Laboratory Techniques for Rabies, 2nd edition*, World Health Organization, Geneva, p. 69.
24. Wiktor,T.J., Macfarlan,R.I., Foggin,C.M. and Koprowski,H. (1984) *Developments in Biological Standardization Joint WHO/IABS Symposium Proceedings*, Vol. **57**, in press.
25. Reichmann,M.E. and Schnitzlein,W.M. (1979) *Curr. Top. Microbiol. Immunol.*, **86**, 123.
26. Sokol.,F., Kuwert,E., Wiktor,T.J., Hummler,K. and Koprowski,H. (1968) *J. Virol.*, **2**, 836.
27. Peters,D., personal communication.
28. Jackson,A.O. and Christie,S.R. (1977) *Virology*, **77**, 344.
29. Jackson,A.O., personal communication.

CHAPTER 5

Pneumoviruses

CRAIG R. PRINGLE

1. UNIQUE FEATURES OF PNEUMOVIRUSES

1.1. General Introduction

The pneumoviruses comprise human respiratory syncytial virus, bovine respiratory syncytial virus and pneumonia virus of mice. Respiratory syncytial (RS) virus, the prototype of the group, is a pleomorphic enveloped virus with an unsegmented negative-stranded RNA genome. It is classified in the genus *Pneumovirus*, which together with the genus *Paramyxovirus* and the genus *Morbillivirus* makes up the family Paramyxoviridae (1). It is uniquely associated with annual epidemics of respiratory illness in infants and is sufficiently distinct in its general and molecular biology and clinical importance to be classified separately from other paramyxoviruses. RS virus is also frequently isolated from agricultural animals (cattle, sheep and goats) and bovine RS virus is associated with respiratory disease in cattle. Antibody to RS virus is frequently present in the sera of domesticated and laboratory animals in contact with man, although there have been no reported isolations. The bovine and human isolates of respiratory syncytial virus are serologically similar, and human isolates have shown no clinically relevant antigenic variation. The serologically unrelated murine pneumonia virus (PVM), which is associated with respiratory disease in mice, has some properties in common with RS virus and is the only other virus sufficiently similar to be included in the genus *Pneumovirus*. Antibody to PVM is present in a high proportion of human sera (Pringle, unpublished data), and historically there is evidence suggestive of a human origin for this pathogen in some laboratory colonies of mice (2).

The principal features of RS virus are listed in *Table 1*. RS virus differs from other paramyxoviruses in the greater number of identified gene products (10 compared with 6 – 7), the gene order (two non-structural protein genes interposed between the 3′ leader sequence and the nucleoprotein gene and another between the putative F and G glycoprotein genes), the lack of both a haemagglutinin and a neuraminidase, and the characteristic dimensions of nucleocapsid (12 – 13 nm diameter compared with 17 – 18 nm) and surface spikes (10 – 12 nm length compared with 8 nm).

These differences, together with the notorious lability of RS virus, justify its separate consideration in this volume. This article is not an exhaustive catalogue of available methods, but rather concentrates on a few reliable procedures for each technique described.

Table 1. General Properties of Pneumoviruses.

Known members of genus:	Human RS virus Bovine RS virus Murine pneumonia virus (PVM)
Nucleic acid:	Single-stranded unsegmented RNA 50S $>5 \times 10^6$ mol. wt.
Proteins[a]:	Two glycosylated envelope proteins: [A 90-K glycoprotein (G), and disulphide bond-linked 48-K and 20-K cleavage products of a 68-K precursor glycoprotein (F?)] Two unglycosylated envelope proteins: a 26-K matrix protein and a 24-K protein of unknown function Three core proteins: a 200-K putative polymerase (L), a 41-K nucleocapsid (N) protein and a 34-K phosphoprotein (P) Three non-structural proteins of unknown function (14 K, 11 K and 9.5 K)
Virion enzymes:	RNA polymerase activity Haemagglutinin (in PVM only) No neuraminidase
Particle density:	1.15 – 1.26 g/cm^2
Particle morphology:	Pleomorphic 80 – 500 nm diameter particles and filaments of 60 – 110 nm diameter up to 5 μm long
Nucleocapsid morphology:	Helical, 12 – 13 nm in diameter
Membrane spike morphology:	10 – 12 nm in length
Genetic interactions:	Eight complementation groups defined No recombination or multiplicity reactivation
Stability:	Thermosensitive. Unstable below pH 3.0. Sensitive to freeze-thawing
Host Range:	Probably restricted *in vivo*
Pathogenicity:	Predominantly associated with respiratory disease, and sporadically with other conditions
Cytopathogenicity:	Cytoplasmic inclusions and rarely nuclear inclusions. Syncytia frequent (except PVM). Propensity to become persistent.

[a]The molecular weights of the proteins have not been determined precisely and slightly different values are used by different authors.

1.2 Nomenclature

Respiratory syncytial virus should be abbreviated as RS virus to distinguish it from the retrovirus Rous sarcoma virus universally abbreviated as RSV. Bovine respiratory syncytial virus should be distinguished from the unrelated bovine syncytial virus, a retrovirus frequently isolated from cattle.

1.3 Virus Isolation

The unusual lability of RS virus and its low titre in secretions and infected tissues require special care in isolation. Isolation from stored specimens is rarely successful. Greatest success is achieved by taking the susceptible culture cells to the patient or infected animal rather than by attempting to transport materials to the laboratory for subsequent inoculation.

Hep-2 or HeLa (Bristol) cells are the most favoured tissue culture cells for isolation of RS virus, but rhesus monkey kidney cells, BS-C-1, Vero, MRC-5, WI-38 and others are sometimes employed. Nasal and oropharyngeal swabs are effective means of isolating RS virus from hospitalised children. Swabs can be transferred to a small volume (2 ml) of buffered salt solution (e.g., Hanks' balanced salt solution with 0.5% gelatine and antibiotics) or broth [e.g., veal infusion with 0.5% bovine serum albumin (BSA) and antibiotics] and the secretions liberated by vigorous swirling. The sampling fluid should be kept on ice and inoculated onto susceptible cells immediately or as soon as possible after collection and certainly within a few hours. The antibiotics should include penicillin (400 units/ml), streptomycin (200 μg/ml), neomycin (100 μg/ml) and mycostatin (100 units/ml).

The important factors in successful isolation of RS virus are avoidance of freeze-thawing, maintenance of uniform low temperature, rapid transfer to susceptible cells and use of foetal animal serum or BSA in culture medium. The sera of adult animals of most of the domestic animals (cattle, sheep, pigs, goats, horses) used to supplement tissue culture medium usually contain neutralising antibody to RS virus at appreciable titres and should never be used. Inoculated cultures should be incubated at 36–37°C and observed for at least 3 weeks with several medium changes. Further blind passages may be required before cytopathic effect (cpe) is observed. A syncytial type of cpe is characteristic but it is not always observed and is dependent on virus strain and type of host cell.

1.4 **Storage**

Pneumoviruses should be stored between −70°C (temperature of solid CO_2 and −198°C (temperature of liquid N_2). Infectivity decays at an appreciable rate at −20°C and this temperature is unsuitable for long term (>3 months) storage. Lyophilisation or addition of glycerol will prolong survival at −20°C. The rate of decay of infectivity varies according to the nature of the material stored, membrane-associated infectivity surviving better than released virus.

2. GENERAL TECHNIQUES

2.1 **Propagation of RS Virus**

2.1.1 *Cell Susceptibility*

A wide range of cell types is susceptible to infection by RS virus. The heteroploid human Hep-2 cell line is the most frequently used in diagnostic laboratories because these cells are easily maintained.

Human diploid cells such as the embryonic lung cell lines WI-38, HeL-1 (3), L-4, L-49 (4), HLDC-1, -4, -6 and the embryonic brain cell lines HBC-1 and HBC-4 (5) may be more suitable for isolation of respiratory syncytial virus than the heteroploid Hep-2 cell line.

Human RS virus can grow in a wide range of non-human cells, including feline kidney, mink kidney and the marsupial Potoroo kidney cell lines (6). The relative susceptibility of different cells is probably dependent to some extent on the passage history of the virus strain employed. For instance, three cycles of growth of the RSN-2 strain from single plaques on WI-38 cells yielded a stock of virus with a 50-fold higher efficiency of plating on WI-38 cells relative to a stock propagated simultaneously in BS-C-1 cells (7).

RS virus has also been adapted for growth in primary cultures of cells from a cold-blooded animal – the tortoise *Testudo gracea* (8) – but it does not multiply in cells of the amphibian *Xenopus laevis* (XTC-2 cells) or in invertebrate (mosquito) cells (Pringle, unpublished observations). Unlike many paramyxoviruses, RS virus does not multiply in eggs or chicken embryo cells. Also, unlike pneumonia virus of mice, it cannot be propagated in BHK-21 cells.

Like the majority of viruses grown in cultured cells, RS virus is prone to accumulate defective interfering (DI) particles during multiplication under conditions of undiluted passage (i.e., repeated cycles of infection at high multiplicity). A simple colorimetric analysis for detection and quantitation of DI particles is available (see Section 3.3.2).

Eagle's minimum essential medium (MEM), preferably supplemented with non-essential amino acids, is a satisfactory maintenance medium for RS virus-infected cells. Actively growing cells usually give better yields of virus. Other media, such as Leibowitz's medium which does not require buffering with gaseous carbon dioxide, are also satisfactory provided cell growth is not adversely affected. RS virus will grow at temperatures of incubation over the range 25–39°C at least, optimum yields being obtained in the region of 37°C.

2.1.2 *Cytopathic Effect*

The characteristic cytopathic effect of RS virus is syncytium formation (*Figure 1*). In syncytia the nuclei are frequently aggregated around an empty central region. Syncytium formation is by no means universal, however, and is dependent on many factors including virus strain, cultural conditions and host cell type. For instance, infection of the BS-C-1 cell line of African green monkey kidney cells results in the formation of foci of aggregated cells. Under agar overlay these foci appear as deeply stained clumps of cells not unlike the foci of transformed cells induced by oncogenic viruses. However, the RS virus-induced foci are surrounded by a lighter stained region (*Figure 2*). This lighter stained region appears to be the consequence of migration or absorption of cells into the central mass. The foci contain enlarged and swollen cells which progress to syncytia. Despite the resemblance to transformation foci, none of these cells remain viable.

2.1.3 *Enhancement Treatments*

Exposure of virus or cells to DEAE-dextran enhances sensitivity to RS virus infection. At an optimal concentration of 40 μg/ml, a 2-fold enhancement of virus yield and an 11-fold increase in the number of cells infected has been reported for bovine RS virus. Nisevich *et al.* (9) reported a 100-fold increase at 15 μg/ml for human RS virus.

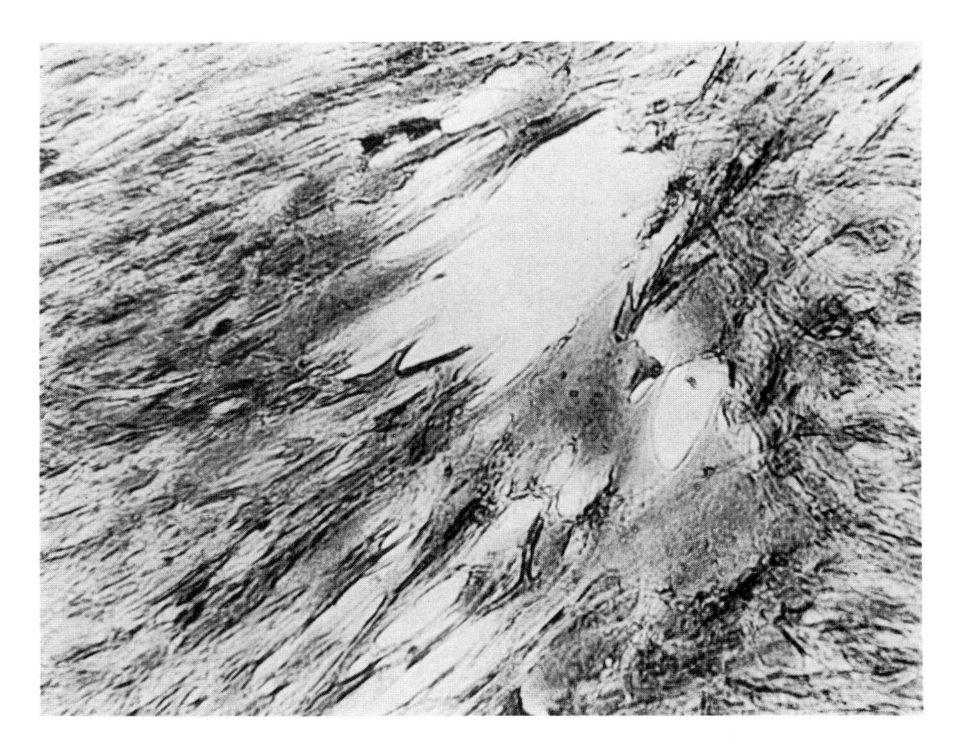

Figure 1. Monolayer of MRC-5 cells showing RS virus-induced syncytial cpe.

2.1.4 *Stabilisation of Infectivity*

RS virus infectivity is stabilized in the presence of high concentrations of sucrose (~40%) or magnesium ions. Fernie and Gerin (10) reported that 1 M $MgSO_4$ increased the stability of RS virus 30-fold, e.g., at +4°C the half-life of RS virus was increased to 12 weeks. Addition of 1 M $MgSO_4$ facilitates concentration and purification of RS virus for physical and chemical purposes, but is not an appropriate additive to medium for isolation of RS virus or routine infectivity assay because the high osmotic pressure will adversely affect the host cells.

2.2 **Assay of Infectivity**

The infectivity of RS virus samples can be assayed by plaque counting. The following method described for BS-C-1 cells is suitable for assays with most types of cells.

(i) Prepare monolayers by seeding 1 x 10^6 cells/dish into 60 mm Petri dishes (and *pro rata* for tissue culture dishes of other dimensions). Incubate overnight at 37°C. Use monolayer immediately confluence is achieved.
(ii) Remove the incubation medium and inoculate 0.2 ml of 10-fold dilutions of virus-containing samples. The diluent may be either a buffered salt solution, such as Dulbecco's phosphate buffered saline (PBS), or preferably Eagle's MEM, both containing 0.5% foetal calf serum (FCS) and antibiotics (penicillin 100 units/ml; streptomycin 100 μg/ml).

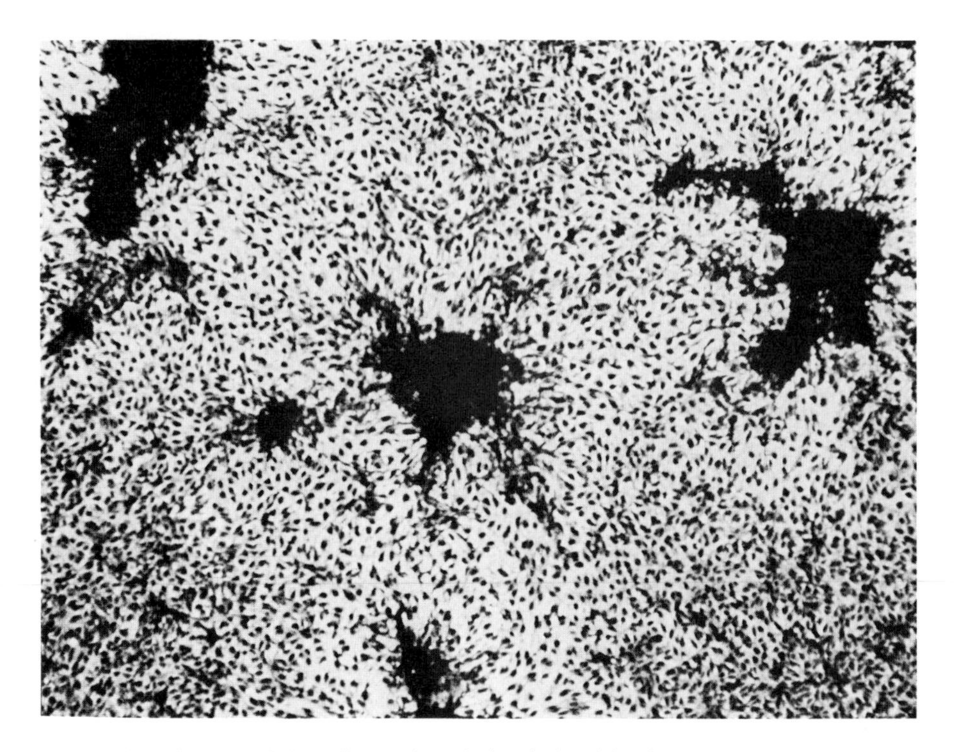

Figure 2. Monolayer of BS-C-1 cells showing RS virus-induced focal cpe.

(iii) Allow 1 h at 31°C for adsorption.

(iv) Add overlay medium at 5 ml per 60 mm Petri dish or *pro rata*. Prior removal of the inoculum is not necessary. The overlay medium can be Eagle's MEM containing 1–2% FCS, antibiotics and 0.9% agar. This medium can be prepared by mixing equal volumes of double-strength Eagle's medium at 37°C with melted 1.8% Noble agar. The agar will remain molten at 43–46°C and will be suitable for application to cell monolayers without adverse effects.

(v) Incubate at 31–39°C for 3–5 days according to temperature; the higher the temperature the shorter the period of incubation. For permanent preparations which can be scored at leisure, add 2 ml of 1% glutaraldehyde in PBS to each 60 mm dish and leave at room temperature for a minimum of 4 h for fixation. Remove fixative and agar overlay and add an appropriate staining solution (e.g., 0.075% Giemsa, 0.1% crystal violet, or similar). Stain for 5 min, wash thoroughly in running tap water and air dry. A low power binocular microscope will be needed for greatest accuracy of plaque counting. Alternatively the infected monolayers can be stained with 0.01% neutral red in PBS or in solid 0.9% agar-containing medium. It is usual to omit phenol red indicator from the overlay medium to enhance contrast, but not essential. The liquid neutral red stain gives a faster result (visible plaques in 2 h), but the preparations are impermanent and the stained monolayers are susceptible to photochemical damage on prolonged exposure to light. Staining under solid neutral red-containing overlay is slower (staining requiring ~4 h), but is suitable for assays where it may be necessary to recover progeny virus from individual plaques.

2.3 **Haemagglutination**

No haemagglutinins have been detected in RS virus-infected tissue culture fluids. Human type O, African green monkey, rhesus monkey, guinea pig, rat, hamster, mouse, goose, sheep, pig, hen and chick erythrocytes have been tested with negative results.

PVM, however, can agglutinate mouse and hamster erythrocytes. The technical details of the haemagglutination assay are described by Compans *et al.* (11).

2.4 **Haemadsorption**

Haemadsorption to RS virus-infected cells has not been observed. Red blood cells from African green monkey, chimpanzee, gelada, guinea pig, rat, hamster, mouse, duck, goose, pigeon, hen and chick have been tested with negative results. However, mouse erythrocytes can adsorb to PVM-infected cells and the technical details have been described by Compans *et al.* (11).

2.5 **Immunofluorescence Methods**

Immunofluorescent staining with polyclonal serum is a convenient and accurate means of enumerating infected cells or establishing the site of virus multiplication by examination of fixed and sectioned tissue. The sub-cellular localisation of viral antigens can be determined using immunofluorescent staining in conjunction with monoclonal antibodies specific for the polypeptide components of the virion.

Immunofluorescent staining can be employed also as a means of rapid diagnosis of virus infection. This aspect of immunofluorescent staining has been reviewed comprehensively (12,13) and the practical details relating to the use of the technique for detection of RS virus in clinical material have been described by Gardner *et al.* (14). Immunofluorescence is a more sensitive technique than virus isolation for confirmation of RS virus infection because, during the convalescent phase of an RS virus illness, infected cells become coated with anti-viral antibody.

The indirect method of immunofluorescent staining is more sensitive than the direct method but non-specific fluorescence is a hazard. Non-specific fluorescence can be eliminated by exhaustive pre-absorption of antisera and the anti-globulin conjugate with acetone-treated tissue culture cells, or by absorption with human and animal liver powders. Acetone treatment is carried out as follows.

(i) Harvest about $10^8 - 10^9$ cells by scraping into suspension.
(ii) Concentrate the cells by centrifugation and resuspend in 0.01 M PBS; this procedure should be repeated three times.

(iii) Re-suspend the cell pellet in acetone (10 times the pellet volume) for 5 min, and recover the cells by centrifugation.
(iv) Re-suspend the cells in 0.01 M PBS and subject them to six cycles of pelleting and resuspension to remove all traces of acetone. The cells can now be stored in 0.01 M PBS at $-20°C$ until required.

The following is a typical protocol for detection of RS virus antigen in infected cells by indirect immunofluorescence.

(i) Seed plastic dishes (50 mm) containing aqua regia-cleaned and ethanol-dehydrated coverslips (13 mm diameter) with 10^6 BS-C-1 cells per dish and, after 20 – 24 h of incubation at 37°C, wash and infect with RS virus (input multiplicity, ~1 p.f.u./cell).
(ii) After 2 h of absorption, wash the infected coverslips thoroughly with MEM plus 2.5% FCS and incubate at either 31 or 39°C.
(iii) Harvest the cover slips 40 – 48 h after infection, fix in pre-cooled acetone (4°C) for 5 min, air dry for 30 min, and store at $-20°C$.
(iv) Pre-absorb the anti-RS virus serum with acetone-treated BS-C-1 cells at 4°C and use at a pre-determined dilution in the range 1:10 – 1:50. Rabbit anti-globulin for several species, conjugated with fluorescein isothiocyanate (FITC), can be obtained commercially. The conjugates should be pre-absorbed with acetone-treated mouse liver powder and BS-C-1 cells at 4°C.
(v) Use the conjugates at their optimum staining times. Expose the coverslips to antiserum for 1 h at 37°C and wash, then add appropriate conjugates for a further 45 min at 37°C.
(vi) After a final wash, mount the coverslips in buffered glycerol saline (pH 8.3) and examine with a fluorescence photomicroscope. See *Figure 3* for an example of the resolution possible by this method.

The specificity of the staining reaction can be verified by (i) absence of staining of uninfected BS-C-1 cells, (ii) absence of staining with FCS (or with antisera to other viruses), and (iii) loss of staining after the absorption of immune serum

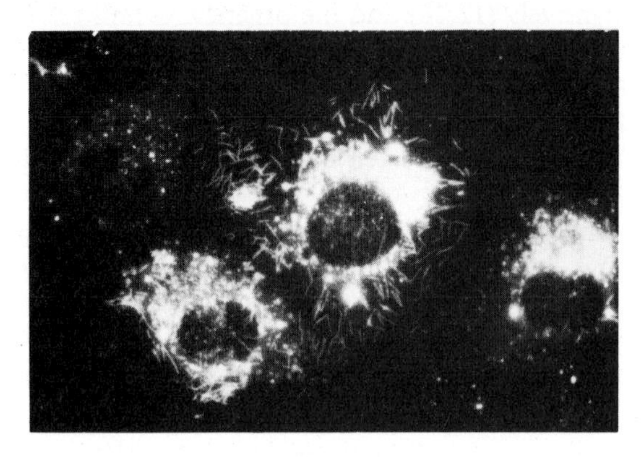

Figure 3. Infected BS-C-1 cells stained with anti-RS virus serum and anti-bovine globulin conjugated with FITC showing extensive early filament production (x 480). Photograph supplied by P.V.Shirodaria.

with BS-C-1 cells infected with RS virus.

Immunoperoxidase staining by standard procedures provides equivalent sensitivity and resolution to immunofluorescent staining, and offers the advantage that a microscope with u.v. optics is not required.

2.6 Enzyme-linked Immunosorbent Assay (ELISA)

The ELISA technique is another immunological procedure for rapid detection and quantitation of RS virus. The following is a typical method.

(i) Coat 96-well plates with either RS virus-infected or uninfected cells or gradient-purified RS virus. The cell antigens can be prepared by either of the following methods.
  (a) Cells are grown in 96-well plates and alternate rows are infected with RS virus leaving the remaining rows in the same plate with uninfected cells. When cytopathic effect is observed, the cells are washed, fixed with ethanol-acetic acid and stored at −20°C.
  (b) Gradient-purified RS virus is added to 96-well plates at approximately 1 μg/well and dried overnight at 37°C. (RS virus persistently infected cells, if available, can also be used as a source of antigen in place of lytically infected cells.)

(ii) Before the ELISA assay, coat all wells with 5% BSA in PBS. Negative controls (pre-immune serum and acetone fluid) should be included in each plate. A standard anti-RS virus serum of known titre should also be included as the positive control. Add the virus-containing samples to the plates (50 μl/well) and incubate overnight at 4°C. Wash the plates then add sheep $F(ab')_2$ anti-mouse Ig conjugated to horseradish peroxidase (New England Nuclear or equivalent) at the recommended dilution (1/500). Incubate the plates for 1 h at 37°C and then wash. Add the substrate, *o*-phenylenediamine, and incubate the plates for 30 min at 37°C. After termination of the assay by addition of 4.5 M H_2SO_4, score the wells using a Multiskan plate reader (Flow Laboratories).

Hierholzer *et al.* (15) have described a procedure whereby protein bands can be quantified by an ELISA reaction following Western blotting (see Section 3.7.1). The advantage of this procedure is that it does not require the use of radioisotopes.

3. ANALYTICAL PROCEDURES

3.1 Radiolabelling, Radioimmunoassay and Radioimmunoprecipitation

3.1.1 *Radiolabelling of Intracellular RS Virus Proteins*

RS virus polypeptides can be labelled effectively using [^{35}S]methionine + [^{35}S]-cysteine as follows.

(i) Labelling is best carried out by adding 2.5 μg/ml actinomycin D at 18 h after infection of cells at a p.f.u./cell ratio of greater than 1.

(ii) After 2 h, replace the incubation medium with medium lacking methionine and cysteine and containing 50 μCi/ml L-[^{35}S]methionine and 25 μCi/ml L-[^{35}S]cysteine.

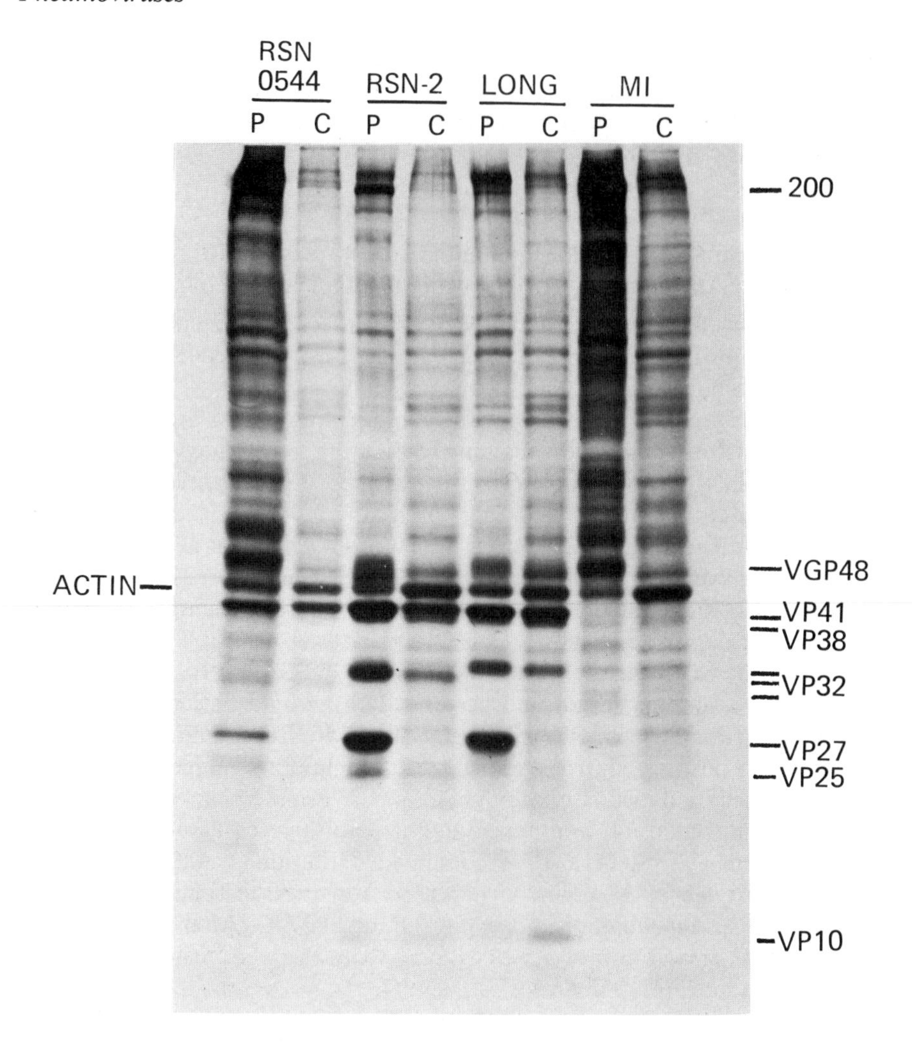

Figure 4. Autoradiograph of [^{35}S]methionine-labelled polypeptides of the RSN-0544, RSN-2 and Long strains of RS virus showing differences in mobilities of the VP32 polypeptide. (C) Cell pellet fraction; (P) polyethylene glycol precipitate of culture supernatant.

(iii) Incubate for 16 h at 37°C or until cpe is extensive.

(iv) Wash the monolayers, then solubilise the cells and extract the proteins by addition of lysis buffer [10 mM Tris-HCl pH 7.4, 0.1 mM EDTA, 0.1 mM phenylmethylsulphonyl fluoride (PMSF), 0.5% (v/v) Nonidet P-40 (NP-40) and 1% (w/v) sodium cholate].

(v) After 2 h at room temperature, transfer to −20°C or use for electrophoresis after boiling for 2 min.

Figure 4 illustrates separation of [^{35}S]methionine-labelled RS virus proteins on a 10% acrylamide slab gel. The viral glycoproteins can be labelled metabolically in a similar manner using 50 μCi/ml D [1-^{3}H]glucosamine or [^{3}H]mannose.

Alternatively, radioiodination of the RS virus envelope glycoproteins can be carried out by surface iodination of infected cells by the lactoperoxidase method (16). The chloramine-T method (17) can be used for iodination of purified RS virus nucleocapsid.

3.1.2 *Radioimmunoassay (RIA)*

A procedure for direct RIA using [^{125}I]goat anti-mouse γ globulin as the second antibody and methanol-fixed persistently-infected mouse cells as the substrate has been described by Cote *et al.* (18).

3.1.3 *Radioimmunoprecipitation (RIP)*

Radioimmunoprecipitation (RIP) is carried out by standard procedures as described for example by Fernie and Gerin (19) or Ward *et al.* (20). The following method is suitable for RS virus (R.M.Elliott, personal communication).

(i) Infect cells in 50 mm Petri dishes at a p.f.u./cell ratio of greater than 1. At times after infection pulse-label by replacing the incubation medium with 1 ml of PBS containing 30 μCi of [^{35}S]methionine + 25 μCi of [^{35}S]-cysteine (or 100 μCi [^{3}H]mannose for labelling glycopolypeptides) and incubate at 31°C for 1 h.
(ii) Remove the radioactive solution and wash the monolayers with cold PBS.
(iii) Scrape the cells into suspension and pellet by centrifugation.
(iv) Resuspend the cell pellet in 200 μl lysis buffer (0.15 M NaCl, 1% sodium deoxycholate, 1% Triton X-100, 0.1% SDS, 1 mM PMSF, 0.01 M Tris-HCl, pH 7.4) and incubate on ice for 30 min. Vortex-mix for 10 s.
(v) Remove nuclei and other debris by centrifugation at 10 000 *g* for 5 min.
(vi) Pre-clear lysate by incubation with 50 μl 10% formalin-fixed *Staphylococcus aureus* Cowan strain (or A protein conjugated beads) for 30 min on ice.
(vii) Remove staphylococci by centrifugation. Add 10 μl hyperimmune serum (or normal serum as control) to 50 μl lysate. Mix and leave for 3 h on ice.
(viii) Add 100 μl of the staphylococcal preparation, mix and leave for 30 min.
(ix) Collect immune complexes by centrifugation for 20 s at 10 000 *g*.
(x) Wash three times rapidly with 500 mM LiCl, 100 mM Tris-HCl, pH 8.5.
(xi) Re-suspend final pellet in 50 μl protein dissociation mix (0.125 M Tris HCl, pH 6.8, 4% SDS, 10% mercaptoethanol, 20% glycerol, 0.1% bromophenol blue) and boil for 2 min. Load on gel or store at −20°C.

3.2 Assay of Defective Interfering Particles

Defective interfering (DI) particles and non-defective RS virus cannot be separated physically, but the existence of DI particles has been inferred from the u.v.-resistant, neutralising antibody-sensitive interfering activity of undiluted passaged virus. A colorimetric assay for quantification of DI particles in RS virus preparations has been described by Treuhaft (21). This assay is based on the uptake of neutral red by cells protected by DI particles and surviving challenge with standard virus. The colorimetric assay is claimed to be more sensitive than yield reduction assay. The protocol for the colorimetric assay is as follows.

(i) Expose monolayers of 1.5 x 10^5 Hep-2 cells in 16 mm wells to DI-containing material for 2 h at 37°C.
(ii) Remove the DI inoculum and replace with 1–3 p.f.u./cell of standard virus (low multiplicity passaged material) and absorb for a further 2 h at 37°C.
(iii) Add 1 ml maintenance medium to each well and continue to incubate at 37°C for 72 h.
(iv) Replace the maintenance medium with 0.5 ml neutral red (33 mg/ml) in Earle's balanced salt solution, and incubate the cells for a further 2 h at 37°C.
(v) Remove the dye and wash the monolayer twice with PBS.
(vi) Extract the dye in the monolayer with 1 ml 50% ethanol in 0.1 M NaH_2PO_4, and measure the amount extracted spectrophotometrically at 540 nm. Compare the amounts of dye extracted from an uninfected cell control and a standard virus infected control.

DI particles can be quantified, assuming a Poisson distribution, by the following relationship:

DI particles/ml = 1/dilution yielding 63% protection x number of cells exposed x 1/adsorption volume.

3.3 **Electron Microscopy**

Pneumoviruses are extremely pleomorphic, and the virion and its nucleocapsid are fragile and affected adversely by the usual procedures employed in virus concentration. Consequently accurate particle counting is impractical.

Transmission electron microscopy is useful for virus identification and for the study of virus morphology and morphogenesis. Scanning electron microscopy is useful for studying the surface morphology of infected cells and it can be used for quantification of surface antigen, making use of the phenomenon of specific bacterial adherence (the SABA/SEM technique).

3.3.1 *Virus Morphology*

Negative stains which have proved satisfactory include phosphotungstic acid (2%; pH 6.0–3%; pH 7.2), sodium phosphotungstate (2%; pH 6.0), potassium phosphotungstate (2%, pH 6.5), sodium silicotungstate (2%, pH 7.0) and uranyl acetate. Concentration of virus should be carried out by the gentlest procedure available avoiding centrifugal force (e.g., polyethylene glycol precipitation, or reduction of fluid volume by Amicon filtration). A droplet of concentrated virus is applied to a carbon-coated Formvar grid. Fixation can be achieved by adding an equal volume of 3% glutaraldehyde in PBS, pH 7.2, mixing and leaving at room temperature for 5 min. The grid is washed in PBS and then negative stained. (The fixation step may be omitted.) Nucleocapsids can be visualised by lysing infected cells with distilled water or hypotonic buffer. Infected cell debris can also be sedimented directly onto the grid prior to negative staining.

3.3.2 *Virus Morphogenesis*

(i) Fix infected cells (in a 50 mm Petri dish) by addition of 2.5% glutaraldehyde to a monolayer *in situ*.
(ii) After fixation for 30 min wash the monolayer twice with PBS prior to addition of 1% osmium tetroxide (10 drops/50 mm Petri dish).
(iii) After 15 min wash the monolayer twice with PBS and finally scrape the cells into suspension in 2 ml PBS.
(iv) Pellet the cells and dehydrate by transfer through 30–90% alcohol (5 min each step; if required the cells can be held in 70% alcohol overnight). The 90% alcohol stage is followed by two treatments of 15 min each in 100% alcohol.
(v) Replace the alcohol with a 1:1 mixture of alcohol and propylene oxide.
(vi) Add fresh alcohol/propylene oxide after 15 min for a further 30 min, then replace it by a 1:1 mixture of alcohol/propylene oxide and EPON, followed by EPON alone for 1 h.
(vii) Layer the compacted cell pellet on top of a little EPON in a BEEM capsule and place the capsules, after topping up with EPON, at 60°C for 48 h for polymerisation. The embedded cells are now ready for sectioning in an ultramicrotome by standard procedures.
(viii) Pick up thin sections on uncoated grids and after 24 h stain by placing the grids, section side down, in a drop of a saturated solution of uranyl acetate in methanol for 5 min.
(ix) Wash the grids with distilled water and dry on a piece of filter paper.
(x) Place a drop of 0.1 M sodium hydroxide on the section before placing the inverted grid on a drop of lead acetate for a further 5 min. The grids after thorough washing and drying on filter paper are now ready for examination in the electron microscope.

3.3.3 *Scanning Electron Microscopy*

The following method is satisfactory for preparation of cells for scanning electron microscopy.

(i) Fix coverslips sparsely seeded with cells in 2.5% glutaraldehyde with phosphate buffer for 1–2 h.
(ii) Place the cells in 1% osmium tetroxide with phosphate buffer for 1 h.
(iii) Dehydrate through a gradient of 30, 50, 70, 90 and 100% ethanol (5 min in each), and transfer the coverslips to a solution of 50% acetone in ethanol for 15 min.
(iv) Replace this with 100% acetone (two changes) before drying at the critical point of CO_2 using a critical point drying apparatus (e.g., as supplied by Polaran Equipment Ltd., Watford, Herts. or equivalent).
(v) The coverslips can now be mounted on aluminium stubs using 'Electrocday 915' (Acheson Colloids Company, Prince Rock, Plymouth) and coated with gold in a Polaron E5000 diode sputtering system or equivalent. These preparations are now ready for examination by scanning electron microscopy.

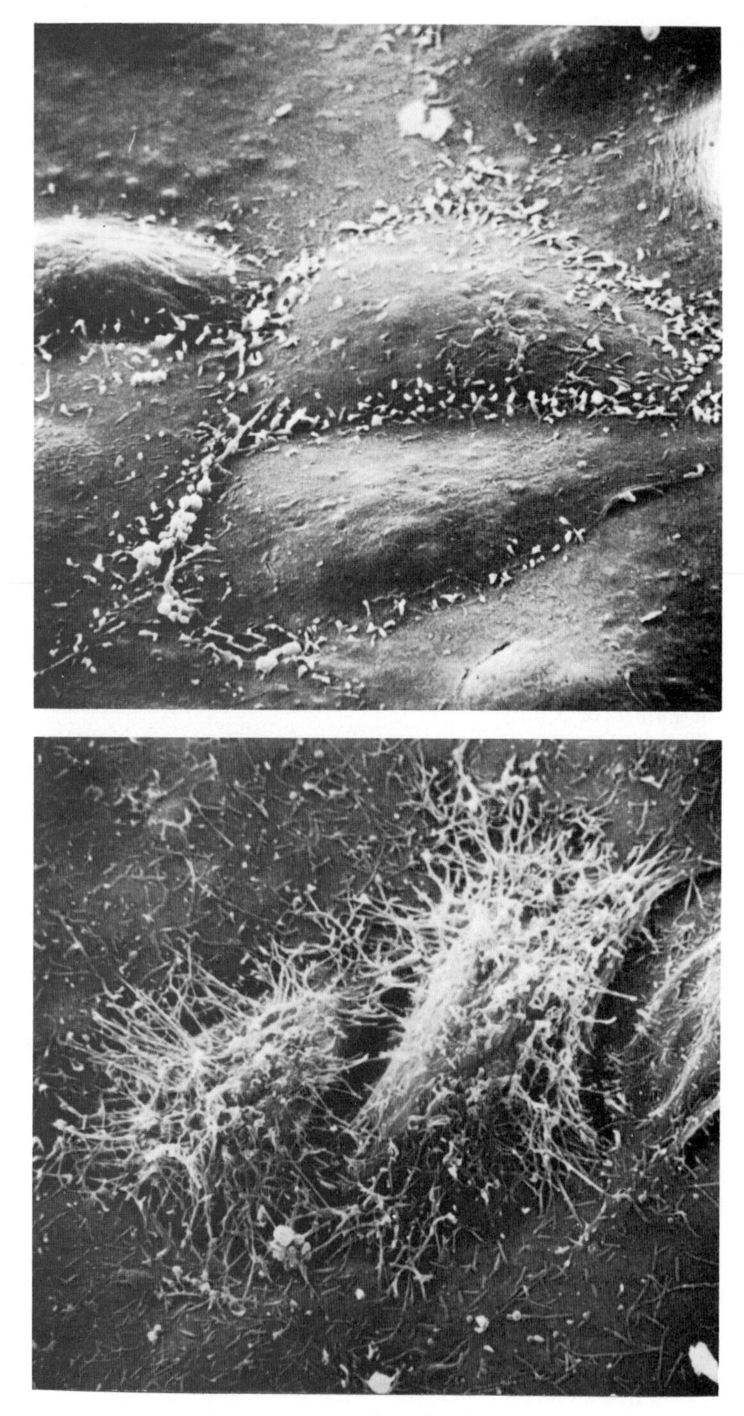

Figure 5. Stereoscan electron micrographs of **(a)** uninfected BS-C-1 cells showing microvilli (x 1820); **(b)** RS virus-infected BS-C-1 cells showing the profusion of filaments (x 1370). (Reproduced from Journal of Virology, with permission.)

Figure 5 illustrates the profuse process formation associated uniquely with infection of cells by pneumoviruses (7,22).

3.3.4 *Localisation and Quantification of Virus Antigen by Adherence of Staphylococci*

The Cowan strain of *S. aureus*, by virtue of expression of the A protein, adheres to the Fc portion of immunoglobulins. The spatial localisation of viral antigen in the plasma membrane can be revealed by the specific adherence of *S. aureus* to infected cells following their exposure to specific anti-viral serum (23). Staphylococci have advantages over latex beads, or polymethyl methacrylate or polystyrene spheres by virtue of their ease of preparation and uniform diameter (~1 μm). With many viruses staphylococci adhere uniformly over the surface of the infected cell, but in the case of pneumoviruses they adhere preferentially to the filaments uniquely associated with these viruses (see *Figure 5b*). The binding is highly specific and confirms the results of immunofluorescent staining which revealed linear staining with the predominant restriction of surface antigen to filamentous structures (see *Figure 3*).

Binding of staphylococci to infected cell surfaces can be carried out as follows.

(i) Grow cells on coverslips to near confluence and infect. The cells of choice are the marsupial Potoroo kidney (PTK-2) cell line, because of their high susceptibility and absence of surface feature.
(ii) After an appropriate period of incubation in maintenance medium, wash the infected monolayers with PBS to remove all traces of serum, and fix for 10–15 min in 2.5% glutaraldehyde in PBS at room temperature.
(iii) Treat the monolayers with specific anti-viral serum for 1 h at 37°C.
(iv) Remove the antiserum by thorough washing with PBS. A suspension of *S. aureus*, Cowan strain, prepared as described by Pringle and Parry (23), or as purchased, is added and incubated at 37°C for 5 min (the timing is critical and should be verified by experiment).
(v) The non-adhering staphylococci are removed by rapid washing in PBS, and the coverslips post-fixed in 2% PBS-buffered osmium tetroxide for 60 min at room temperature. The cover slips can then be processed for scanning electron microscopy as described in Section 3.3.3.

Enumeration of the number of adherent staphylococci can provide an accurate measure of the accumulation of viral antigen in the cell membrane, provided appropriate controls are included as described by Pringle and Parry (24).

3.4 Concentration of Virus

RS virus can be concentrated simply by polyethylene glycol precipitation as follows.

(i) Make up a 36% (w/v) solution of polyethylene glycol 6000 in distilled water.
(ii) Chill and dilute polyethylene glycol into virus-containing fluid to give final concentration of 6%.
(iii) Leave at +4°C for 2 h.

(iv) Collect precipitate by centrifugation (4000 *g* for 10 min) at +4°C.
(v) Drain off supernatant, surface wash pellet twice with buffer and resuspend in a small volume (0.5 ml per 100 ml starting material).

3.5 Purification and Radiolabelling of RS Virus

Clinical isolates of RS virus are highly cell associated, but several laboratory strains release sufficient amounts of virus for successful purification. The Long and A2 strains are the most suitable for virus purification.

3.5.1 *Gradient Centrifugation*

There are many possible protocols the following one has been used successfully for purification of the A2 strain (25).

(i) Infect Hep-2 monolayer cultures with RS virus at a multiplicity of 1 p.f.u./cell. Adsorb for 2 h at 37°C, then add incubation medium (Eagle's medium plus 5% heat-inactivated FCS).
(ii) At 10 h post-infection, add 0.3 μg/ml actinomycin D. To label RNA add fresh Eagle's medium with 20 μCi/ml of [5-^{3}H]uridine from 16–20 h. To label proteins add 5 μCi/ml of [^{35}S]methionine.
(iii) Harvest the supernatant when cytopathic effect and syncytium formation are evident (24–40 h). Remove cellular debris by centrifugation at 11 000 *g* for 20 min. Concentrate by centrifugation at 65 000 *g* for 90 min. Keep chilled.
(iv) Resuspend pellet in NTE (0.01 M Tris HCl, pH 7.4, 0.1 M NaCl, 0.001 M EDTA) by sonication.
(v) Centrifuge through a 10–50% linear sucrose gradient for 1 h at 150 000 *g*. Collect the fractions containing a visible band, or locate by infectivity or radioactivity.
(vi) Dilute pooled samples with NTE and sonicate. Reband the virus by velocity sedimentation in a 20–60% sucrose velocity gradient at 150 000 *g* for 2 h.
(vii) Concentrate by re-centrifugation (65 000 *g* for 90 min).

This type of procedure will provide adequate preparations of semi-purified radiolabelled RS virus. More rigorous purification can be achieved by more complex procedures (e.g., see ref. 10), and the addition of 1 M $MgSO_4$ will stabilise infectivity which is normally very sensitive to centrifugation (see Section 2.5.4).

3.5.2 *Isopycnic Banding of RS Virus*

This can be achieved by longer periods of centrifugation in sucrose (16 h at 73 000 *g*) or in 15–45% (w/v) metrizamide (190 000 *g* for 6 h). The latter has advantages as a density medium because of its lower viscosity and lower osmolarity than sucrose. Solutions of metrizamide should always be prepared fresh in 10 mM Hepes pH 7.8 before use (26).

3.6. Radiolabelling of Virion RNA

Preparation of intact virion RNA is difficult to achieve. The following procedure is recommended by Huang and Wertz (25).

(i) Purify the virus as described in Section 3.5.1 and isolate RNA by phenol extraction. Recover the precipitate by ethanol precipitation.
(ii) Purify the extracted RNA by velocity sedimentation in a 15–30% sucrose/SDS gradient for 15 h at 19 000 r.p.m.
(iii) Radiolabel metabolically as outlined in Section 3.5.1 using [^{3}H]uridine or ^{32}P.

Alternatively, RNA extracted from unlabelled virions can be labelled at the 3′ end using RNA ligase and cytidine 3′,5′-bis [^{32}P]phosphate, as described by Wertz and Davis (27). 5′-end labelling can be carried out after removal of the 5′-terminal phosphates of the viral RNA by calf intestinal phosphatase, followed by inactivation of the enzyme by proteinase K. The RNA is then phenol extracted and ethanol precipitated, and labelled by reaction with [γ-^{32}P]ATP in the presence of polynucleotide kinase from T4-infected *Escherichia coli*. The labelled RNA is then separated from residual ATP by Sephadex G50 column filtration. A suitable protocol for RS virus can be found in Schubert *et al.* (28).

3.7. Analysis of RNA and Viral Proteins by Polyacrylamide Gel Electrophoresis

3.7.1 *Proteins*

Radiolabelled polypeptides synthesised *in vivo* or *in vitro* can be analysed on 6–15% gradient or 10% single concentration polyacrylamide gels by standard methods as described in Marsden *et al.* (29).

Proteins separated by polyacrylamide gel electrophoresis can be transferred to nitrocellulose paper ('Western blot transfer') for biochemical or immunological analysis. The application of this technique to RS virus has been described by Hierholzer *et al.* (15). Transfer from acrylamide gel to nitrocellulose paper is achieved by electrophoresis as follows.

(i) After electrophoretic separation of proteins, place the acrylamide gel in a transblot electrophoresis cell (e.g., no. 170-3905, Bio-Rad Laboratories) along with a nitrocellulose sheet (e.g., BA85, 0.45 μm thick, Schleicher and Schuell) and 0.025 M Tris, 0.193 M glycine buffer, pH 8.35, with 20% methanol.
(ii) Electrophorese at 60 V for a minimum of 4 h at 4°C.
(iii) Cut the nitrocellulose sheet into strips and place in capped tubes.
(iv) Incubate the strips with specific anti-RS antibody (e.g., human convalescent serum diluted 1/100 in PBS with 0.3% Tween 20) for 2 h at room temperature.
(v) Wash three times with PBS/Tween.
(vi) Incubate with conjugate (peroxidase-labelled goat anti-human IgG) diluted 1:1000 for 3 h at room temperature.
(vii) Wash three times with PBS/Tween.
(viii) Develop with a solution of 50 mg of 3,3′-diaminobenzidine and 100 μl of 3% hydrogen peroxide per 100 ml of PBS, pH 7.2.
(ix) Wash in water and dry on filter paper (Whatman no. 2).

(x) Quantification can be achieved by spectrophotometry after treatment with plastic solvents (37.8% aqueous dimethylformamide or 40% aqueous N-methylpyrrolidone) to render the nitrocellulose strips transparent.

3.7.2 *RNA*

Radiolabelled RNA from purified virions or the cytoplasm of RS virus-infected cells can be analysed by electrophoresis in 1.5% agarose-6 M urea gels as follows (P.Collins, personal communication).

(i) To prepare one gel 3 mm thick by 50 cm long by 17 cm wide, make up separately (a) 120 ml of 4.5% agarose (3 x concentration), using a microwave oven or melting the agarose by autoclaving; (b) 1.5 x urea in citrate buffer. Dissolve 129.6 g urea in 9 ml 1 M citrate stock pH 3 and make up to 240 ml with distilled water. (N.B. Urea swells so add only ~135 ml of distilled water initially.)
(ii) Warm the 1.5 buffer and mix with the 3 x agarose.
(iii) Pour the gel in the cold in a flat bed apparatus in three stages with a 45 min delay between each stage; gel feet, gel legs and finally gel slab.
(iv) Place the comb in the gel immediately.
(v) Use after 2–3 h. Remove the comb and rinse with urea buffer. Refill the wells with urea buffer and load the samples in sample buffer into electrophoresis buffer.
(vi) Run the gel in the cold at 5 V/cm for 18 h. No pre-run is necessary. The stock solutions are prepared as follows:

 1 M citrate stock pH 3.0: mix 1 M citric acid and 1 M sodium citrate until pH reaches 3.0.

 Electrophoresis buffer: 1:40 dilution of 1 M citrate stock = 0.025 M citric acid pH 3.0.

 Sample buffer: 0.025 ml 1 M citrate stock, 3.6 g 6 M urea, 2.0 g 20% sucrose, 0.1 ml stock 0.005% bromophenol blue, make up to 10 ml with water.

 6 M urea buffer: 3.6 g urea, 0.25 ml 1 M citrate pH 3, make up to 10 ml with distilled water.

3.8 **Purification of Messenger RNA for In Vitro Translation**

Messenger RNA for translation *in vitro* can be prepared as follows.

(i) Infect 50 x 10^6 cells with RS virus at a p.f.u./cell ratio of five or greater. Incubate at 37°C for 10–12 h.
(ii) Wash the monolayers and scrape into buffer (0.01 M Hepes, pH 7.6, 0.01 M NaCl, 0.001 M $MgCl_2$). Leave at +4°C for 10 min to swell the cells.
(iii) Disrupt the cells by Dounce homogenisation (15 strokes) and remove cell debris by centrifugation at +4°C.
(iv) Re-suspend pellet in 2 ml buffer, homogenise again to clarify. Pool the supernatants from the first and second homogenisations.

(v) Add $CsCl_2$ and N-lauryl sarcosine at the rate of 1 g and 0.02 g per ml of supernatant. Mix thoroughly and heat at 51°C for 1–2 min.
(vi) Layer 7 ml over 2 ml of 5.7 M CsCl, 0.1 M EDTA, 2% sarcosine in a SW40 tube and add buffer to fill the tube. Spin in SW40 rotor at 25 000 r.p.m. for 16 h at 25°C (or equivalent).
(vii) Resuspend the pellet in 0.5 ml sterile distilled water and ethanol precipitate. Ethanol precipitation should be carried out twice adding 1.2 ml ethanol, plus 25 μl 4 M NaCl and 10 μl 10% SDS.
(viii) Re-suspend the final precipitate in NTE buffer (0.01 M Tris-HCl, pH 7.4, 0.1 M NaCl, 0.001 M EDTA) + 0.2% SDS and chromatograph on an oligo d(T)-cellulose column. Elute bound material and ethanol precipitate, adding rabbit liver tRNA as carrier. Pellet, resuspend in buffer without SDS and re-precipitate with 2 volumes of ethanol plus 0.2 M potassium acetate. Re-suspend in 250–500 μl of 0.01 M Hepes, pH 7.6.

Cell-free protein synthesis can be achieved using a messenger-dependent rabbit reticulocyte lysate system as described by Preston (30) or using a commercially available test. Uninfected cell mRNA should be prepared simultaneously as a control.

Cell-free translation of individual messages can be achieved by hybrid selection from total mRNA using defined recombinant cDNA plasmids as described by Collins and Wertz (31).

3.9 Molecular Cloning and Oligonucleotide Sequencing

Ten RS viral genes have been identified by cDNA cloning, mapping and translation of RS virus mRNA. The cDNAs were reverse transcribed from mRNAs from virus-infected Hep-2 cells and cloned using the *E. coli* plasmid pBR322. The methodology involved is beyond the scope of this chapter. Reference should be made to the original papers of Collins and Wertz (31), Collins *et al.* (32) and Venkatesan *et al.* (33).

The nucleotide sequence of the RS virus nucleocapsid gene and the inferred amino acid sequence of the nucleocapsid protein have been published by Elango and Venkatesan (34), and those of the M gene and M protein by Satake and Venkatesan (35). The methodology of oligonucleotide sequencing as applied to RS virus is described fully in these papers.

3.10 Purification of Proteins

3.10.1 *Viral Polypeptides*

Purified viral polypeptides can now be obtained by affinity chromatography using specific monoclonal antibodies, or by expression of cloned genes in prokaryotic (for unprocessed products) or eukaryotic (for processed products) cells. Protocols specific for RS virus have not been described so far.

3.10.2 *Nucleocapsid Protein*

Nucleocapsid protein can be obtained from purified nucleocapsids. Nucleocapsids can be purified from either virions or infected cell extracts. A method for purification from cell extracts (which gives more material) is given below.

(i) Infect 50 x 10^6 cells at a p.f.u./cell ratio of one or greater.

(ii) Harvest cells at 36 – 48 h post-infection by treatment with 10 mM Tris-HCl pH 7.4 containing 0.18 M NaCl, 0.25 mM EDTA and 0.1 mM PMSF.

(iii) Collect cells by centrifugation (5000 *g* for 15 min) and re-suspend in 1 ml lysis buffer [10 mM Tris-HCl pH 7.4, 0.1 mM EDTA, 0.1 mM PMSF, 0.5% (v/v) NP-40] at +4°C for 20 min.

(iv) Disrupt the cells by Dounce homogenization.

(v) Add NaCl to 0.1 M and remove the cells, debris and nuclei by centrifugation at 600 *g* for 2 min.

(vi) Pellet the nucleocapsids by centrifugation at 100 000 *g* for 30 min.

(vii) Wash once in lysis buffer and again in 10 mM Tris-HCl, 0.1 mM EDTA, pH 7.4.

(viii) Purify the nucleocapsids from the re-suspended pellet by isopycnic centrifugation on a 15 – 55% (w/w) potassium tartrate gradient in 10 mM Tris-HCl, 0.1 mM EDTA, pH 7.4 buffer.

3.10.3 *Virion Glycoproteins*

Purification of the virion glycoproteins involving solubilisation with non-ionic detergent in low ionic strength buffer has been described by Ueba (36).

(i) Pellet the gradient-purified virions (see Section 3.5) through 30% sucrose-NTE buffer at 50 000 *g* for 60 min, and re-suspend in 0.01 M phosphate buffer (pH 7.2) with 10% (v/v) Triton X-100.

(ii) Stir at room temperature for 30 min and centrifuge at 200 000 *g* for 60 min.

(iii) Extract the supernatant with n-butanol.

(iv) Suspend the extracted proteins in 0.01 M phosphate buffer with 1% Triton X-100. Fractionate by rate-zonal centrifugation in a linear 5 – 25% sucrose gradient containing 1% Triton X-100 at 100 000 *g* for 18 h. Collect the fractions from the top and dialyse against 0.01 M phosphate buffer.

(v) Remove Triton X-100 by n-butanol extraction.

(vi) Identify the glycoprotein-containing fractions by polyacrylamide gel electrophoresis.

4. SPECIALISED BIOLOGICAL PROCEDURES

4.1 **Production of Monoclonal Antibodies**

Monoclonal antibodies to several RS virus proteins have been obtained by standard procedures. The following is a representative protocol.

(i) Inoculate BALB/c mice intraperitoneally with 10^7 RS virus-infected cells per mouse. Repeat 10 days after the first inoculation.

(ii) Prepare cell suspension from the spleen of immunised mice in RPM1 1640 medium buffered with sodium bicarbonate containing 15 μg penicillin/ml, 15 μg gentamicin/ml, 1 mM sodium pyruvate, 2 mM L-glutamine and 15% FCS.

(iii) Fuse with murine myeloma cells, e.g., BALB/c NS1/1 maintained by growth in the above medium, by mixing 3 x 10^7 spleen cells and 3 x 10^7 myeloma cells. Centrifuge the cells at 200 *g* for 10 min and equilibrate at 37°C for 1 min. Add slowly 1 ml of 50% polyethylene glycol 4000 containing 5% DMSO to initiate fusion.

(iv) Incubate for 90 sec at 37°C and stop fusion by slow addition of 20 ml of medium.

(v) Sediment the cells by centrifugation and wash once with medium. Distribute to flat-bottomed 96-well microtitre plates at 2 x 10^5 cells/well in HAT medium (i.e., medium plus 100 μM hypoxanthine, 16 μM thymidine, 0.4 μM aminopterin).

(vi) Incubate in the presence of macrophage feeder layers prepared 1 day before the fusion, each well containing 10^4 peritoneal macrophages from a normal mouse.

(vii) Add 50 μl of fresh HAT medium to each well after 6 days of incubation.

(viii) Test for antibody activity 9–12 days after fusion by an ELISA test (see Section 10) with RS virus antigen to determine the specificities of the secreted antibodies.

(ix) Clone the hybrids producing specific antibody by limiting dilution in 96-well microtitre plates with macrophage or spleen cells as feeder layers. Repeat the cloning a second time. Hybrids can be considered stable when 95% of the progeny clones produce antibody.

4.2 Isolation of Temperature-sensitive Mutants

Temperature-sensitive mutants are useful for the analysis of gene function and as biochemical tools. Temperature-sensitive mutants can be isolated as follows.

(i) Infect 10^6 BS-C-1 cells in a 50 mm Petri dish or 30 ml screw-capped bottle with wild-type RS virus (i.e., a triple-plaque purified stock of RS virus able to form plaques at 39°C) at a p.f.u./cell ratio of 0.01 and absorb for 1 h at 31°C.

(ii) Wash off inoculum with two changes of 3 ml medium and add 3 ml Eagle's medium containing 2.5% FCS and mutagen (50 μg/ml 5-fluorouracil, or 2.5 μg/ml N-methyl-N′-nitro-N-nitrosoguanidine, or 50 μg/ml 5-azacytidine).

(iii) Incubate at 31°C until the cpe is visible in infected control cultures without mutagen.

(iv) Plate out dilutions of the supernatant from mutagenised cultures, after clarification but without prior freeze-thawing. Incubate at 31°C under 0.9% agar-containing overlay for 5–7 days at 31°C.

(v) Transfer well-separated plaques to fresh monolayers using Pasteur pipettes inserted through the agar and prepare individual stocks.

(vi) Screen for plaque formation at 31°C and 39°C by inoculation and incubation of duplicate plates at these temperatures. Alternatively the plaques from mutagenised virus can be screened directly by stabbing isolates into individual sectors of duplicate plates prepared by addition of a 2 mm thick layer of 0.9% agar-containing incubation medium. After absorption for 30 min at 31°C, a second overlay of 5 ml per plate is added. Incubate one of each pair of sectored plates at 31°C and one at 39°C. Recover virus from any sector showing cpe at 31°C but not at 39°C and grow stocks for confirmatory testing.

Temperature-sensitive mutants can be classified into functional groups by complementation tests. The factors affecting complementation have been described in detail (37,38).

As with all unsegmented negative strand RNA viruses, recombination between temperature-sensitive mutants whether in the same or different complementation groups has never been observed.

4.3 **Establishment of Persistently Infected Cultures**

Persistent infection of cultured cells with minimal or no cytopathology can be achieved by several means. Passage of virus at high multiplicity of infection and serial transfer of surviving cells has been sufficient to achieve persistent infection. Other methods include infection of naturally restrictive cells in the presence of antibody or with interferon treatment. Incubation of cells infected with temperature-sensitive mutants of RS virus to suppress cytopathogenicity can also produce persistently infected cultures which can be propagated at sub-restrictive temperature once persistence has been established. In this case the outcome is dependent on the properties of the initiating virus. Temperature-sensitive mutants of complementation group B which exhibit only limited antigen synthesis at restrictive temperature are unsuitable, producing either abortive or cytolytic infections. Mutants showing appreciable antigen synthesis at restricted temperature (complementation group D) consistently yield persistently infected cultures which can be propagated indefinitely (39). Persistently infected cultures provide a convenient source of antigen for ELISA tests (see Section 2.10).

5. REFERENCES

1. Kingsbury,D.W., Bratt,M.A., Choppin,P.W., Hanson,R.P., Hosaka,Y., ter Meulen,V., Norrby,E., Plowright,W., Rott,R. and Wunner,W. (1978) *Intervirology,* **10**, 137.
2. Horsfall,F.L.,Jr. and Curnen,E.C. (1946) *J. Exp. Med.,* **83**, 43.
3. Anderson,J.M. and Beem,M.O. (1966) *Proc. Soc. Exp. Biol. Med.,* **121**, 205.
4. Dreizin,R.S., Ponomareva,T.I., Rapoport,R.I. and Petrova,E.I. (1967) *Acta Virol. (Praha),* **11**, 533.
5. Larionov,A.S. and Soloneva,A.I. (1968) *Vopr. Virusol.,* **13**, 164.
6. Parry,J.E., Shirodaria,P.V. and Pringle,C.R. (1979) *J. Gen. Virol.,* **44**, 479.
7. Faulkner,G.P., Shirodaria,P.V., Follett,E.A.C. and Pringle,C.R. (1976) *J. Virol.,* **20**, 487.
8. Schindarov,L., Galabov,A., Runerski,N. and Wassileva,W. (1969) *Zentralbl. Bakteriol.,* **210**, 313.
9. Nisevich,L.L., Konstantinova,L.A. and Stakhanova,V.M. (1972), *Vopr. Virusol.,* **17**, 344.
10. Fernie,B.F. and Gerin,J.L. (1980) *Virology,* **106**, 141.
11. Compans,R.W., Harter,D.H. and Choppin,P.W. (1967) *J. Exp. Med.,* **126**, 267.
12. Gardner,P.S. and McQuillin,J. (1974) *Rapid Virus Diagnosis,* published by Butterworths, London.
13. Gardner,P.S. (1977) *J. Gen. Virol.,* **36**, 1.
14. Gardner,P.S., McQuillin,J. and McGuckin,R. (1970) *J. Hyg. Camb.,* **68**, 575.
15. Hierholzer,J.C., Coombs,R.A. and Anderson,L.J. (1984) *J. Virol. Methods,* **8**, 265.
16. Marchalonis,J.J., Cone,R.E. and Santer,V. (1971) *Biochem. J.,* **124**, 921.

17. Greenwood,F.C., Hunter,W.M. and Glover,J.S. (1963) *Biochem. J.*, **89**, 114.
18. Cote,P.J., Fernie,B.F., Ford,E.C., Shih,J.W. and Gerin,J.L. (1981) *J. Virol. Methods*, **3**, 137.
19. Fernie,B.F. and Gerin,J.L. (1982) *Infect. Immun.*, **37**, 243.
20. Ward,K.A., Lambden,P.R., Ogilvie,M.M. and Watt,P.J. (1983) *J. Gen. Virol.*, **64**, 1867.
21. Treuhaft,M.W. (1983) *J. Gen. Virol.*, **64**, 1301.
22. Parry,J.E., Shirodaria,P.V. and Pringle,C.R. (1979) *J. Gen. Virol.*, **44**, 479.
23. Pringle,C.R. and Parry,J.E. (1980) *J. Virol. Methods*, **1**, 61.
24. Pringle,C.R. and Parry,J.E. (1982) *J. Gen. Virol.*, **58**, 207.
25. Huang,Y.T. and Wertz,G.W. (1982) *J. Virol.*, **43**, 150.
26. Wunner,W.H., Buller,R.M.L. and Pringle,C.R. (1976) in *Biological Separations in Iodinated Density Gradient Media*, Rickwood,D. (ed.), IRL Press, London, p. 159.
27. Wertz,G.W. and Davis,N.L. (1981) *Nucleic Acids Res.*, **9**, 6487.
28. Schubert,M., Keene,J.D. and Lazzarini,R.A. (1979) *Cell*, **18**, 749.
29. Marsden,H.S., Crombie,I.K. and Subak-Sharpe,J.H. (1976) *J. Gen. Virol.*, **31**, 347.
30. Preston,C.M. (1977) *J. Virol.*, **23**, 455.
31. Collins,P.L. and Wertz,G.W. (1983) *Proc. Natl. Acad. Sci. USA*, **80**, 3208.
32. Collins,P.L., Huang,Y.T. and Wertz,G.W. (1984) *J. Virol.*, **49**, 572.
33. Venkatesan,S., Elango,N. and Chanock,R.M. (1983) *Proc. Natl. Acad. Sci. USA*, **80**, 1280.
34. Elango,N. and Venkatesan,S. (1983) *Nucleic Acids Res.*, **11**, 5941.
35. Satake,M. and Venkatesan,S. (1984) *J. Virol.*, **50**, 92.
36. Ueba,O. (1980) *Acta Med. Okayama*, **34**, 245.
37. Gimenez,H.B. and Pringle,C.R. (1978) *J. Virol.*, **27**, 459.
38. Pringle,C.R., Shirodaria,P.V., Gimenez,H.B. and Levine,S. (1981) *J. Gen. Virol.*, **54**, 173.
39. Pringle,C.R., Shirodaria,P.V., Cash,P., Chiswell,D.J. and Malloy,P. (1978) *J. Virol.*, **28**, 199.

CHAPTER 6

Growth, Purification and Titration of Influenza Viruses

THOMAS BARRETT and STEPHEN C.INGLIS

1. GENERAL INTRODUCTION

The influenza viruses are enveloped single-stranded RNA viruses which cause serious epidemic disease in humans and a number of animal species such as horses, fowl, seals and swine. There are three types of influenza virus, which are classified on the basis of the complement fixing properties of their nucleoprotein core, and which differ biologically and biochemically. The type A virus, by virtue of its capacity for gross antigenic change (antigenic shift), is the cause of major pandemics of influenza in humans, and for this reason has been studied in the greatest detail. Most of the methods available for growth, purification and titration of influenza viruses have therefore been developed for this virus. Type B viruses, although they do not show antigenic shift, appear in other respects to be very similar biochemically and biologically to the type A viruses and so in general may be propagated and titrated using similar methods. Type C viruses, by contrast, show very little biochemical similarity with types A and B and accordingly have rather different growth properties. Since this group has been much less well studied than the others, this chaper will concentrate on methods developed for the A and B viruses, particularly those with which we have had direct experience. These methods are however usually applicable also to influenza C viruses. This chapter is primarily designed as a practical guide and so, in general, theoretical aspects of the methods have not been discussed. Where necessary, reference has been made to some of the biological features of the virus, but for a complete account of the influenza viruses, the reader is referred to a number of comprehensive recent reviews (1 – 3).

2. PROPAGATION SYSTEMS

2.1 Embryonated Eggs

2.1.1 *Introduction*

Fertile hen eggs provide a very convenient and inexpensive system for the propagation of influenza viruses. Almost all strains of the virus, of human or animal origin, will grow to some extent in eggs, and for most egg-adapted laboratory strains the titres achieved are relatively high [$\sim 10^4 - 10^5$ haemagglutinating units (HAU)/egg or $10^9 - 10^{10}$ plaque-forming units (p.f.u.)/egg]. In addition, the egg provides an enclosed environment for virus growth and so no special sterile conditions are required for their use.

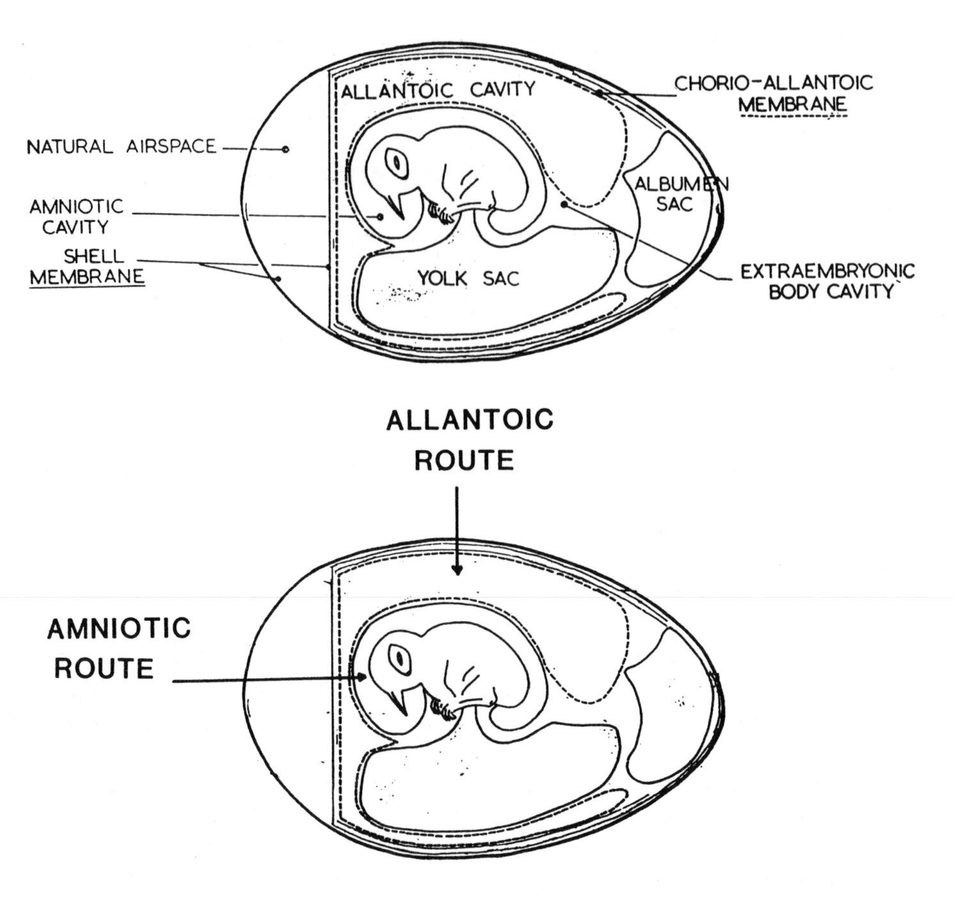

Figure 1. Diagrammatic section through an embryonated hen egg at 11 – 16 days of incubation. The arrows in the bottom diagram indicate the route of virus inoculation for either the amniotic or the allantoic route of virus inoculation. Holes are drilled in the shell in the appropriate position after sterilisation of the area with alcohol, and virus is introduced into the embryo by injection with a fine gauge needle.

The virus is usually inoculated into the fluid-filled allantoic cavity of the egg (see *Figure 1*) from where it can infect the cells of the chorioallantoic membrane (which lines the allantoic cavity). Progeny virus is released into the allantoic fluid and so may be recovered easily by aspiration of the fluid. Some strains of virus, particularly new field isolates and also type C viruses, do not grow well in the chorioallantoic cells, but will replicate if inoculated directly into the amniotic sac; the virus then has access to a variety of different embryonic tissues, and the resulting progeny is released into the amniotic fluid. This fluid may then be harvested separately.

2.1.2 *Preparation of Eggs for Inoculation*

The virus grows best in 10 – 12 day old embryonated eggs, and generally it is convenient to purchase freshly-fertilised eggs from a local farm and to incubate them

in the laboratory for the required time. It is not essential to have spe
ment for this although a proper egg incubator is a valuable asset. The e
incubated successfully at 37°C in a humidified incubator provided the
ed regularly (several times a day). The type of egg used is not considered important, but in certain cases it may be desirable to eliminate the possibility that the eggs are already infected with other microorganisms, in which case special pathogen-free eggs may be purchased from certain suppliers.

Before virus inoculation, the eggs should be checked for infertility. This may be done simply by examining the eggs against a bright light in a darkened room (candling). From 4 or 5 days after fertilisation, the embryos should be clearly visible.

2.1.3 *Preparation of Virus Inoculum*

During routine passage of laboratory stocks, the virus inoculum will usually consist of a sample of infected allantoic fluid or tissue culture medium. This is usually diluted to about 10^3-10^4 p.f.u./ml ($10^{-2}-10^{-1}$ HAU/ml) in Hank's balanced salt solution (see Section 8) before use, since inoculation of large amounts of virus into the egg encourages the formation of defective-interfering (DI) virus (4). The diluent should be supplemented with gelatin (0.5%) if the diluted virus is to be stored before inoculation. Obviously care should be exercised to ensure that the virus inoculum is free of bacterial contamination. However, if this is impossible, the antibiotic gentamycin should be added to the inoculum (to a final concentration of 50 μg/ml).

A detailed description of the preparation of clinical material for growth and isolation of influenza viruses is beyond the scope of this chapter and has been presented elsewhere (5). The essential features of the protocol however involve homogenisation to disperse the material, low-speed centrifugation to clarify the material, and inoculation of an undiluted sample of the resulting supernatant *via* the amniotic route.

2.1.4 *Virus Inoculation*

(i) *The allantoic route.*

(1) Turn the eggs on their side, and swab the exposed surface with alcohol in order to sterilise the area around the inoculation point.
(2) Drill a small hole in the shell (see *Figure 1*) using a dentist's drill fitted with a cutting disc. The hole should be deep enough so that a syringe needle can easily be inserted through the shell, but ideally should not penetrate the shell membrane.
(3) Using a syringe with a 25 gauge needle, inoculate 0.1 ml of the virus inoculum into the allantoic cavity by pushing the needle just below the surface of the shell.
(4) Seal up the hole in the egg with a little melted paraffin wax.
(5) Incubate the inoculated eggs, pointed end downwards, at 37°C in a humidified incubator. It is not necessary to turn the eggs during the incubation period.

(ii) *The amniotic route.*

(1) Sterilise the blunt end of the egg with alcohol and drill a small hole as indicated in the diagram (*Figure 1*).

(2) Position the egg against a bright lamp in a darkened room so that the embryo is visible.

(3) Using a syringe with a 23 gauge, 4 cm needle, introduce the inoculum through the pre-drilled hole and, with a rapid stab aimed at the embryo, insert the needle into the amniotic sac. Deliver 0.1 ml of the inoculum and withdraw the needle carefully.

(4) Seal up the hole with hot wax and incubate the egg as described in the previous section.

2.1.5 *Incubation Times*

The optimal incubation time for virus yield varies depending on the virus strain, and so should be checked. For some avian influenza A strains such as fowl plague, virus yield reaches a maximum at 24–26 h, but human A strains and influenza B and C strains usually require 48–72 h.

2.1.6 *Harvesting of Virus*

Before attempting to harvest infected fluid from the eggs, it is advisable to chill them at 4°C for 4–18 h or at –20°C for 30 min. This step kills the embryo (if the virus has not already done so) and causes contraction of the surrounding blood vessels. The risk of contaminating the infected fluid with red blood cells (which could adsorb the virus and so reduce the virus yield) is therefore greatly reduced. If contamination of the fluid with red cells in unavoidable, the resulting loss of virus can be minimised by incubating the infected fluid at 37°C for 30 min. Under these conditions, the virus-associated neuraminidase activity should destroy the red cell receptors to which the virus is bound, and so promote virus release.

(i) *Harvesting allantoic fluid.*

(1) Support the eggs, blunt end uppermost, on a tray, and sterilise the exposed area with alcohol.

(2) Break open the shell at the blunt end using either sterile forceps or a proprietary egg punch. Remove the shell around the air sac so that the chorioallantoic membrane is easily accessible.

(3) With sterile blunt forceps, tear the chorioallantoic membrane and peel it towards the side of the egg.

(4) Use the forceps to push the embryo and the yolk sac gently out of the way, and suck out the allantoic fluid with a blunt-ended pipette. An average-sized egg should yield about 7–10 ml of fluid, and the fluid should be a pale yellow colour. The presence of egg yolk in the fluid can interfere with subsequent virus purification and so care should be taken to avoid disruption of the yolk sac during harvesting. If the harvested virus is to be used for biochemical studies, for example analysis of virion-associated enzyme activities, the fluid should be collected on ice.

(5) Clarify the allantoic fluid by centrifugation at 10 000 *g* for 10 min and store the resulting supernatant either at 4°C or at −70°C (see Section 6 on virus storage).

(ii) *Harvesting amniotic fluid.*

(1) Sterilise the surface of the egg with alcohol and break open the blunt end as before. Try to remove as much of the shell as possible so that the contents of the egg may be decanted easily.
(2) Tear away the chorioallantoic membrane and empty the contents of the egg into a sterile Petri dish. The amniotic sac should remain intact.
(3) With a syringe and a 25 gauge needle, puncture the amniotic sac and carefully remove as much of the fluid as possible (0.2−0.4 ml is a reasonable yield).
(4) Clarify and store as for allantoic fluid.

2.1.7 *Expected Yields*

The extent of virus growth in eggs depends on the particular strain used. Wild strains will generally produce very low titres initially (for example ~100 HAU/ml of amniotic fluid), but after several passages in eggs, should grow to much higher titres. Egg-adapted laboratory strains vary considerably in their yield, but most will grow to about 2000−5000 HAU/ml and occasionally up to 20 000 HAU/ml of allantoic fluid. Egg adaptation is clearly due to genetic change, and so it must be borne in mind that detailed genetic and biochemical analyses of virus strains which have been passaged in eggs may give information which is not relevant to the original virus strain.

2.2 **Tissue Culture Systems**

2.2.1 *Introduction*

While embryonated hen eggs are the most widely used system for bulk growth of influenza viruses, analysis of the fine detail of influenza virus genetics requires the virus to be grown productively in cell culture. The most commonly used and reliable culture system consists of monolayers of primary chick embryo fibroblast cells (CEF), but primary monkey kidney cells, or the Madin-Darby bovine or canine kidney cell lines (MDBK or MDCK) are also suitable. Although many types of cell culture may be susceptible to virus infection, they are usually non-permissive for virus multiplication. In some cases the cell may support only a partial replication cycle (abortive infection) which does not allow the formation of new virus particles. In other cases, however, progeny virus may be assembled and released from the cell but in a form which is non-infectious.

One factor which is known to be essential for virus infectivity is proteolytic cleavage of the virus haemagglutinin (HA). The haemagglutinin molecule on the surface of the virus is cleaved by endogenous cellular enzymes to form two smaller molecules HA1 and HA2 (7,8). In some cell systems which lack a suitable endogenous protease, influenza virus can be induced to grow productively and form plaques by the addition of trypsin to the overlay medium (9). Influenza viruses differ in their susceptibility to cleavage by these enzymes and recent studies have shown that the primary protein structure at the HA1-HA2 junction is important in determining this characteristic (10). The amino acid sequence at the cleavage site of the HA molecule is very similar to that found at the cleavage site of the F protein of Sendai virus and it has been suggested that they act in an analogous way to enable virus penetration to occur (11). While cleavage of the HA is essential for infectivity it is not the only factor which determines the pathogenicity of the virus. This property appears to be controlled by a combination of virus genes.

2.2.2 *Experimental*

Monolayer cultures of cells growing on plastic dishes or flasks are convenient for carrying out virus infection, and best results are usually obtained using cells which have just reached confluence. Serum contains substances which reduce virus infectivity and so the tissue culture medium should be removed and the monolayers rinsed with phosphate-buffered saline (PBS) before addition of the virus inoculum, which will generally consist of allantoic fluid diluted in PBS to give the required multiplicity of infection. Allow the virus to adsorb to the cells for 30–60 min at room temperature; under these conditions, virus should attach to the cells but not begin replication. Remove the inoculum and replace with culture medium containing a reduced concentration of serum (2% maintenance medium). When it is particularly important that all unadsorbed virus be eliminated, e.g., when measuring the amount of virus released by cells, the monolayers should be treated with virus-specific antibody to reduce the background from unadsorbed virus. In this case, wash the monolayers three times with warm PBS after removal of the inoculum. Add maintenance medium and incubate the cells at the appropriate temperature for 60 min. Then replace the medium with 2 ml of PBS containing virus antiserum (60 HAI units/ml). After 15 min, wash the monlayers three times with warm PBS, cover with fresh maintenance medium and incubate at the required temperature.

Where trypsin is required for virus replication it is normally added to a final concentration of 10 μg/ml. Trypsin normally used for passage of tissue culture cells is suitable. It should be stored as a frozen concentrated stock and diluted just before use. It is important to check each batch of trypsin since the activity can vary from preparation to preparation. Each batch should be checked on uninfected cells at various concentrations and a concentration of not more than half that required to remove the cells used. The normal cytopathic effect (cpe) observed with influenza virus is a rounding up of infected cells and their subsequent detachment from the tissue culture dish; this is what happens with uninfected cells when the typsin concentration is too high.

The speed of virus replication in tissue culture is obviously dependent on the particular virus strain and type of cell, but generally during a productive infection, virus release begins from 4 to 6 h after infection and reaches a maximum by 16–24 h. This process may be monitored easily by titrating samples of the tissue culture medium for haemagglutinating activity (see Section 3.1). A typical growth curve obtained in this way is shown in *Figure 2*.

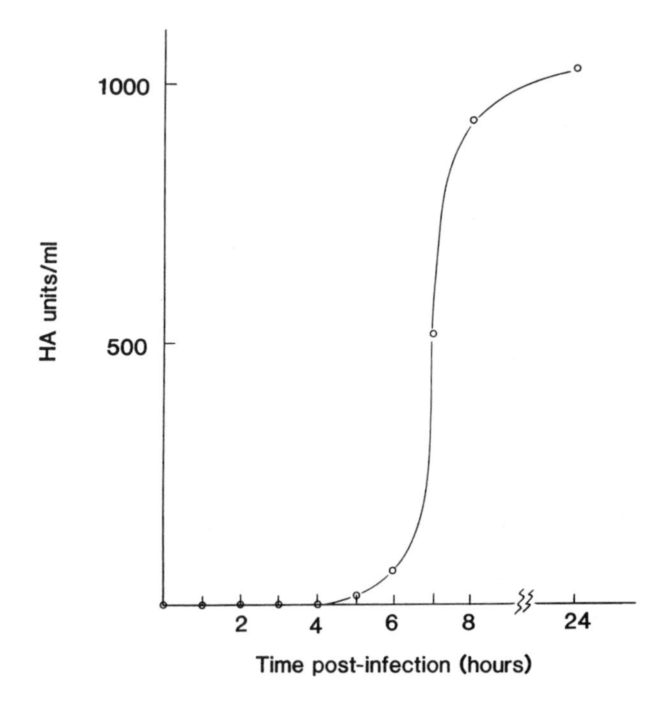

Figure 2. Growth of influenza A virus in primary chicken embryo fibroblast (CEF) cells measured by the release of virus haemagglutinating activity. A 15 cm dish of confluent CEF cells was infected at a multiplicity of approximately 50 p.f.u./cell and 0.25 ml samples of medium were removed at the times indicated for direct determination of haemagglutinating activity.

2.3 **Organ Culture**

Influenza virus has been reported to grow in a variety of different types of organ culture, but such systems are clearly more useful for experimental purposes rather than for virus propagation. A detailed description of organ culture systems is beyond the scope of this chapter, but there are reports of successful cultivation of virus in human adult tissues such as nasal mucosa, uterus, conjunctiva and bladder, and foetal tissues such as trachea, oesophagus, intestine, bladder, conjunctiva and lung. A variety of animal tissues have also been used such as avian trachea and colon, and ferret nasal turbinates, uterus, bladder, oviduct, oesophagus, conjunctiva, pharynx and respiratory tissue. For a discussion of the available systems, see Sweet and Smith (2).

3. METHODS FOR VIRUS TITRATION

3.1 **Titration of Influenza Viruses by Haemagglutination**

3.1.1 *Introduction*

Like many other animal viruses, influenza viruses can agglutinate red blood cells. The basis for this effect is the interaction of one of the virus surface proteins, the haemagglutinin, with receptors on the surface of the red cells. Since virus par-

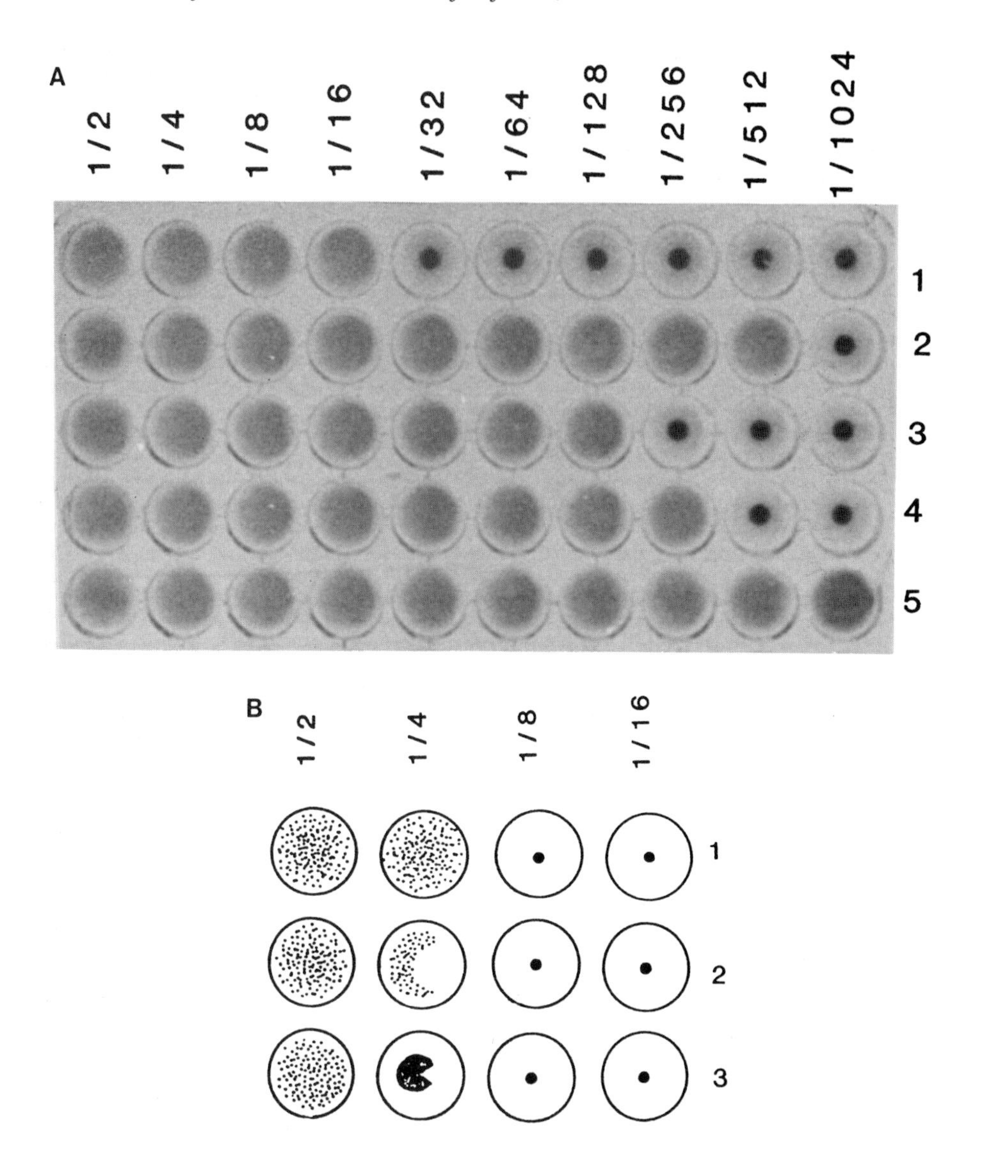

Figure 3. Titration of influenza virus by the haemagglutination assay. The assay is set up by making serial doubling dilutions of the test sample in round-bottomed wells as described in Section 3.1. The end-point is taken as the smallest amount of virus which can agglutinate 50% of the red cells in the standard assay. In row 1 of **A**, virus dilutions up to 1/16 give complete agglutination while the higher dilutions give no agglutination. The end-point is taken as the dilution which would contain 1 haemagglutinating unit (HAU) i.e., the amount of virus needed to agglutinate the standard volume of 1% red blood cells. In the case of row 1 in **A** this is taken as the half-way point between 1/16 and 1/32, i.e., a dilution of 1/24. In the same way the end-point dilutions for rows 2, 3 and 4 can be calculated as 1/768, 1/192, 1/384. This means in effect that 0.25 ml (the volume of virus solution in the well) of a 1/24 dilution of the original virus stock contains 1 HAU. The concentration of virus may then be calculated through multiplying by the appropriate factor, in this case giving a value of 96 HAU/ml. In row 5 the virus concentration is greater than 1024 HAU and a complete dilution series has not been carried out. In such cases a 1/10 dilution of the original virus should be carried out before repeating

the HA titration. The end-point in the titration assay is usually clear cut as seen in **A**. However, sometimes intermediate patterns of agglutination are seen and **B** illustrates the other types of end-point which may be encountered. In row 1 the normal end-point dilution (as illustrated in **A**) is shown, i.e., 1/6. The end-point in the example in row 2 is taken as 1/3 of the way between the 1/4 dilution and the 1/8 dilution, i.e., 1/5. In row 3 it is taken as the 1/4 dilution. This gives values of 24 (row 1), 20 (row 2) and 16 (row 3) HAU/ml.

ticles contain many HA molecules; in the form of spike-like projections, they are capable, provided they are present in sufficient quantity, of forming a complete network of linked red cells. Network formation can be detected very simply in a 'U'-shaped cup or tube; unagglutinated cells precipitate as a small pellet in the bottom of the tube, but agglutinated cells settle out as a continuous sheet (see *Figure 3*). This reaction provides an extremely convenient means for estimating the amount of virus in a particular preparation.

A fixed amount of red cells is mixed with an equal volume of different dilutions of the unknown virus suspension in a series of cups, and incubated undisturbed for 30–45 min. The extent to which the virus suspension can be diluted before agglutination activity is lost can then be measured. The amount of virus which can agglutinate 50% of the red cells in this situation is defined as 1 haemagglutinating unit (HAU), and so knowing the extent to which the virus may be diluted before this point is reached, the virus titre, in HAU/ml, may be calculated. The unit of haemagglutinating activity is defined by the conditions of the assay, such as the volume of the reaction and the concentration of red cells, and it must be noted that different laboratories may carry out the test in slightly different ways. The assay conditions described here are those in routine use in the authors' laboratory.

The HAU is a useful unit for comparative purposes, but clearly does not measure directly either the overall amount of virus present or the number of infectious particles. However, for any particular virus strain, a relationship between the HAU and other units of virus measure may be established empirically and, provided the virus has not changed its genetic character, used subsequently to estimate other parameters. For most laboratory strains of influenza virus, the following approximate relationship holds:

$$\begin{aligned} 1 \text{ HAU} &\equiv 2 \times 10^6 \text{ egg infectious particles } (\text{EID}_{50}) \equiv 2 \times 10^5 \text{ p.f.u.} \\ &\equiv 10^7 \text{ physical particles by electron microscopic count} \end{aligned}$$

3.1.2 *Preparation of Red Cells*

A variety of different red cell types may be used for the assay, for example human type 'O' or guinea pig cells, but most often fowl red cells are used. The following protocol is convenient for preparation of a 1% suspension.

(i) Filter the blood through two layers of muslin.

(ii) Centrifuge the filtered blood at 1500 r.p.m. in a bench centrifuge for 5 min.

(iii) Remove the supernatant and surface layer of white cells with a Pasteur pipette attached to a water pump. Wash the cells three times in saline solution by resuspension and re-centrifugation at 1500 r.p.m. as before.

(iv) Resuspend the cells in fresh saline and pack cells by centrifugation at 2500 r.p.m. for 10 min. Discard the supernatant and dilute aliquots of packed cells 100-fold in saline. Shake well and store at +4°C.

3.1.3 *Experimental*

(i) In a series of cups, prepare doubling dilutions of the unknown virus suspension in saline so that the final volume in each well is 0.25 ml. Set up a control well with saline alone.
(ii) Shake the 1% suspension of red blood cells thoroughly and add 0.25 ml to each cup. Swirl to mix and leave the cups undisturbed on a sheet of white paper for 45 min.

The pattern of agglutination at the end-point generally falls into one of three types, and this may be used to calculate an approximate titre (see legend to *Figure 3*).

An alternative and more economical assay procedure is to use a scaled-down test (0.05 ml each of virus and red cells) which can be carried out in a microtitre tray. It should however be borne in mind that in this case a unit of haemagglutinating activity is defined as the amount of virus *in 0.05 ml* which can agglutinate 50% of the red cells *in 0.05 ml* of a 1% suspension of red cells.

3.1.4 *Factors Which May Affect the Test*

A number of factors may interfere with the assay.

(i) It is important that the blood should not be stored for too long before use. After 5–7 days, haemolysis will begin to occur, making assays at first inaccurate and eventually impossible.
(ii) Since a lower concentration of red blood cells in the assay will give a falsely high virus titre, it is important that the red cell suspension should be shaken thoroughly before use.
(iii) The assay should not be left for too long before reading the result. As well as the haemagglutinin protein, influenza viruses contain on their surface a neuraminidase enzyme which can remove sialic acids from the ends of oligosaccharide chains. Since the HA protein binds to its receptors *via* sialic acids, the action of neuraminidase serves to break the bonds between virus and cells. In time therefore a fully agglutinated sheet of red cells will begin to collapse. Some virus strains have a very high neuraminidase activity which, at room temperature, may remove virus receptors too quickly to allow proper agglutination. This problem may be avoided by carrying out the assay at 4°C.
(iv) Poor results may be obtained if the assay trays are not properly cleaned. Trays should be immersed after use in 4% sodium hydroxide solution for 1 h before thorough rinsing in distilled water.
(v) The virus suspension may contain substances which inhibit agglutination, either specifically or non-specifically. This is a particular problem when titrating tissue culture medium, since serum is known to contain such inhibitors. Many of these inhibitors may be inactivated by heating the serum to 56°C for 30 min before use in the experimental medium, but for a full

discussion, see Hoyle (12). The reverse may also occasionally be true; some batches of serum contain factors which cause agglutination of red cells in the absence of virus. This possibility should be controlled by titrating some uninfected medium as well as the virus containing sample.

(vi) It should be remembered that many other viruses also haemagglutinate red cells, and so might interfere with the assay. The identity of the virus under test can of course be checked by haemagglutination inhibition test (see Section 3.1.5) using an antiserum which is known to react with the virus.

3.1.5 *The Haemagglutination Inhibition (HAI) Test*

The ability of influenza viruses to bind to and agglutinate red blood cells can also be exploited to measure the concentration of antibodies against the virus in a serum sample. A series of dilutions of the serum are reacted with a fixed amount of virus (enough to allow complete red cell agglutination in normal circumstances), before addition of red blood cells; the extent to which the serum can be diluted and still inhibit red cell agglutination therefore indicates the amount of antibody present.

(i) Set up a series of 2-fold dilutions of the serum under test using saline as the diluent. Depending on whether the assay is to be carried out in standard or microtitre haemagglutination trays, the final volume of each dilution should be 0.25 ml or 0.05 ml, respectively.

(ii) Add 4 HA units of the test virus to each well (in a final volume of 0.05 ml or 0.01 ml depending on the assay volume). Incubate the virus and antibody at room temperature for 30 min.

(iii) Add the appropriate amount of 1% red blood cells (0.25 ml or 0.05 ml), mix and allow to stand undisturbed for 30 min.

(iv) Read the assay and determine the concentration of antibody in the serum from the highest dilution which is still able to inhibit agglutination.

Note: appropriate controls should be included to check that the added virus is the correct concentration, that the serum itself does not cause agglutination of the red cells and that the red cells will settle out in the absence of virus.

3.2 **Titration by the Egg Infectious Dose**

3.2.1 *Introduction*

The amount of infectious virus in a particular suspension may be estimated by measuring the highest dilution that can still initiate infection in an embryonated egg. The smallest amount of virus which is capable of doing this, on 50% of occasions, is known as the egg infectious dose (EID_{50}). The assay is the most sensitive of all methods for measuring small amounts of infectious virus, since for most strains 1 p.f.u. of virus is equivalent to about 10 EID_{50}. One EID_{50} represents about 5 – 10 virus particles as determined by counting in the electron microscope (see Section 3.5).

3.2.2 *Experimental*

(i) Prepare serial 10-fold dilutions of the unknown virus suspension in Hank's balanced salt solution (see Section 8), and inoculate 0.1 ml samples of each dilution into the allantoic (or if appropriate the amniotic) cavity of a series of 11-day-old embryonated eggs (see Section 2.1.4). The number of eggs used for each dilution depends on the accuracy required, but generally 10 eggs are sufficient. Incubate for the required period (see Section 2.1.5).

(ii) Harvest the allantoic fluid from each egg and carry out a haemagglutination test on the resulting samples. It is not necessary for this purpose to set up complete dilution series for each egg sample. Since a positive or negative result for virus growth is all that is required, a spot test is sufficient. This may be performed simply by diluting 25 μl of the test sample directly into 0.25 ml of saline in a cup or tube, and adding 0.25 ml of red cells.

3.2.3 *Calculation of Virus Titre*

A number of methods may be used to calculate the virus titre (in EID_{50}). One of the most useful is that described by Thompson (13). In the example shown in *Table 1*, 0.1 ml samples of each of a series of 10-fold dilutions of a virus stock (Y EID_{50}/ml) have been inoculated into 10 eggs. For each dilution, the number of eggs showing virus growth (as judged by a haemagglutination test) has been determined, and this number is expressed as a proportion of the total inoculated (the proportional infectivity). For each dilution, a 'mean proportional infectivity' is calculated by averaging the proportional infectivity for that particular dilution with those obtained for the two adjacent dilutions. These 'moving averages' (p) of the infectivities will usually form an increasing sequence without any reversals

Table 1. Calculation of Virus Titre.

Dilution	*Dose*	*Log[dose]*	*No. of eggs infected*	*Proportion of eggs infected*	*Moving average (p)*
10^{-10}	Y x 10^{-11}	$-11+\log[Y]$	0/10	0	
10^{-9}	Y x 10^{-10}	$-10+\log[Y]$	0/10	0	$0.33(=\bar{p}_0)$
10^{-8}	Y x 10^{-9}	$-9+\log[Y]$	1/10	0.1	$0.266(=\bar{p}_1)$
10^{-7}	Y x 10^{-8}	$-8+\log[Y]$	7/10	0.7	$0.600(=\bar{p}_2)$
10^{-6}	Y x 10^{-7}	$-7+\log[Y]$	10/10	1.0	$0.900(=\bar{p}_3)$
10^{-5}	Y x 10^{-6}	$-6+\log[Y]$	10/10	1.0	

In this case the moving average of the infectivities for dilutions 10^{-9}, 10^{-8} and 10^{-7} $(=\bar{p}_1)$ $=(0 + 0.1 + 0.7)/3 = 0.266$ and the moving average for dilutions 10^{-8}, 10^{-7} and 10^{-6} $(=\bar{p}_2)$ $= 0.1 + 0.7 + 1.0)/3 = 0.600$. The log[dose] which would give a moving average of $\bar{p}=0.5$, represents 1 EID_{50} and may be calculated by interpolation as follows:

In the example, an average infectivity of $\bar{p}_1=0.266$ corresponds to a log[dose] of $-9+\log[Y]$. The log[dose] giving an average infectivity of 0.5 will be a proportionate distance

$$\frac{0.500 - 0.266}{0.600 - 0.266} = \frac{0.234}{0.334} = 0.7$$

between the values of $-9 + \log[Y]$ and $-8 + [\log Y]$; i.e.

$$-9 + \log[Y] + 0.7(\log[10]) = -8.3 + \log[Y] = 1\ EID_{50}$$

$$\text{therefore } \log[Y] = 8.3$$

$$Y = 1.99 \times 10^8$$

(whereas the same may not be true of the unaveraged infectivities). These values are used to calculate, by interpolation, the virus dose which would give a mean proportional infectivity of 0.5. This dose therefore represents 1 EID_{50}.
The accuracy of results obtained by this method may be assessed by statistical analysis as described by Thompson (13).

3.3 **Titration of Virus using Egg-bits**

3.3.1 *Introduction*

The use of egg-bits, i.e., pieces of 11-day embryonated egg shell with chorioallantoic membrane attached, rather than whole eggs, is an economical way of measuring infectivity of influenza virus samples. This method (14), developed to titrate influenza virus before tissue culture systems became available, is still a very useful method for titrating virus strains which will not grow in tissue culture, or if suitable tissue culture cells are not available.

The use of bits of chorioallantoic membrane attached to the shell is important since the shell helps to keep the tissue intact and also acts as a strong buffering system to keep the pH at a constant 7.5.

3.3.2 *Preparation of Egg-bits*

(i) Open embryonated (11 day) eggs with scissors at the albumin (pointed) end and tip out the embryo. If the chorioallantoic membrane is removed with the embryo the egg should be discarded.
(ii) Rinse the shell with membrane attached twice with standard medium (see Section 3.3.6) and cut strips of shell about 0.6 cm wide running the length of the egg. Cut these in turn to form square pieces.
(iii) Store the pieces in a Petri dish containing standard medium. Each egg can provide up to 100 6 mm^2 pieces.

3.3.3 *Virus Inoculation*

(i) Carry out the assays in small test-tubes (5 x 1.25 cm) or more conveniently in the trays used for HA assay.
(ii) Sterilise the trays by soaking them in 95% ethanol. Drain the excess ethanol and wrap the trays individually in clean brown wrapping paper. Put the trays in a warm air oven overnight to evaporate off the residual ethanol, they are then ready for use.
(iii) Arrange the egg-bits in the HA trays (1 x 6 mm^2 bit/well) using 0.3 ml standard medium in each well. The egg-bits may be kept for several hours at room temperature in standard medium or for 24 h at 37°C with agitation without significant effect on subsequent virus growth.
(iv) Prepare serial 10-fold dilutions of the test virus and add virus inoculum of 0.025 ml to the wells.

3.3.4 *Incubation of Samples*

(i) If a large batch of virus samples are to be assayed, stack the trays one on top of the other separated by 0.3 cm spacers to ensure efficient aeration of the wells.

(ii) Fix the battery of trays into a suitable frame and surround by moist lint.

(iii) Seal the assembly of trays using wrapping film or aluminium foil, place in a warm room and shake vigorously for 2–3 days depending on the virus being used. A horizontal shaker with a speed of 120 oscillations/min and a thrust of 8 cm is recommended. Efficient shaking is essential both to avoid build up of poisonous metabolites and to ensure sufficient aeration of the cells.

3.3.5 *Reading the Assay*

At the end of the incubation period remove the egg-bits from the wells with fine forceps and add 0.25 ml of standard 1% red blood cells to each well. Briefly shake the trays and then leave undisturbed on a white background for 30 min. Positive haemagglutination indicates infection. The titre of virus can then be determined from the dilution series as for the whole egg assay.

3.3.6 *Preparation of Standard Medium*

Weigh out the following:

NaCl	8.0 g
KCl	0.6 g
$CaCl_2$	0.8 g
$MgCl_2$	0.05 g
Glucose	0.3 g
Gelatin (acid-free)	2.0 g
Chloramphenicol	0.1 g

Neutral red (25 ml 0.01% stock) and water are then added to make the volume up to 1 litre. The pH is adjusted to approximately 7.0 by the addition of a few drops of normal NaOH. The neutral red acts as an indicator and the colour should be yellowish-orange at pH 7.0. The complete solution is sterilised by autoclaving at 115°C for 30 min.

3.4 **Plaque Assay**

3.4.1 *Introduction*

Serial 10-fold dilutions are prepared in PBS supplemented with gelatin (0.5%). Plaque assay can be carried out in any suitable tissue culture vessel but is most conveniently performed in 6-well multiwell dishes with a cell diameter of 4 cm. This enables a complete assay to be carried out in one container, saving space in incubators and lessening the chance of confusion from badly labelled dishes. Where it is important to keep the temperature constant with a high degree of accuracy, as for example when working with temperature-sensitive (*ts*) mutants at the non-permissive temperature, these multiwells can be stacked and sealed in a plastic bag and submersed in a water bath with an accurate thermostat. This greatly reduces the frequency of failed assays due to temperature drops when incubator doors are opened.

3.4.2 *Infection of Monolayers*

It is important that cell monolayers are evenly confluent before starting the plaque assay.

(i) Remove the medium, and wash out traces of serum by rinsing the monolayers with warm PBS. An inoculum of 0.2 ml is sufficient to cover the monolayer in a 4–5 cm dish.
(ii) Add the virus inoculum (starting with the most dilute sample) drop by drop and spread it from the centre of the dish. If the virus is added at the side of the dish, surface tension tends to prevent it spreading to the centre and all the plaques will be crowded at the edge of the dish.
(iii) Allow the virus to adsorb at room temperature for 30 min. It is good practice to shake the dishes to encourage the even spread of virus once or twice during the adsorption period.

3.4.3 *Addition of Overlay Medium*

At the end of the adsorption period remove the inoculation medium (although this is not essential), and cover the cells with 2 ml of maintenance medium containing between 0.7 and 1.25% agar. It is very important to ensure that the temperature of the overlay medium is not too high or it will kill the cells. It is equally important to ensure that the agar is not too cool as it will set in the pipette. In CEF cells a temperature of 45°C is suitable. To prepare overlay medium, heat 2 x concentrated medium to 45°C and add an equal volume of 2 x concentrated molten agar, previously cooled to about 60°C. Allow the complete overlay medium to cool to 45°C.

For rapid preparation of molten agar a microwave oven is very convenient. It is important to avoid bubbles in the overlay medium as these will confuse the final results since they can resemble plaques, or obscure areas where plaques have formed. Where plaque assays are to be carried out in cells more sensitive than CEFs, the agar can be substituted by low gel temperature agarose which does not set until the temperature is below 37°C. Overlay medium containing carboxymethylcellulose (CMC) can also be used (see Section 8). This avoids problems with the temperature of the overlay medium but can cause 'rocketing' of the plaques if the dishes are disturbed during the assay.

3.4.4 *Staining of Plaques*

Normally plaques take 2–3 days to develop to the point where they may be detected by staining. Where cells have been incubated with an agar overlay, staining plaques may be carried out using neutral red.

(i) Prepare overlay medium, containing agar, as above and add neutral red (made up at 1% in water) to a final concentration of 0.1%.
(ii) Add 2 ml of this medium to each 4 cm dish as above.
(iii) Return the dishes to the incubator for 3–4 h, when the plaques should become visible.

If it is not possible to count all the plaques at once plaques will remain visible for at least 48 h and should not enlarge if sealed in a box in an atmosphere of 5% CO_2 and stored at +4°C. An alternative staining method, if a more permanent record is required, is to peel off the agar overlay and then fix the cell monolayer in formol-saline [10% v:v formaldehyde (38%) in PBS] for 2–3 min. The monolayers are then stained with a few millilitres of toluidine blue (1% in water) for 10 min. The stain is rinsed off in tap water and the dishes drained and dried.

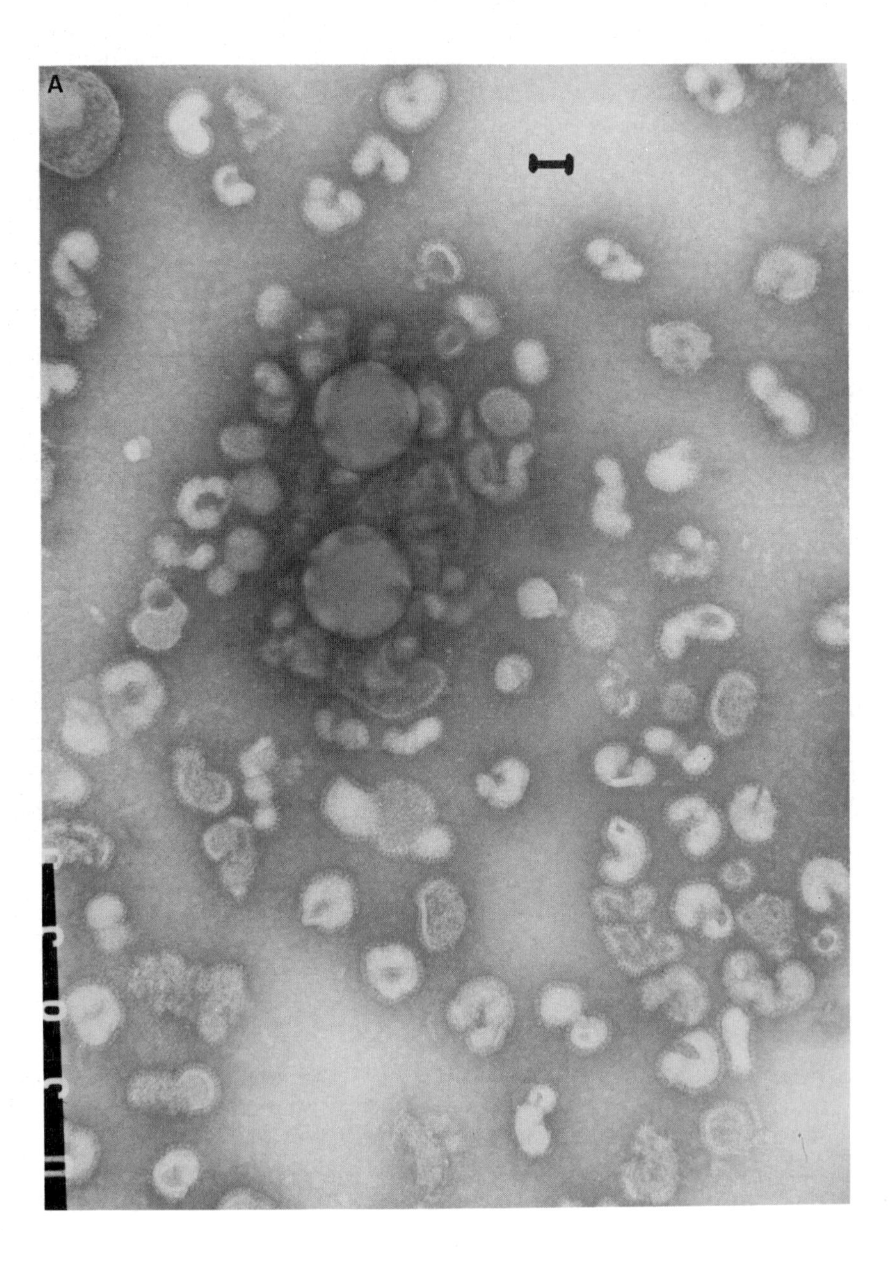

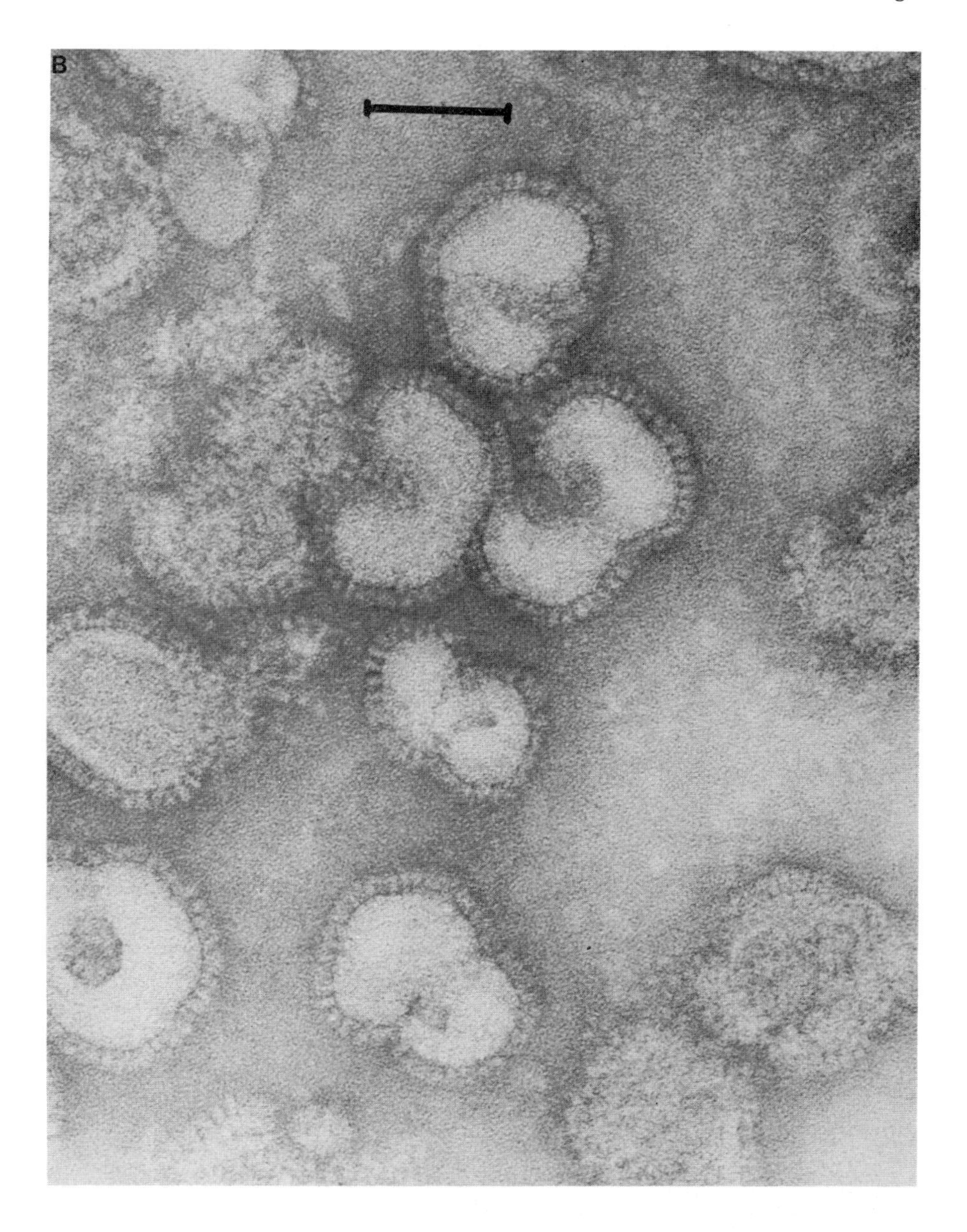

Figure 4. Electron micrograph of purified influenza A virus. Virus was purified by density gradient centrifugation as described in Section 4.3.3. The virus sample was negatively stained with phosphotungstic acid. Latex beads of known concentration were included with virus sample and can be seen in **A**. Virus particles have a mean diameter of 100 nm and the envelopes show distinct projecting spikes which are composed of the virus haemagglutinin and neuraminidase proteins. The spikes can be seen more clearly in **B** which shows a greater magnification. The bar in both **A** and **B** represents 100 nm.

This method of staining is also suitable for plaque assays which have been carried out in the presence of CMC medium.

3.4.5 *Plaque Purification of Virus*

To grow viruses from a plaque, choose a dish which contains few well-separated plaques. Remove the required plaques by stabbing each with a sterile Pasteur pipette. This will remove an agar plug to which some virus-infected cells have attached. Flush the agar plug into 50 – 100 μl of sterile Hank's balanced salt solution. This can be used directly to inoculate into fertile hens eggs to amplify the virus. Alternatively the virus can be plaqued again directly on tissue culture cells (plaque to plaque purification) and then amplified in fertile hens eggs after the desired number of plaque to plaque purifications. Normally three cycles of plaquing are required to plaque purify the virus. The plaque purified virus is grown up in fertile hens eggs to produce a 'master stock' (egg pass 1) and working stocks are prepared as outlined in Section 6.

3.5 **Electron Microscopy**

Previous sections have dealt with methods for measuring virus by assaying its biological properties such as the ability to infect tissue culture cells and eggs, or to bind to red blood cells, but these tests clearly do not measure the number of virus particles in a preparation. The relationship between the unit of infectivity or of haemagglutination and the virus particle is to a large extent dependent on the virus strain, and may also vary from preparation to preparation, and so it is often desirable to perform a particle count using the electron microscope. The most usual method for doing this is to mix a sample of the virus with an equal volume of a known concentration of latex beads which are approximately the same diameter (100 nm) as the virus particles. The mixture is then spread on an electron microscope grid after negative staining (for example with phosphotungstic acid) and the relative numbers of virus particles and latex beads in a particular set of fields are counted (see *Figure 4*). Knowing the concentration of latex beads in the original mixture, the concentration of virus particles may be calculated (15).

4. VIRUS PURIFICATION

4.1 **Introduction**

Various methods have been developed to purify influenza virus from infected allantoic fluid or cell medium. The method of choice depends on the degree of purity required. The purity of the final virus preparation can be established by polyacrylamide gel electrophoresis (see *Figure 5*) but, as with all enveloped viruses which bud from the cell membrane, some degree of host-protein contamination is inevitable even if the most rigorous care is taken in virus purification. In addition, the virus can be pleiomorphic and this can lead to difficulties when banding the virus on density gradients. However, with a virus which grows to high titre in either eggs or tissue culture, a purified virus preparation can be obtained in which host protein contamination is negligible. The first stage in

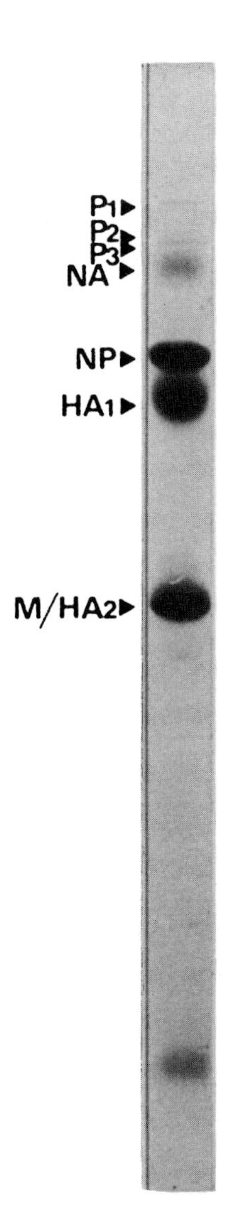

Figure 5. Polyacrylamide gel electrophoresis of the structural proteins of purified influenza A virus. Avian influenza A/FPV (Dobson strain) was purified by density gradient centrifugation as described in Section 4.3.3 and separated by electrophoresis on a 15% polyacrylamide gel.

purification is to remove all cell debris from the infected allantoic fluid or medium by low-speed centrifugation, usually at 2000 – 10 000 *g* for 10 min. All operations should be carried out on ice to minimise loss of virus infectivity and to avoid proteolytic breakdown of virus proteins.

4.2 Concentration of Virus

Since the virus is normally in a large volume of allantoic fluid or tissue culture medium it must first be concentrated to a manageable volume before any subsequent purification steps can be undertaken. There are many ways of concentrating the virus-infected medium and the choice will depend to a large extent on the equipment available in the laboratory.

4.2.1 *Centrifugation*

Since large volumes are usually involved (>1 litre) in most cases, a large capacity ultracentrifuge rotor is required. Rotors are available with capacities of 1–2 litres which are capable of pelleting virus from infected medium. A force of 50 000 *g* for 90 min is sufficient to pellet influenza virus, but for smaller volumes, i.e., after subsequent purification steps, or if small volumes of labelled virus are being purified, the centrifugation time may be reduced to 30 min using a force of 200 000 *g* in a smaller rotor. The virus pellets are drained well to remove excess fluid and resuspended in a small volume of NTE buffer using a syringe and a large gauge needle (19 G). The resuspended virus should then be passed several times through a fine needle (25 G) or sonicated for 1 min in a water bath sonicator to break up virus aggregates. The volume of NTE in which the virus is resuspended will depend on the size of the pellets and the volume required for the subsequent purification steps. An approximate volume is 2–5 ml for the pellet from 1–2 litres of infected fluid.

4.2.2 *Polyethylene Glycol Precipitation*

Polyethylene glycol (PEG) precipitation is commonly employed to precipitate virus from infected medium. It is useful when a large capacity high-speed rotor is not available. For this procedure solid PEG (mol. wt. 6000) is added to the virus-containing fluid to a final concentration of 8%(w/v) and stirred on ice or at 4°C for 1 h. The virus precipitate is collected by centrifugation at 10 000 *g* for 20 min and resuspended in a small volume of NTE buffer. Resuspension of the virus-PEG pellet is difficult since it forms a glutinous mass. The pellet is best resuspended using a tissue homogeniser and then either sonicated or passed through a fine needle (25 G). Any large particles which fail to resuspend should be removed by centrifugation at 10 000 *g* for 10 min before further purification is carried out.

4.2.3 *Mechanical Concentrator*

Several commercial concentration devices are available which can quickly reduce 1–2 litres of infected fluid to 50–100 ml. These devices are useful since problems incurred in resuspending virus pellets are avoided. However, the final volume is too great for some of the further purification steps described below.

4.2.4 *Red Cell Adsorption*

This very simple and rapid method of virus concentration takes advantage of the ability of influenza viruses to bind to red blood cells (see Section 3.1). It can also be used as a one step purification method for the virus (see Section 4.3.4).

4.3 **Purification Methods**

Several methods for purification of concentrated virus are given below. The choice of method employed will depend on the availability of equipment and the degree of purity required in the final virus preparation.

4.3.1 *Centrifugation onto a Sucrose Cushion*

This rapid one-step method for virus purification is suitable for small quantities of infected fluid, e.g., when radiolabelling of virus is being carried out. It can also be used for a crude purfication of virus concentrated from a larger quantity of fluid. The virus-infected fluid, for example tissue culture medium, is layered over a step gradient consisting of 5 ml of 15% sucrose above the same volume of 60% sucrose in NTE buffer and centrifuged at 200 000 *g* for 30 min. The interface region between the 15% sucrose layer and the 60% sucrose cushion is removed and the virus can be pelleted out of this solution after dilution (>4-fold) in NTE buffer (see Section 4.2.1).

4.3.2 *Controlled-pore Glass Bead Chromatography*

This method (16) allows the rapid purification of influenza virus from concentrated virus.

(i) Fill a suitable column with controlled-pore glass beads (CPG 700-120, Sigma Chemical Company) with a mean diameter of 72.9 nm and a particle size distribution between 125 and 177 μm (80 – 100 mesh). Glass beads of these dimensions will exclude material of greater than 70 nm (influenza particles have a mean diameter of ~100 nm) and so will effectively separate the virus particles from smaller host cell contaminants. A column 0.9 cm in diameter and at least 100 cm long is required for effective separation of volumes of about 5 ml. For larger volumes (20 – 50 ml) a column of 2.5 x 100 cm is required.

(ii) Before pouring the column, wash the beads in distilled water three times and then in running buffer twice. To ensure an even settling of the beads, agitate the column constantly while the beads are being poured. It is convenient to assemble the column on a horizontal rotary shaker with a wide funnel attached to the top of the column.

(iii) Fill the column to 4/5 capacity with NTE loading buffer and expel air bubbles. Then with the column flow adaptor opened, add the slurry of glass beads slowly through the funnel. When all the slurry has been added and the column is full, close the flow adaptor and allow the beads to settle while the column is still shaking. The column can then be removed and set up in a cold room and equilibrated with more NTE buffer.

(iv) Run the column at a flow-rate of 1 – 5 ml/min depending on the column size, by attaching a peristaltic pump.

(v) Monitor the progress of the virus through the column using a suitable spectrophotometer or u.v. monitor.

(vi) Collect fractions of 2 – 10 ml, depending on the column size. The virus peak elutes first from the column and can easily be identified by its haemagglutinating activity. A sharp peak of virus, well separated from cell contaminants, should be obtained, from which the virus can be pelleted after suitable dilution in NTE. Most (70 – 90%) of the haemagglutinating activity can be recovered in the pellet.

Once they are set up, the columns can be re-used after washing with NTE. After several cycles of fractionation, the beads should be removed and acid washed before re-use.

4.3.3 *Density Gradient Purification*

For this procedure virus concentrate is subjected to rate-zonal sedimentation on linear 15 – 60% (w/v) sucrose gradients in NTE.

(i) Use 10 ml or 30 ml gradients depending on the volume of virus to be purified.
(ii) Centrifuge the gradients at 110 000 *g* for 1 h at 4°C. If the virus is present in sufficient quantity, it may be observed as a visible white diffuse band about half way down the gradient.
(iii) Collect the virus band using a syringe with a needle with the tip bent at right angles. If there is very little virus and it is difficult to see, visualise it by illuminating it from above against a black background. Mark the area of the gradient containing virus and harvest this area of the gradient when the light is removed. Alternatively fractionate the gradient stepwise, and monitor individual fractions for haemagglutinating activity.
(iv) Purify the virus further by a second gradient centrifugation, either by another cycle of rate-zonal sedimentation on a 15 – 60% sucrose gradient or preferably by equilibrium density gradient sedimentation on a 20 – 45% (w/v) potassium tartrate gradient.
(v) Load the virus directly onto the tartrate gradient after at least 1:1 dilution in NTE or re-pellet at 200 000 *g* for 30 min at 4°C and resuspend in a small volume of NTE for subsequent gradient centrifugation.
(vi) Centrifuge tartrate gradients at 100 000 *g* for at least 5 h; an overnight spin is generally most convenient.
(vii) Collect the virus band as before, dilute 6-fold, and pellet at 100 000 *g* for 30 min at 4°C.
(viii) Suspend the virus pellet in a suitable buffer for storage (see Section 6).

4.3.4 *Adsorption to Red Blood Cells*

Virus can be concentrated from clarified allantoic fluid or medium by the addition of formalinised red blood cells at a concentration of 5% (v/v). The virus-red blood cell suspension is stirred gently at 4°C for 1 h. Erythrocytes with adsorbed virus are pelleted at 2000 *g* for 10 min and washed twice in ice-cold saline. The efficiency of adsorption can be monitored by an HA test performed on the fluid before and after adsorption to the red cells. If adsorption is less than 90% efficient the operation can be repeated with a second lot of red blood cells. Adsorbed virus is eluted from the erythrocytes by stirring in 10 ml standard saline at 37°C for 30 min. The red cells are then removed by centrifugation at 2000 *g* for

10 min and the virus-containing supernatant removed. Efficient elution depends on the virus neuraminidase removing the sialic acid receptors from the red blood cells to release virus. Purified virus can be recovered from the supernatants, after dilution more than 6-fold in NTE buffer, by centrifugation at 200 000 *g* for 30 min at 4°C. The virus is then resuspended in a suitable buffer for storage.

Preparation of formalinised red blood cells

(i) Filter blood through two layers of muslin and pellet the cells by centrifugation at 2000 r.p.m. for 10 min in a bench centrifuge.
(ii) Remove the supernatant and white cell layer and wash the pellet three times in a volume of PBS approximately equivalent to that of the supernatant by resuspension and re-centrifugation as before. After the last wash centrifuge the cells for 20 min to form a tightly packed pellet.
(iii) Resuspend each 25 ml of pelleted cells in 200 ml PBS and transfer to a large measuring cylinder.
(iv) Fill a piece of dialysis tubing with 50 ml of formalin (38% formaldehyde), which has previously been adjusted to pH 5.5 – 6.0, and then place the tubing in the measuring cylinder with the red blood cells. Agitate the red cells slowly, avoiding frothing, using a magnetic stirrer for 3 h at room temperature. This allows the formalin to diffuse slowly into the red cells.
(v) After 3 h puncture the dialysis tubing and allow the formalin to drain into the red cells. Continue stirring for 12 h.
(vi) Filter the cells through muslin and wash them five times with PBS. Use a glass rod to resuspend the cells.
(vii) Resuspend the packed cells in an equal volume of 0.2 M Tris-glycine buffer pH 8.3 and incubate at 37°C for 2 h. Add solid sodium borohydride to 1 mM (0.038g/l).
(viii) Pellet the cells by centrifugation at 2000 r.p.m. and wash twice more in PBS.
(ix) Finally resuspend the cells in PBS to give a 50% suspension of red cells. Store in the presence of 0.01% sodium azide at 4°C. The cells can be stored indefinitely at this stage.
(x) Wash the appropriate amount of red cells in Tris-glycine buffer before use.

5. ASSAY OF VIRION-ASSOCIATED ACTIVITIES

5.1 Virion Transcriptase

5.1.1 *Introduction*

The genome of influenza virus is negative-sense RNA, i.e., it must first be transcribed into positive-sense mRNA before virus-coded proteins can be synthesised in infected cells. To accomplish this the virus particle contains an RNA-dependent RNA polymerase. This virus-associated RNA polymerase can transcribe full-length virus mRNAs *in vitro* which are terminated at the correct position and are polyadenylated. The enzyme is greatly stimulated by the addition of specifc dinucleotides (17) or capped mRNAs, e.g., globin mRNA (18), which act as primers in the reaction. If capped mRNAs are added to the system *in vitro*, the 5′ cap structure along with the adjacent 10 – 14 nucleotides is transferred to the 5′ end of the virus-coded transcripts (19), accurately reflecting the situation which occurs *in vivo* (20,21).

5.1.2 *Experimental*

Virion transcriptase activity can be demonstrated in purified virus by the addition of suitable buffer, ribonucleoside triphosphates, divalent cations and a detergent to permeabilise the virus envelope. A typical assay mix (100 μl) would contain the following:

Tris-HCl, pH 8.2	50 mM
KCl	150 mM
$MgCl_2$	8 mM
Dithiothreitol	5 mM
Nonidet P-40	0.5%
ATP	2.0 mM
GTP	0.2 mM
CTP	0.4 mM
[^{3}H]UTP	0.4 mM
Purified virus	0.5 – 1.0 mg/ml.

(i) Assemble the reaction mixtures on ice, and then incubate them at 31°C (the optimum temperature for the reaction).

(ii) When primer is to be added, pre-incubate the reaction mixtures with the primer for 10 min without $MgCl_2$; then initiate the assay by the addition of the $MgCl_2$.

(iii) To determine the time course of RNA synthesis, remove aliquots (10 μl) at intervals and add to 20 μl of N perchloric acid containing 0.125 M sodium pyrophosphate.

(iv) Add calf thymus DNA (10 μg) and leave the samples on ice for 10 min.

(v) Collect the precipitated RNA on Whatman GF/C glass fibre filters and then wash them with 0.88 M HCl, 0.125 M sodium pyrophosphate and finally with ethanol before drying.

(vi) Determine the associated radioactivity by liquid scintillation counting using a toluene-based scintillant.

The rate of the reaction will depend on the virus used and whether or not specific dinucleotides or capped mRNAs are added to the reaction. Typical reaction curves for the unstimulated and stimulated reactions are shown in *Figure 6.*

5.2 **Neuraminidase**

5.2.1 *Introduction*

Influenza virus particles contain on their surface a neuraminidase enzyme in addition to the haemagglutinin protein. The neuraminidase enzyme takes the form of a mushroom-like spike which projects outward from the virion envelope and which is composed of a tetramer of polypeptide chains. The enzyme is capable of cleaving the terminal N-acetyl neuraminic acid residues from the sugar chains of

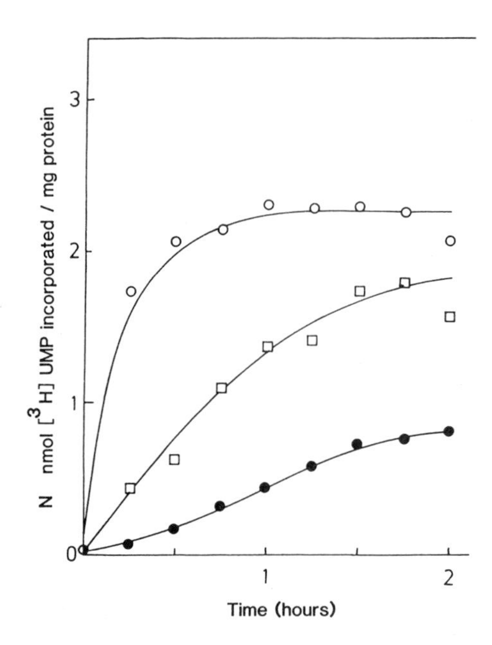

Figure 6. Assay of the RNA-dependent RNA polymerase activity of purified influenza A virus. Virus purified by density gradient centrifugation as described in Section 4.3.3 was assayed for RNA polymerase activity in the presence or absence of added primers. Reactions (100 μl) were set up as described in Section 5.1.2. The assay was carried out at 31°C and 10 μl aliquots were removed at the times shown and the acid-precipitable counts determined. From these values the amount of [^{3}H]UTP incorporated into acid-insoluble material was calculated. The ApG primer was added at a concentration of 0.3 mM and the globin mRNA primer at 100 μg/ml. These are the optimal amounts of primer for this virus (influenza A/FPV/Rostock strain). ●—●, unstimulated reaction; □—□, mRNA-primed reaction; ○—○, ApG-primed reaction.

glycoproteins, and its function appears to be to prevent aggregation of virus during maturation and release (22). The enzyme activity may be assayed by incubating purified virus with a suitable substrate such as fetuin which may be prepared from foetal calf serum (23,24).

5.2.2 *Experimental*

See Section 5.2.3 for assay reagents.

(i) Prepare the sample for testing. Purified virus is best but it must be borne in mind that glycerol and sucrose may interfere with the assay. Some neuraminidases do not work well in the presence of Ca^{2+} whereas others require Ca^{2+} for their activity. Virus in allantoic fluid may be used if it is first dialysed to remove interfering substances.

(ii) Prepare a series of tubes containing between 5 and 40 μg of N-acetylneuraminic acid [in 0.2 ml of 0.2 M sodium phosphate buffer (pH 6.0)] in order to obtain a calibration curve for the test.

(iii) Make up an assay mixture of 0.2 ml containing 0.1 ml of the sample to be tested (virus preparation or virus diluent as a control) and 0.1 ml of fetuin (10 mg/ml) in 0.2 M sodium phosphate buffer, pH 6.0 (the pH optima for different neuraminidases are known to vary, but pH 6.0 is generally satisfactory). Mix the reagents well and incubate them at 35°C for 0.5 – 2 h depending on the enzyme activity. The temperature optima for neuraminidases of different strains may vary (some enzymes, for example those of the N1 strain, are unstable at 37°C) and so should be checked for the particular strain under test.

(iv) Cool the reaction mixtures to room temperature and add 0.1 ml of periodate reagent. Mix well and incubate at room temperature for 20 min.

(v) Add 1.0 ml of arsenite reagent. Mix until evolved iodine redissolves.

(vi) Add 2.5 ml of thiobarbituric acid reagent. Mix and place in a boiling water bath for 15 min. A red-pink colour develops which fades on cooling.

(vii) Add 4 ml of butanol reagent [N-butanol containing 5% concentrated HCl (v/v)] and mix vigorously to extract the colour. Centrifuge at about 1500 r.p.m. in a bench centrifuge for 10 min at room temperature to separate the phases. Read on a spectrophotometer at 549 nm against the control (about 0.6 O.D. is ideal) and compare the results with those obtained with the calibration curve.

5.2.3 *Reagents for the Assay*

Periodate reagent:	sodium periodate	4.28 g
	orthophosphoric acid	62 ml
	H_2O	38 ml

Dissolve periodate in water and then add the acid. Store in an amber bottle in a cool dark place (not at 4°C). Stable.

Arsenite reagent:	sodium arsenite	10.00 g
	sodium sulphate (anhydrous)	7.10 g
	H_2O	100 ml
	conc. H_2SO_4	0.28 ml

Filter through Whatman no.1 paper. Stable.

Thiobarbituric acid reagent:	Thiobarbituric acid	1.2 g
	sodium sulphate (anhydrous)	14.2 g
	H_2O	200 ml

Dissolve in boiling water bath. A precipitate forms on storage.

5.3 **Fusion Activity**

5.3.1 *Introduction*

The haemagglutinin molecule of influenza virus is responsible for binding the virus to cell receptors and mediates fusion of the virus and host cell membranes. This fusion activity requires acidic pH. Bound virus is internalised by endocytosis in coated pits and the virus can later be detected in larger endosomes and in lysosomes where the acidic pH necessary to trigger fusion is found (25). The exact mechanism whereby fusion of virus and cell membranes is accomplished is not understood, but cleavage of the haemagglutinin molecule to form the HA1 and HA2 subunits is essential. The N-terminal amino acid sequence of the HA2 molecule is very highly conserved and is very hydrophobic. It is thought that at neutral pH this region of the molecule is buried in the three-dimensional structure of the molecule. At low pH the molecule has been shown to undergo a dramatic conformational change and it is thought that this exposes the N-terminal region of HA2 and allows its fusion activity to occur (26). This fusion activity can be shown *in vitro* by the ability of the virus to lyse red blood cells in a pH-dependent manner (27,28).

5.3.2 *Experimental*

(i) Set up a series of centrifuge tubes containing 20 μl of virus-infected allantoic fluid (~1000 HAU/ml) and 400 μl of 1% red blood cells. Purified virus can also be used in the assay. Keep these on ice for 30 min to allow the virus to bind to the red cells.

(ii) At the end of this period add 40 μl of 0.4 M buffer (with pH ranges 5.2–6.2, see below) to each tube and continue incubation at 37°C for 20–30 min with periodic agitation.

(iii) Pellet the cells at 2000 *g* for 10 min or at 10 000 *g* for 30 sec (micro-centrifuge) and measure the absorbance at 540 nm relative to that of water. The pH value at which 50% lysis occurs can be calculated and used to compare the pH fusion profiles of other influenza viruses.

Assays should be performed in duplicate in steps of 0.2 pH units and uninfected allantoic fluid should be assayed in the same way to determine the extent of lysis in the absence of virus. Human (type O, Rh –) or guinea pig blood works well in the assay but not chicken blood. Assays are most conveniently carried out in micro-centrifuge tubes. The buffers used can be varied depending on the pH range to be studied but in the range 5.2–6.2, 2-(n-Morphilino) ethanesulphonic acid (MES) and its salt is suitable.

6. STORAGE OF VIRUS STOCKS

Influenza virus is best stored in concentrated form as diluted virus is very unstable. The virus will retain its infectivity for up to 1 week at 4°C. For long-term storage, samples of allantoic fluid or infected medium are dispensed into small vials and snap-frozen in a methanol-dry ice bath. Rapid freezing of the virus has a negligible effect on virus infectivity. Virus can also be stored in a freeze-dried form at room temperature for an indefinite period. This is the most convenient way to transport virus samples. Purified virus for biochemical analysis can be stored at –20°C for about 6 months without significant loss of enzymatic activity in a buffer containing 60% glycerol, 100 mM NaCl, 10 mM Tris-HCl, pH 7.4 at a concentration of about 5 mg/ml. In this buffer virus can also be stored at 4°C for up to 2 weeks. Purified virus can be stored indefinitely in liquid nitrogen.

When dealing with a new or plaque-purified virus isolate, the following storage scheme will help to minimise the risk of introducing genetic changes through multiple passage.

(i) Grow up the initial virus isolate in fertile hen eggs to produce a 'master stock' (egg pass 1) of allantoic fluid. Hold a number of aliquots of this stock in different freezers in case of mechanical or electrical failure.

(ii) Grow a 'submaster' stock (egg pass 2) from this by egg inoculation, and once again store a number of aliquots at $-70°C$.

(iii) Finally, grow samples of this stock in eggs to give a 'working stock' (egg pass 3) which is used for experimental purposes.

This scheme can generate a virtually unlimited amount of virus which is always the same small number of passes away from the original plaque isolate. If variants of the virus arise during egg passage there is always a master stock available from which the original virus can be recovered. This system is essential when working with virus mutants since wild-type contamination of stocks can accidentally occur.

7. RADIOLABELLING VIRUS

7.1 **Labelling of RNA**

Virus can be labelled most easily in tissue culture cells with either ^{32}P supplied as carrier-free inorganic phosphate or [^{3}H]uridine. Virus also can be labelled, although less efficiently, in de-embryonated eggs using [^{32}P]orthophosphate (29).

(i) To label the virus in tissue culture with ^{32}P, first starve the cells of phosphate by incubating them overnight in phosphate-free medium. The phosphate-free medium should be supplemented with lactalbumin hydrolysate and glucose (0.05% final concentration of each), but serum should be omitted.

(ii) Carry out infection as normal but use physiological saline (0.9% NaCl) rather than PBS to wash the cells and to dilute the virus inoculum.

(iii) Add phosphate-free medium ($\sim$10 ml for a 15 cm diameter plate of cells) as described above (but containing 0.5 mCi/ml [^{32}P]orthophosphate).

(iv) Harvest medium containing the labelled virus at 24 h post-infection and purify using the method of choice (see Section 4).

For [^{3}H]uridine labelling, incubate the cells overnight in serum-free medium and infect in the usual way, and then add [^{3}H]uridine (0.1 mCi/ml) to the cells in serum-free medium. Harvest the virus-containing medium at 24 h post-infection. If the virus normally grows to a high titre in the cells used for the experiment, the virus produced from one 10 cm dish should just be visible on a 12 ml gradient if gradient purification is the method chosen to purify the virus. If virus production is low a parallel gradient with a larger amount of unlabelled virus can be run and the equivalent fraction harvested from the radioactive gradient. Otherwise the gradient must be fractionated and the radioactivity of each fraction measured.

If radioactively-labelled intracellular virus RNA species are required then the above procedures are followed and the RNA extracted from the cells at the required time after infection. The virus-specific RNA segments can be conveniently separated on a 6% polyacrylamide gel containing 7 M urea in the TBE buffer system (30). In this system the three largest segments (segments 1–3) will run together. To separate these a low percentage acrylamide gel system described by Loening (31) is required.

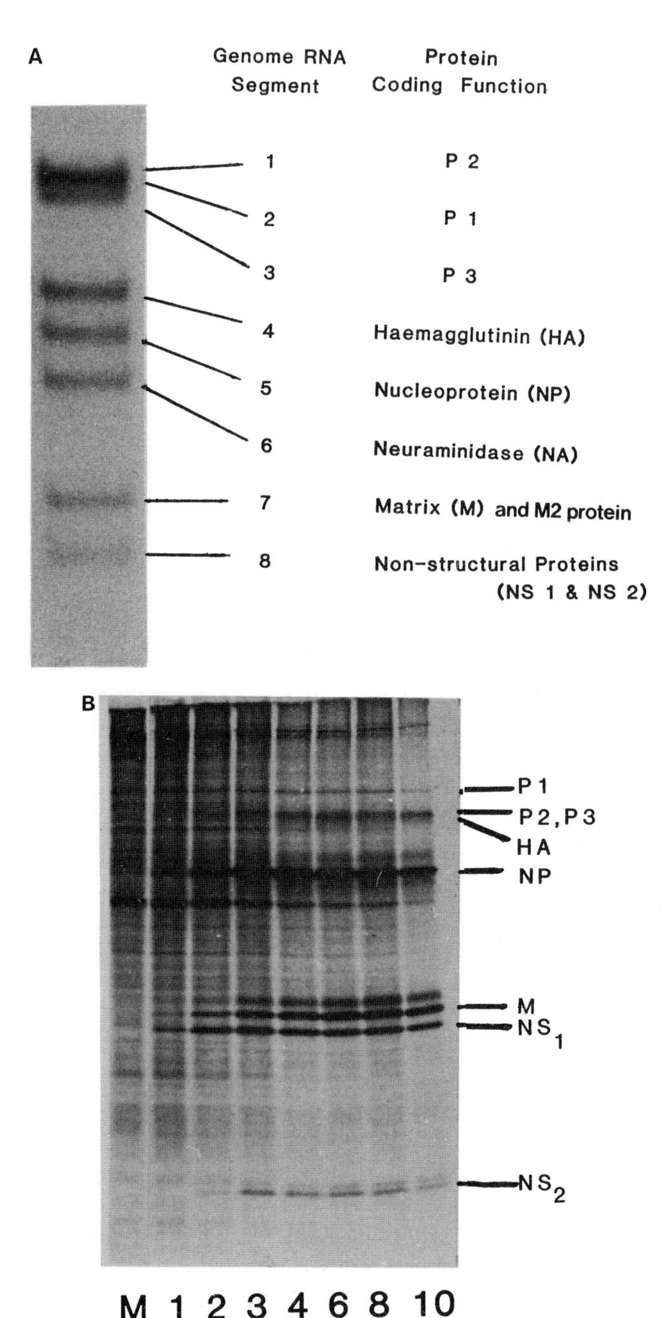

Figure 7. Electrophoresis of virus RNA and protein radioactively labelled as described in Section 7. **Panel A:** virus was labelled for 24 h with [^{32}P]orthophosphate and the labelled RNA extracted from gradient-purified virus. **Panel B:** CEF cells were pulse-labelled for 30 min with [^{35}S]methionine at various times after infection with influenza A/FPV/34 (Rostock strain), disrupted by boiling in SDS-containing gel buffer and electrophoresed on 15% polyacrylamide gels. **M:** mock-infected cells. The numbers under the other tracks refer to the time post-infection at which the pulse label with [^{35}S]methionine was carried out.

7.2 Labelling of Protein

The most convenient radiolabel for labelling influenza virus proteins is [^{35}S]methionine. Where labelled virions are required, cells should be incubated in methionine-free medium overnight before infection, in order to maximise the efficiency of labelling. After infection methionine-free medium containing 200 μCi/ml [^{35}S]methionine is added and the virus-containing medium is harvested at 24 h.

It is more often desirable to study the intracellular production of virus proteins, for example, to monitor the time course of infection or test the effects of inhibitors of replication. In this case it is not necessary to pre-starve the cells in methionine-free medium. Infection is carried out normally and at the desired time post-infection the medium is removed, the cells washed twice with PBS and then labelled by the addition of methionine-free medium containing 10 – 20 μCi/ml [^{35}S]methionine. A labelling period of 30 min is normally sufficient to label all the virus-specific proteins strongly, but this is also long enough for proteolytic cleavage of the HA molecule (into the HA1 and HA2 subunits) to occur. If this is not desired shorter pulses of label (5 – 10 min) should be employed and the concentration of [^{35}S]methionine increased accordingly. During short labelling periods (i.e., up to 30 min) it is advisable to add a strong buffer (e.g., Hepes pH 7.5 to 20 mM) to keep the pH constant since there is insufficient time for the CO_2-$NaHCO_3$ buffer system in the medium to equilibrate to pH 7.5. At the end of the labelling period the cells can be harvested directly or given a 'chase' by adding fresh growth medium supplemented with unlablled methionine (10 mM) after removal of the labelled medium.

Wastage of expensive [^{35}S]methionine may be minimised by carrying out infections for labelling in 24- or 96-well multiwell dishes of cells. In these wells, 50 or 100 μl, respectively, of labelled medium is all that is required and many different labelling experiments can be carried out with a small amount of [^{35}S]methionine.

When the cells are to be harvested, they should be rinsed, while still attached to the culture vessel, with saline. They may then be removed by scraping or flushing in saline and pelleted by centrifugation at 2000 *g* for 5 min or for 30 sec in a micro-centrifuge. If the cells are to be analysed by SDS-polyacrylamide gel electrophoresis, they are then resuspended in a volume of gel loading buffer equivalent to the amount of labelled medium used. An alternative and more convenient harvesting procedure involves the addition of polyacrylamide SDS-gel loading buffer directly to the cell monolayers (at least 200 μl of buffer/10^6 cells) followed by removal of the cell lysate. If the samples are very viscous due to a high concentration of DNA it is best to shear the DNA either by sonication in a sonicator bath or by passing the sample through a fine syringe needle. All virus-specific proteins can be conveniently separated using a 15 – 17.5% polyacrylamide-SDS gel using the discontinuous gel system described by Laemmli (32).

8. MEDIA AND BUFFERS

8.1 Media for CEF Cells

Standard and deficient media can be obtained from many commercial sources.

The media used in this laboratory are obtained from Biocult Ltd., Paisley, Scotland.

Growth medium: Medium 199 (obtained as a 10 x concentrated stock) is made 10% (v/v) with newborn calf serum, 1% (w/v) with penicillin (100 units/ml), 0.01% (w/v) with streptomycin, 0.02% (w/v) with kanamycin, 0.00025% (w/v) with fungizone and 0.06% with sodium bicarbonate. Sterile double-distilled water is added to adjust the final volume.

Maintenance medium: Maintenance medium for virus growth in CEF cells is made as above but with a final concentration of 2% (v/v) newborn calf serum.

Phosphate-free medium:
Phosphate-free medium is supplemented with 0.5% (w/v) glucose, 0.5% (w/v) lactalbumin hydrolysate, 2 mM glutamine and antibiotics as described above. Serum is omitted.

Methionine-free medium:
This medium is supplemented with glutamine and antibiotics as described above. Hepes buffer (pH 7.4) is added to 20 mM.

Carboxymethylcellulose (CMC) medium:
CMC is prepared by mixing 16 g CMC powder (low viscosity sodium salt) with 300 ml PBS. The mixture is left overnight at 4°C and a further 200 ml of PBS added before autoclaving for 20 min. 50 ml of this solution is added to 200 ml of overlay medium.

8.2 **Buffers**

PBS: This solution is prepared by dissolving commercially available tablets (Dulbecco 'A' tablets, Oxoid Ltd.) in the appropriate volume of double-distilled water. The solution is then autoclaved. The final concentration of salts is as follows: 0.137 M NaCl, 0.027 M KCl, 0.032 M Na_2HPO_4, 0.015 M K_2PO_4 (pH 7.3).

Saline: 0.9% (w/v) NaCl.

NTE buffer: 100 mM NaCl, 10 mM Tris HCl, pH 7.5, 1 mM EDTA.

Hanks balanced salts:

NaCl	0.8 g
KCl	0.4 g
$MgSO_4.7H_2O$	0.2 g
$CaCl_2.2H_2O$	0.185 g
Na_2HPO_4 (anhydrous)	0.0475 g
KH_2PO_4	0.06 g
$NaHCO_3$	0.35 g

Glucose	1.0 g
Phenol red	0.017 g
H_2O	to 1 litre

The $MgSO_4$, $CaCl_2$, KH_2PO_4, and Na_2HPO_4 are each dissolved separately in about 100 ml of distilled water and added slowly to the bulk solution with constant stirring. Phenol red is added last and the solution made up to the final volume. The solution is dispensed into convenient aliquots and autoclaved.

9. ACKNOWLEDGEMENTS

We are grateful to many of our colleagues in the Division of Virology for much helpful advice in the preparation of this chapter. In particular we would like to thank Dr. C.R. Penn for providing material for *Figures 5* and *6*, Mr. R.C. Patterson of the Veterinary School, University of Cambridge, for preparing the electron micrographs and Dr. F. Tomley for critical reading of the manuscript.

10. REFERENCES

1. McCauley,J. and Mahy,B.W.J. (1983) *Biochem J.,* **211,** 281.
2. Sweet,C. and Smith,H. (1980) *Microbiol. Rev.,* **44,** 303.
3. Palese,P. and Kingsbury,D.W. (1983) *Genetics of Influenza Viruses,* published by Springer-Verlag, Wien/New York.
4. Von Magnus,P. (1951) *Acta Pathol. Microbiol. Scand.,* **28,** 278.
5. Grist,N.R., Bell,E.J., Follett,E.A.C., and Urquhart,G.E.D. (1979) *Diagnostic Methods in Clinical Virology,* published by Blackwell Scientific Publications.
6. Kilbourne,E.D. (1983) in *Genetics of Influenza Viruses,* Springer-Verlag, Wien/New York, p. 1.
7. Lazarowitz,S.G. and Choppin,P.W. (1975) *Virology,* **68,** 440.
8. Klenk,H.-D., Rott,R., Orlich,M. and Blodorn,J. (1975) *Virology,* **68,** 426.
9. Appleyard,G. and Maber,H.B. (1974) *J. Gen. Virol.,* **25,** 351.
10. Bosch,F.X., Garten,W., Klenk,H.-D. and Rott,R. (1981) *Virology,* **113,** 725.
11. Gething,M.-J., White,J.M. and Waterfield,M.D. (1978) *Proc. Natl. Acad. Sci. USA,* **75,** 2737.
12. Hoyle,L. (1968) *The Influenza Viruses,* Virology Monographs, published by Springer-Verlag, Wien/New York.
13. Thompson,W.R. (1947) *Bacteriol. Rev.,* **11,** 115.
14. Fazekas de St. Groth,S. and White,D.O. (1958) *J. Hyg. (Lond.),* **56,** 535.
15. Williams,R.C. and Backus,R.C. (1949) *J. Am. Chem. Soc.,* **71,** 4052.
16. Heyward,J.T., Klimas,R.A., Stapp,M.D. and Obijeski,J.F. (1977) *Arch. Virol.,* **55,** 108.
17. McGeoch,D.J. and Kitron,N.K. (1974) *Virology,* **15,** 686.
18. Plotch,S.J., Bouloy,M. and Krug,R.M. (1979) *Proc. Natl. Acad. Sci. USA,* **76,** 1618.
19. Robertson,H.D., Dickson,E., Plotch,S.J. and Krug,R.M. (1980) *Nucleic Acids Res.,* **8,** 925.
20. Krug,R.M., Broni,B.A. and Bouloy,M. (1979) *Cell,* **18,** 329.
21. Caton,A.J. and Robertson,J.S. (1980) *Nucleic Acids Res.,* **8,** 2591.
22. Palese,P., Tobita,K., Ueda,M. and Compans,R.W. (1974) *Virology,* **61,** 397.
23. Warren,L. (1959) *J. Biol. Chem.,* **234,** 1971.
24. Aminoff,D. (1961) *Biochem. J.,* **81,** 384.
25. Marsh,M. (1984) *Biochem. J.,* **218,** 1.
26. Skehel,J., Baley,P., Brown,E., Martin,S., Waterfield,M., White,J., Wilson,I. and Wiley,D. (1982) *Proc. Natl. Acad. Sci. USA.,* **79,** 968.
27. Maeda,T. and Ohnishi,S. (1980) *FEBS Lett.,* **122,** 283.
28. Huang,R.T.C., Rott,R. and Klenk,H.-D. (1981) *Virology,* **110,** 243.
29. Dimmock,N.J., Carver,A.S., Kennedy,S.I.T., Lee,M.R. and Luscombe,S. (1977) *J. Gen. Virol.,* **36,** 503.
30. Peacock,A.C. and Dingman,C.W. (1967) *Biochemistry (Wash.),* **6,** 1818.
31. Loening,U.E. (1968) *J. Mol. Biol.,* **38,** 355.
32. Laemmli,U.K. (1970) *Nature,* **227,** 680.

CHAPTER 7

Double-stranded RNA Viruses

M.A. McCRAE

1. CLASSIFICATION OF DOUBLE-STRANDED RNA-CONTAINING VIRUSES

Viruses which have double-stranded (ds) RNA as their genetic information are ubiquitous in nature, infecting organisms from the bacterial, animal and plant kingdoms. The majority of these viruses have been classified into a single virus family, the Reoviridae (1), which is composed of six genera (*Table 1*). A few virus isolates have been made, e.g., housefly virus (2) and chum salmon virus (3), which although they have not as yet been assigned to any of the reovirus genera, have features which suggest that on further characterisation they will find a place in the Reoviridae. By contrast, there are a number of ds RNA-containing viruses, with infectious bursal disease virus (IBDV), infectious pancreatic necrosis virus (IPN), bacteriophage $\phi 6$ and yeast killer factor particles being the best studied examples, where differences from *bona-fide* Reoviridae members make it unlikely that they will be included in this virus family. A proposal has been made to group some of these latter viruses into a new virus family termed the Birnaviridae (P. Dobos, personal communication).

Table 1. Classification of the Reoviridae Family.

Genus	*Host range*	*No. of genomic segments*	*Examples*
Orthoreovirus	Vertebrates	10	Mammalian reoviruses Avian reoviruses
Orbivirus	Vertebrates	10 and 12	Bluetongue African horse sickness Colorado tick fever
Rotavirus	Mammals and birds	11	Bovine rotavirus SA11 Wa
Cypovirus	Insects	10	Cytoplasmic polyhedrosis virus
Phytoreovirus	Plants and insects	12	Wound tumour virus Rice dwarf virus
Fijivirus	Plants and insects	10	Fiji disease virus Maize rough dwarf virus

2. VIRUS ISOLATION AND TISSUE CULTURE GROWTH

The three serotypes of mammalian reoviruses were isolated in the late 1950s and early 1960s (4). Their ability to adapt directly to high titre growth in a wide range of mammalian cell lines has meant that they have been subjected to the most detailed analysis of any of the Reoviridae. Consequently, the majority of procedures used in working with members of the other genera represent derivatives of protocols developed with the mammalian reoviruses. Therefore, this article will focus on procedures for use with mammalian reoviruses and, in particular, the Dearing strain of reovirus type 3 (the most widely studied reovirus). Where appropriate, protocols for other ds RNA viruses which differ significantly from those for reovirus will be given.

2.1 **Maintenance of Cells for Virus Growth**

The majority of mammalian cell lines can be infected successfully with mammalian reoviruses. This wide host range, coupled to high stability of the virus, means that great care is required when handling reovirus to ensure that adventitious contamination of tissue culture stocks does not occur. The most widely used cells for work on reovirus are mouse L-cells, which have the advantage that if maintained appropriately they can be changed at will between monolayer and suspension culture. For growth of virus, L-cells are most conveniently maintained in suspension culture in a sterile flat-bottomed flask at 37°C. Joklik's modified minimal essential medium supplemented with 5% foetal calf serum is used, the high level of phosphate in the medium being sufficient to maintain the correct pH without additional CO_2 gassing. Cell density should be kept within the range $5 \times 10^5 - 1 \times 10^6$ cells/ml; this can be done by splitting the culture 1:2 every 24 h. If maintained continuously in suspension culture for long periods (>4–5 weeks) L-cells lose their ability to settle out and form an attached cell monolayer in plastic dishes. The convenient attribute of being able to switch the cells rapidly between the two culture states can be maintained if approximately once a month the suspension culture is transferred to rolling tissue culture bottles. After rolling overnight at 37°C the unattached cells are discarded and those attached to the glass or plastic substrate trypsinised and returned to suspension culture. It is important to note that L-cells are very sensitive to trypsinisation and care should be taken not to over-trypsinise the cells if they are being passaged in monolayer culture. If cells have been trypsinised to set up a suspension culture, the residual trypsin should be removed by centrifugation of the cells before they are diluted out in fresh medium for suspension culture growth. When establishing the initial suspension, culture cells should be diluted out to 1×10^6 cells/ml, allowed to grow for 6–8 h and then split 1:2.

2.2 **Preparation of Virus Inocula**

In common with many other viruses, reovirus will rapidly generate defective interfering particles which will considerably depress virus yield if propagated at high input multiplicities of infection (m.o.i.). Consequently it is important to ensure, both in maintaining virus stocks and in the preparation of inocula, that virus is always propagated with an input m.o.i. of 0.1 or less. Reovirus also exhibits a temperature-dependent cytotoxicity (4) which is evident during growth at 37°C and more pronounced at 39°C. Therefore virus growth is best carried out at 34°C or even 31°C to ensure that observed cytopathic effects (cpe) are not due solely to cytotoxicity. For the preparation of virus inocula, infect two 150 cm^2 flasks containing confluent monolayers of mouse L-cells at a m.o.i. of no greater than 0.1. Allow virus growth to proceed at 34°C until complete cpe is obtained (2–4 days). The majority of the virus yield (80–90%) will remain associated with cellular debris and so the total contents of each flask should be harvested, titrated on L-cells and stored at −70°C until use. The virus yield from two 150 cm^2 flasks will be sufficient to infect several eight-litre cultures of cells for virus growth.

2.3 Large-scale Growth of Virus

This is normally carried out in 4 litre batches (or multiples thereof) of mouse L-cells in suspension. The following procedure for maximum virus yield, is as described by Smith *et al.* (5).

(i) Concentrate L-cells at a density of 1 x 10^6 cells/ml by centrifugation (low speed refrigerated centrifuge 2000 r.p.m. for 20 min), resuspend in 1/10 volume of Puck's balanced salts, add virus to a final m.o.i. of 1–10 and stir the suspension at 34°C to allow virus adsorption to occur.

(ii) After 1 h, dilute the suspension to its original volume in fresh medium and allow virus growth to proceed at 34°C to reduce viral cytotoxicity.

Whilst undoubtedly maximising virus yields, this procedure is laborious, difficult to perform aseptically on large volumes of cells and expensive in terms of media requirements. Consequently a simpler alternative is recommended, in which the suspension culture to be infected is shifted directly to 34°C and the cells are not concentrated for infection. The steps in the procedure are given below.

(i) Place the suspension culture at 34°C, and allow the cells to equilibrate to the lower temperature for 1–2 h.

(ii) Dilute the inoculum virus to 10–20 ml in MEM, and add directly to the culture.

(iii) Allow virus growth to proceed for 36–42 h at 34°C before transferring the whole culture to 4°C and allowing the infected cells to settle. During virus growth at 34°C it is quite normal for the colour of the suspension to change from red to orange. Growth of the type member of reovirus serotype 2, i.e., Jones strain, is slower than that of the other two serotypes and so, to maximise yield, growth should be allowed to proceed at 34°C for 48–56 h.

3. VIRUS PURIFICATION

3.1 Harvesting Infected Cells and First Virus Extraction

The majority of reovirus particles (~ 90%) produced during virus growth remain associated with fragments of lysed cells and are not released freely into the surrounding medium. This fact can be used to simplify the problem of harvesting virus from large volumes of suspension culture. Thus, after virus growth, allow the infected cellular debris to settle out by standing the suspension culture at 4°C for at least 24 h. (The culture can be left at 4°C for up to 10 days without seriously affecting final virus yield.) The majority of the suspension medium can then be carefully removed using a suction line. Whilst the majority of virus particles will remain associated with the settled cellular debris, the medium will still contain considerable amounts of infectious virus and should be handled accordingly. After removing most of the overlaying medium, decant the infected cellular debris, and concentrate by low speed centrifugation (2000 r.p.m., 4°C for 20 min). Resuspend the cell pellet from this spin in approximately 100 ml per 4 litres of original suspension in reovirus resuspension buffer (50 mM Tris-HCl pH 8.0, 10 mM NaCl, 1.5 mM β-mercaptoethanol).

The initial virus extraction involves homogenisation and phase separation steps and relies on the fact that reovirus does not contain any lipid in its two concentric capsid shells. Thus the resuspended cellular material is mixed with 1/4 volume of trifluoro-trichloroethane. (This compound is used as a commercial refrigerant and can be purchased under the trade names Arcton 113 or Freon 113). Homogenise the mixture for 3 min at half maximum speed using a Virtis 23 homogeniser.

Two points should be made concerning this homogenisation step. Firstly other homogenisers such as the Waring blender can be used but, if employed, care must be taken to avoid a pressure build up due to the volatility of the Freon in the homogenisation vessel. Secondly, this homogenisation step inevitably produces considerable aerosols and so it should be done well away from areas where other tissue culture work is going on and preferably in a well-vented fume cupboard.

Following homogenisation, separate the two phases by low-speed centrifugation (2000 r.p.m., 4°C for 20 min). This produces a reddish-pink opaque upper aqueous phase, which should be decanted and stored at 4°C, and a white gelatinous lower Freon phase which contains the lipid-bearing cellular material. Add approximately 50 ml of resuspension buffer to the Freon phase, and repeat the homogenisation and centrifugation steps with the resulting aqueous phase being pooled with that from the first extraction. Continue re-extraction of the Freon phase until it is apparent from the opacity of the aqueous phase that little more material is being extracted (4–5 extractions are usually quite sufficient). Back-extract the combined aqueous phases exactly as for the forward extractions using a 1/8 volume of Freon. Concentrate the virus present in the final aqueous phase by high-speed centrifugation (24 000 r.p.m., 4°C, 1 h Beckman SW 28 rotor). Resuspend the resulting virus pellets in a small volume (~3–5 ml each) of resuspension buffer and store overnight at 4°C.

3.2 Gradient Purification of Virus

Two alternative procedures can be followed for purifying the virus. The standard virus purification method of Smith *et al.* (5) involves a sedimentation velocity banding of virus on a sucrose gradient followed by density equilibrium banding on caesium chloride.

(i) For this method, load the resuspended virus pellets onto a 10–40% sucrose gradient (made up in 50 mM Tris-HCl buffer pH 8.0) and centrifuge at 25 000 r.p.m., 4°C for 1 h in a Beckman SW 28 rotor. The pellet material from 1–1.5 litres of original infected cell suspension can be loaded onto a single gradient. The virus band should be in the middle of the gradient and can be conveniently collected by side puncture of the tube.

(ii) Dilute the material in virus resuspension buffer to reduce the sucrose concentration and concentrate the virus by centrifugation (25 000 r.p.m., 4°C, 1 h in a Beckman SW 28 rotor).

(iii) Layer the resuspended virus pellets onto a pre-formed caesium chloride gradient, density 1.2–1.4 g/ml (made up in 50 mM Tris-HCl pH 8 and containing a 5–10% glycerol gradient to stabilise the gradient during pouring) and centrifuge for 2 h at 24 000 r.p.m., 4°C in a Beckman SW 28 rotor. The opalescent-blue virus band should appear in the middle of the gradient and can be collected as before. Also visible in the top third of the gradient should be the virus top component band (virion capsids lacking viral RNA) which varies in relative amount between virus growths and can be collected separately if required.

(iv) Dilute the virus band material in 50 mM Tris-HCl pH 8.0, concentrate by centrifugation, and resuspend in a small volume (~1–2 ml) of 50 mM Tris-HCl pH 8.0. It is important to note that Tris buffer and not virus resuspension buffer should be used for this last dilution and resuspension step since the latter contains β-mercaptoethanol which would interfere with the absorbance measurements used to calculate virus yield.

The second gradient purification procedure is a simplified alternative of the original method described by Smith *et al.* (5). In this, the sucrose gradient step, which does not give a dramatic improvement in virus purity, is omitted and the virus is banded directly in pre-formed caesium chloride gradients.

3.3 Measurement of Virus Yield and Storage of Purified Virus

The yield of virus can be measured in a number of ways to obtain an estimate of the numbers of physical particles and infectious particles. The number of physical particles can be estimated using a formula derived by Smith *et al.* (5) which relates the optical density of a purified virus suspension at 260 nm to both virus protein concentration and numbers of physical particles: 5.42 OD units at 260 nm equals 1 mg/ml purified virus which in turn equals 1.13×10^{13} virus particles. A more accurate measurement can of course be obtained by a particle count in the electron microscope but the above absorbance measurements of yield are quite sufficient for most purposes.

Measurement of infectious virions is carried out by plaque assay. Reovirus will form plaques in a wide variety of cell lines, but mouse L-cells or monkey CV-1 cells have been most widely used. Virus plaques take 4–5 days to develop at 37°C under a 1% agar overlay and can be vital stained using neutral red or stained with 0.01% crystal violet after formalin fixing of the cells and removal of the agar overlay. Typically, particle:p.f.u. ratios in the region of 20–50:1 are obtained

Day no.	*Procedure*
1	Start with 100 ml of suspension L-cells at 1 x 10^6 cells/ml. Split 1:2 in Joklik's MEM + 5% FCS (10 min).
2	Split 200 ml of cells to 400 ml (~10 min).
3	Split 400 ml of cells to 800 ml (~10 min).
4	Split 800 ml of cells to 1.6 litres (~10 min).
5	Split 1.6 litres of cells to 3.2 litres (~10 min).
6	Split 3.2 litres of cells to 6.4 litres (~10 min).
7	Remove cells (100 – 400 ml) from the culture to maintain as stock and transfer remainder at about 12 Noon to 34°C water bath (10 min). At about 4.00 p.m. add sufficient virus to the culture to give an m.o.i. of 1 – 10 (~5 min) and leave to grow at 34°C.
9	Transfer infected cells at about 10.00 a.m. to 4°C and allow cells and cellular fragments to settle out. Minimum time at 4°C, 48 h, maximum 14 days.
11	Remove bulk of medium by suction and concentrate cells by centrifugation. Resuspend cells in approximately 100 ml resuspension buffer and Freon 113 extract.

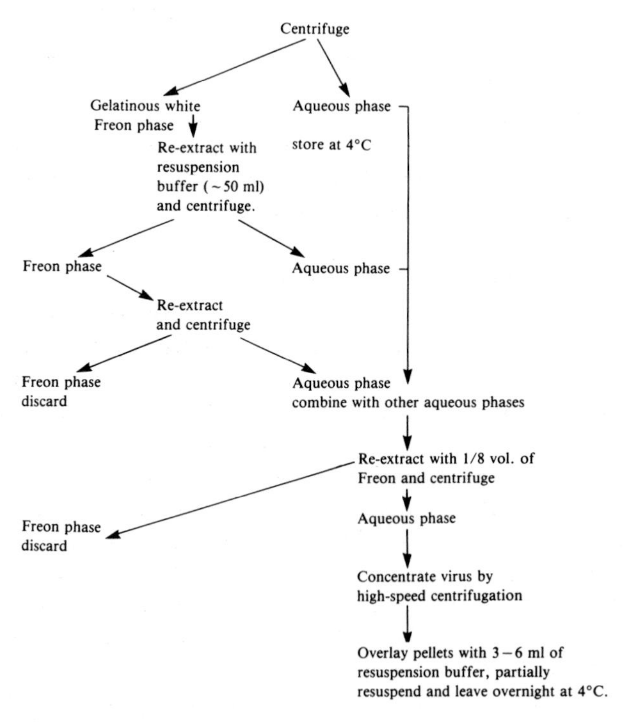

The harvesting of infected cells and Freon extraction should take approximately 3 h to complete. Centrifugation is for 1 h and resuspension should take about 10 min.

12	Complete resuspension of pellets and pour CsCl gradients (1 h). Band virus on CsCl (2 h centrifugation) Collect virus bands and dilute in 50 mM Tris-HCl pH 8.0 (30 min). Concentrate virus by centrifugation. Resuspend in Tris buffer and measure OD_{260} (30 min).

Figure 1. Flow diagram to indicate approximate timings of Reovirus purification procedure.

for reovirus grown and purified as described above and any significant increase in that value indicates the development of defective interfering particles in the inoculum which will almost certainly be accompanied by a reduction in virus yield. As mentioned earlier, reovirus grows extremely well in mouse L-cells and yields as high as 25 mg purified virus/litre of infected suspension cells have been obtained, although yields in the range of 3 – 10 mg/litre of infected suspensions are those normally achieved for reovirus types 1 (Laing) and 3 (Dearing), and 2 – 5 mg/litre for reovirus type 2 (Jones).

Purified virus is normally stored at −70°C. However, reovirus is sufficiently stable to allow it to be stored for long periods (months – years) at 4°C with only marginal loss of infectivity. If virus is stored for long periods at 4°C then azide (0.01%) should be present to prevent bacterial growth.

A flow chart of the virus growth and purification procedure for reovirus type 3 is given in *Figure 1*.

4. GROWTH AND PURIFICATION OF ROTAVIRUSES

Rotaviruses have become recognised in recent years as the major aetiological agents of acute viral infantile gastroenteritis in humans, and in all the major species of domestic livestock (6,7). They are consequently of considerable medical and agricultural importance. In contrast to the mammalian reoviruses, rotaviruses have fastidious requirements for growth in tissue culture which has been, and continues to be, a hurdle to their rapid characterisation.

4.1 Adaptation to Growth in Tissue Culture

Rotaviruses from all animal species have until recently been extremely difficult, and in some cases impossible, to adapt to tissue culture growth. This problem was particularly acute for the human rotaviruses where for some years all attempts to adapt clinical isolates to routine growth in tissue culture met with failure. A procedure involving multiple passaging in piglets before adaptation to tissue culture growth was successfully employed by Wyatt and co-workers (8) to achieve the first successful adaptation of a human rotavirus to tissue culture growth. However, this technique is not feasible for widespread application. Despite these initial problems, significant advances have been made by a number of Japanese groups (9 – 11) who made use of earlier results on enhancement of virus infectivity by protease treatment (12,13). The outcome of this Japanese work has been the development of a tissue culture adaptation procedure that can be applied to virus isolates from all infected species. This procedure consists of two main elements, trypsin treatment of virus inocula and 'blind passage' of isolates for a number of passages without observable cytopathic effect. A protocol that will result in a high (50 – 80%) adaptation efficiency is as follows:

(i) Make a 10% suspension of the infected faecal specimen in Minimal Essential Medium (MEM) and centrifuge at low speed (4000 r.p.m., 30 min, 4°C) to remove particulate material.
(ii) Add an equal volume of medium containing 20 μg/ml of trypsin to the

supernatant and incubate this mixture at 37°C for 30 min before diluting 1:20 in MEM.

(iii) Add 0.2 ml of the trypsin-treated virus to a confluent monolayer of MA 104 cells that have been grown in 100 mm x 10 mm rolling culture tubes in MEM supplemented with 10% foetal calf serum, 100 units/ml penicillin and 100 μg/ml streptomycin. The cell monolayer should be washed three times with MEM before virus addition. (The MA 104 continuous epithelial cell line was derived originally from the kidneys of embryonic rhesus monkeys and has been found by numerous groups to be one of the best cell lines for supporting growth of rotaviruses).

(iv) After adsorption for 1 h at 37°C, remove the inoculum and overlay the cells with MEM containing antibiotics and 0.5 μg/ml trypsin. Incubate the tubes at 37°C in a roller drum and check every 2 days against uninfected controls for signs of a cytopathic effect. If a clear cytopathic effect is observed, harvest the cells and culture fluid by freeze-thawing of the tubes. If however, as is more common, no clear cpe is observed, allow virus growth to proceed for 10 – 12 days before harvesting.

(v) Treat the harvested culture fluids from passage (1) above with trypsin as before and passage again on fresh confluent monolayers of MA 104 cells. It often requires 3 – 10 'blind passages' in culture before a clear virus-induced cytopathic effect is obtained 48 – 72 h after inoculation.

(vi) In order to avoid problems from experimentation with virus mixtures that might have co-adapted to tissue culture growth, three cycles of virus growth from a single infectious virion should be undertaken at the earliest opportunity. Ideally this should be done by plaque purification of the virus; however, some isolates, whilst growing in MA 104 cells, do not form plaques, and in these cases virus purification should be done using the terminal dilution cytopathic effect procedure. Make serial dilutions of the earliest virus passage producing a clear virus-induced cytopathic effect, and plate out onto confluent monolayers of MA 104 cells in 24 or 48 well tissue culture dishes. After 1 h adsorption at 37°C, remove the inocula and overlay the wells with MEM containing 0.5 μg/ml trypsin before incubating the dishes at 37°C for 48 – 72 h. At a dilution in the series where only a low percentage (5 – 10%) of the wells have a cytopathic effect, harvest several wells separately, then dilute and plate out as before. This process should be repeated at least three times before the virus is grown up for biochemical study. If the virus isolate does produce plaques on MA 104 cells, then three cycles of plaque picking should be carried out as a more satisfactory approach to virus purification prior to use in experiments.

Some general points are worth noting concerning the above procedures. Whilst MA 104 cells will grow well in newborn or normal calf serum, the ubiquitous nature of rotaviruses and consequent high levels of anti-rotavirus antibodies in commercial calf serum means that all tissue culture work should be done using foetal calf serum. The growth of rotaviruses and the ensuing cytopathic effect are achieved in the presence of 0.5 μg/ml trypsin. It is essential that uninfected cell controls are incorporated into each experiment to ensure that the cytopathic effects observed are not due to the action of the trypsin on the cells. Concerning the trypsin itself, in order to minimise variability in its use, it should be made up as a 10 mg/ml stock solution, which is aliquoted in single experiment volumes and stored at $-20°C$. It should *not* be subjected to repeated freeze-thawing before use.

5. GROWTH AND PURIFICATION OF ROTAVIRUS FOR BIOCHEMICAL STUDIES

Despite the success of the above tissue culture adaptation protocol, only a small number of virus isolates [SA11 (14) and the UK tissue culture adapted bovine viruses (15) are the best studied isolates of this type] grow sufficiently well in tissue culture to facilitate routine culture of the large quantities of virus required for many biochemical studies. The UK tissue culture adapted bovine rotavirus grows best in the epithelioid African green monkey kidney cell line BSC-1 and a convenient batch size for large-scale virus growth consists of twenty 80 oz roller bottles of confluent BSC-1 cells.

(i) Prepare the virus inoculum by infecting two confluent 150 cm^2 bottles of BSC-1 cells at an m.o.i. of 0.1, allowing virus growth to proceed in the presence of 5 $\mu g/ml$ trypsin until there is a complete cpe ($\sim$2 days).

(ii) Titrate the culture; the yield should be in the region of $1-3 \times 10^9$ p.f.u.

(iii) Dilute this virus in MEM to a titre of 5×16^6 p.f.u./ml and use 10 ml to infect each roller bottle.

(iv) After 1 h of adsorption at 37°C, add 15 ml of MEM containing 7.5 $\mu g/ml$ of trypsin to the roller bottle, and continue incubation at 37°C for 72 h.

After approximately 24 h it is quite normal to observe that the cells have become detached from the glass surface which facilitates their easier harvesting at the end of the virus growth period. We have not found that the inclusion of trypsin into the growth medium has any marked beneficial or detrimental effect on the final yield of infectious virus, but it causes stripping of the infected cells from the glass which greatly simplifies their harvesting. This is important, since the great majority of the virus remains cell associated at the end of the infectious cycle.

(v) At the end of the virus growth period decant the infected cell suspension into 250 ml centrifuge bottles and concentrate the virus and infected cell debris by centrifugation (50 000 *g*, 90 min at 4°C).

(vi) Resuspend the resulting cell/virus pellets in a small volume (20–30 ml) of virus resuspension buffer (50 mM Tris-HCl pH 8.0, 10 mM NaCl, 1.5 mM β-mercaptoethanol, 3 mM $CaCl_2$). The resuspended material can be stored at $-70°C$ at this stage or immediately processed through the virus purification procedure. This is carried out exactly as described for reovirus except that only two re-extractions of the Freon phase are usually required. The final virus yield does vary, but 200–500 μg of purified virus/80 oz bottle is the normal range.

6. BIOCHEMICAL PROCEDURES

6.1 **In Vitro Transcription of Virus mRNAs**

Reovirus was one of the first viruses shown to carry a virion-borne transcriptase, in this case an RNA-dependent RNA polymerase (16,17). This enzyme is easy to activate, extremely stable and can be used to synthesise very large amounts of pure viral mRNA. These properties of this transcriptase system together with an ability to manipulate it to produce mRNAs with various types of 5′-terminal cap structure (18) have led to it being used for a wide range of *in vitro* studies on mRNA function. Production of reovirus mRNA is achieved in a three-step process detailed in the following sections.

6.1.1 *Removal of Outer Virion Shell*

The RNA transcriptase activity of reovirus is not evident in the intact virion but can be unmasked by the removal of the outer of the two concentric virion shells. This activation step has been achieved in various ways, however that now generally used is digestion of the intact particle with the protease chymotrypsin.

(i) Digest the purified virus suspension, at a final concentration of 1 mg/ml, in freshly-made chymotrypsin at 50 μg/ml in a reaction mixture containing 20 mM Tris-HCl buffer pH 8.0 and 0.12 M KCl.

(ii) Incubate with gentle stirring at 37°C for 60–65 min.

(iii) Following digestion, collect the enzymatically active viral cores by centrifugation (30 000 *g* 4°C, 20 min).

Whilst viral cores are a great deal more resistant to chymotrypsin digestion than the intact virion, it is possible to over-digest the mixture leading to the loss of RNA transcriptase activity. Consequently it is important to adhere to the relative virus to protease concentrations and digestion times given above. The amount of starting virus used is obviously dependent to some degree on the quantity of viral mRNA required but it is usual to start with 5–10 mg of purified virus.

In the case of other ds RNA-containing viruses, different procedures are employed for this step. Thus for rotaviruses, 20 min incubation in 3 mM EDTA at 37°C before centrifuging as described above is sufficient to activate the transcriptase, whereas in the case of cytoplasmic polyhedrosis virus the absence of an outer virion shell negates the requirement for this step.

6.1.2 *Synthesis of Viral mRNA*

Following activation, viral cores do not store well for enzymatic activity and so should be used immediately. Resuspend the viral cores collected by centrifugation directly in the synthesis reaction mixture at a final concentration equivalent to 1 mg of purified virus per ml. The synthesis reaction mixture should contain the following per 10 ml:

30 mg ATP
20 mg CTP
20 mg GTP

10 mg UTP
37.5 mg Phospho-enol-pyruvate
1 mg Rabbit skeletal pyruvate kinase
1 ml 1 M Tris – HCl buffer pH 8.0
0.5 ml 10 mM S-adenosyl methionine
37.5 μl 0.4 M EDTA
50 μl Inorganic pyrophosphatase (100 units/ml)
0.1 ml of 30 mg/ml Macaloid. [This is an RNase inhibitor designed to help preserve the integrity of the mRNA following its synthesis and can be substituted with a number of other suitable inhibitors, e.g., Bentonite (final concentration 250 μg/ml), or placental RNase inhibitor (final concentration 500 units/ml).]
200 – 500 μCi [^{3}H]UTP. (The addition of radioactive UTP to the reaction mixture is not essential, but merely serves as a convenient and rapid way of monitoring the progress of the synthesis reaction. If it is not included then 20 mg UTP/10 ml reaction mixture can be used.)

After the viral cores have been gently and completely resuspended in the appropriate volume of the above reaction mixture, the synthesis reaction can be initiated by adding 1 M $MgCl_2$ to give a final concentration in the reaction of 13 mM. The synthesis is allowed to proceed at 37°C with gentle stirring and its progress can be monitored by taking small (10 μl) samples at 1 h intervals and measuring the incorporation of [^{3}H]UTP into acid-insoluble material. After 4 – 6 h the synthesis reaction will stop. This does not represent inactivation of the virion RNA transcriptase but simply exhaustion of enzyme substrates. Therefore after 5 h of synthesis the viral cores should be collected by centrifugation (30 000 *g*, 4°C 20 min), the mRNA-containing supernatant removed, the cores resuspended in fresh synthesis reaction mixture and synthesis allowed to proceed for a further 5 h. (N.B. If Macaloid or Bentonite are used as the inhibitors of RNase activity, their particulate nature will ensure that they are collected together with the viral cores at the end of the first round of mRNA synthesis and so these inhibitors should not be added to subsequent reaction mixtures.)

The rounds of mRNA synthesis can be performed five or six times before there starts to be a significant diminution of the yield of viral mRNA. A convenient schedule to follow for this whole procedure is to begin chymotrypsin digestion of virus early on day 1, carry out two rounds of synthesis during that day, allow the third synthetic round to continue overnight (8 – 10 h) and then carry out two further rounds of synthesis before terminating the process at the end of day 2.

The reaction mixture detailed above results in the synthesis of predominantly fully 5′ capped and methylated mRNA. To produce mRNA that has the 5′ cap but is not methylated the S-adenosyl methionine in the reaction mixture should be replaced by the same concentration of S-adenosyl homocysteine. It is also possible to manipulate the system to produce non-capped, non-methylated mRNA in which case the S-adenosyl homocysteine is used to replace the S-adenosyl methionine, the amount of GTP in the reaction mixture is reduced from 20 mg to 1 mg and the inorganic pyrophosphatase is replaced with 25 μl of 0.2 M sodium pyrophosphate.

6.1.3 *Collection of Viral mRNA*

The supernatants obtained at the end of each round of synthesis contain the viral mRNA which is released from viral cores following its synthesis. Extract the collected supernatants immediately, once with an equal volume of water-saturated phenol, and four times with ether. After removal of residual ether with a nitrogen stream, add 10 M LiCl to the supernatant to a final concentration of 2 M to 'salt-out' the synthesized mRNA and allow to stand overnight at 4°C. This procedure gives a precipitate of viral mRNA which can be simply collected by low-speed centrifugation without precipitating the unincorporated nucleotide triphosphates as would occur if ethanol precipitation were used. Redissolve the collected mRNA in 5 mM Tris-HCl buffer pH 8.0, re-precipitate with ethanol to further concentrate it, and resuspend the precipitate at an appropriate concentration (2–4 mg/ml) in water.

Concerning the yield of mRNA that can be expected when this protocol is followed, as mentioned above the virion-associated transcriptases of ds RNA-containing viruses are very stable and very active and total yields of mRNA in the 4–6 mg range can be expected if the procedure is started with 5 mg of purified virus.

6.2 In Vitro Translation of Viral mRNA

The relative simplicity of producing large quantities of viral mRNA *in vitro* has meant that reovirus mRNA has been used extensively for studying mRNA function and has consequently been translated in a wide variety of *in vitro* translation systems. The high activity, coupled to low endogenous levels of synthesis following treatment with micrococcal nuclease (19), of the *in vitro* translation system produced from rabbit reticulocytes, means that for most purposes it is the system of choice. If only a small amount of *in vitro* translation is being contemplated it is probably simplest and most economical to purchase the system from one of the commercial suppliers currently distributing it, and to use it according to their instructions. Larger scale users will find a brief description of a method for making an *in vitro* translation system from rabbit reticulocytes in *Appendix A*.

A typical *in vitro* translation of viral mRNA can be set up as follows

(i) 50 μl of micrococcal nuclease-treated reticulocyte system.

(ii) 4 μl of amino acid mixture containing 2.5 mM of all amino acids except methionine. (This mixture can be made up as a stock solution by dissolving the appropriate weight of each amino acid in approximately 20 ml of water, neutralising the solution to pH 7.4 with KOH, bringing the solution to a final volume of 25 ml, aliquoting into suitable volumes and storing at −20°C. N.B. The low solubility of some amino acids, particularly tyrosine, means that this stock solution will always be cloudy.)

(iii) 5 μl of 20 x energy generating system. This 20 x stock can be made as follows:

129 mg dipotassium ATP (final concentration 20 mM)
24 mg GTP (final concentration 4 mM)
1 ml 2 M Tris-HCl buffer pH 7.5
7 ml water.

Neutralise the solution to pH 7.5 with KOH, and make up to 10 ml and store at −20°C.
To each 2 ml aliquot of this solution add 145.2 mg of di-Tris creatine phosphate and 2 mg of creatine phosphokinase, check the pH and re-neutralise to pH 7.5 if necessary. This solution can be aliquoted and stored for up to 1 month at −20°C.

(iv) 1 M KCl. The amount actually needed here will vary according to the salt optimum of the particular batch in use, but should be in the range 5 − 10 μl.

(v) 50 mM magnesium acetate. The amount added will vary according to the Mg^{2+} optima but should be in the range 1.5 − 3.0 μl.

(vi) 5 − 25 μCi [^{35}S]methionine.

(vii) 4 μl of mouse liver tRNA at 2 mg/ml. This is not an essential ingredient but does give some improvement in the stimulation achieved (~2-fold).

(viii) 5 − 10 μg viral mRNA. (Saturation of the system is achieved with an mRNA concentration in the 50 − 100 μg/ml range.)

(ix) Filling buffer (5 mM Tris-HCl pH 7.4) to bring the final volume of the reaction to 100 μl.
The reaction is usually started by addition of the nuclease-treated system to the other ingredients and its progress can be monitored by measuring the incorporation of [^{35}S]methionine in hot-acid precipitable material. Ideally the reaction should show a linear increase in [^{35}S]methionine incorporation for 30 − 60 min.

6.3 In Vitro Translation of Viral Genomic ds RNA for Determining Protein Coding Assignments

The development of a simple trick to facilitate *in vitro* translation of the positive-sense strand of viral genomic ds RNA (20) has provided a relatively straightforward method for achieving the RNA-protein coding assignments for any dsRNA-containing virus. The procedure consists of the three steps detailed in the following sections.

6.3.1 *Fractionation of Individual Species of Genomic ds RNA*

This is usually achieved using polyacrylamide gel electrophoresis and obviously the actual electrophoresis conditions used are dependent on the virus under study. As an example, individual fractionation of most segments of the rotavirus genome can be achieved by electrophoresing 100 − 200 μg of genomic ds RNA on a 40 x 20 x 0.0015 cm 6.0% polyacrylamide gel in the Laemmli discontinuous buffer system (21) at 30 mA for 24 h. Following electrophoresis the separated RNA segments can be visualised either under u.v. light after ethidium bromide staining of the gel or preferably by autoradiography if a tracer amount of ^{32}P-labelled viral ds RNA is added to the loaded sample. Following individual excision of the RNA segments from the whole gel, elute the RNA from the polyacrylamide gel piece. This can be done by a myriad of different procedures, but possibly the most reliable and gentle is that of electroelution (22). The elution should be done in 50 mM Tris-acetate buffer pH 7.4 and, following electrophoresis, the contents of the dialysis bag are extracted twice with water-saturated phenol, four times with ether and then the eluted RNA concentrated by ethanol precipitation.

6.3.2 *Denaturation of Individual ds RNA Segments Prior to Translation*

The simple trick used to allow the translation of the positive-sense strand of the normally translationally inert ds RNA is to denature it at low temperature in 90% dimethylsulphoxide (DMSO). The original procedure used involves resuspending the ds RNA in water at a concentration of approximately 20 mg/ml. Add DMSO to this RNA to a final concentration of 90%. It is important to ensure that the RNA is completely dissolved before adding the DMSO and also that it mixes completely with the DMSO. The latter is best achieved by increasing the DMSO concentration gradually with good mixing at each stage. Heat the mixture to 45°C for 20 min to ensure denaturation and then add directly to the *in vitro* translation system. Using the above starting RNA concentration, the addition of 2.5 μl per 100 μl translation assay will be sufficient to saturate it for RNA and still give a low enough final DMSO concentration not to poison the translation system (20).

Two alternatives to the above procedure have been developed. In the first (McCrae, unpublished), denature the ds RNA using DMSO, but immediately after incubation at 45°C, concentrate denatured RNA by ethanol precipitation. Resuspend the resulting desiccated pellet of denatured RNA directly in the *in vitro* translation mixture. This modification of the original procedure simplifies the resuspension of the fractionated ds RNA segments since their concentration is now relatively unimportant at that stage. Some workers (23,24) have used incubation in 10 mM methyl mercury hydroxide as an alternative RNA denaturant.

6.3.3 *In Vitro Translation of Denatured ds RNA and Characterisation of Translation Products*

The *in vitro* translation of denatured ds RNA is carried out exactly as described above for mRNA. The translation products of individual RNA segments are compared by polyacrylamide gel electrophoresis with virion and virus-infected cell polypeptides in order to determine the virus protein encoded by each RNA segment.

6.4 **Molecular Cloning of the Genomes of ds RNA Viruses**

Recombinant DNA technology is useful for complete structural characterisation of viral RNA molecules and, in some cases, for high level production of their encoded proteins. A number of groups have described a general cloning strategy for these viruses based on the use of genomic ds RNA as the starting material (25–28). This strategy can be divided into a number of discrete steps.

6.4.1 *Polyadenylation of Viral Genomic RNA*

To allow oligo(dT) priming to be used for reverse transcription of viral RNA, the non-polyadenylated viral RNA must first be polyadenylated using *Escherichia coli* poly(A) polymerase. Start the procedure using 10 μg of unfractionated genomic ds RNA, which is first denatured in 90% DMSO as described in Section 6.3 and then ethanol precipitated. This denaturation step is not essential for the polyadenylation reaction to work but most workers have found that polyadenylation proceeds more efficiently using a denatured template. The polyadenylation reaction is set up as follows.

(i) Resuspend the dried pellet of denatured ds RNA in 96 μl of water.
(ii) Add 2 μl 100 mM $MnCl_2$ (freshly made).
(iii) Add 100 μl 2 x polyadenylation reaction mixture. (1 x polyadenylation reaction mixture = 50 mM Tris-HCl buffer pH 8.0; 100 μM ATP; 40 μCi/500 μl of [^{3}H]ATP; 10 mM $MgCl_2$; 50 μg/ml bovine serum albumin (nuclease free); 0.25 M NaCl.)
(iv) Add 2 μl (10 units) *E. coli* poly(A) polymerase.
(v) Incubate the mixture at 37°C for 15 min, during which time an average of 20–25 A residues are added to each molecule, and terminate by phenol extraction.
(vi) Pass the phenol-extracted material over a gel filtration column (Sephadex G50) and recover the excluded polyadenylated RNA by ethanol precipitation.

6.4.2 *Reverse Transcription of Polyadenylated Genomic RNA*

The advantage of employing denatured genomic RNA for reverse transcription lies in the fact that both cDNA strands are synthesised in a single reverse transcription reaction thereby obviating the need for the normal first and second strand synthetic reactions required when mRNA is used.

(i) First, denature the polyadenylated RNA in 90% DMSO and ethanol precipitate.
(ii) Resuspend the dried pellet and add the following constituents of the reverse transcription reaction:
 40 μl water;
 40 μl 5 mM methyl mercury hydroxide;
 incubate at 20°C for 3 min;
 120 μl reverse transcriptase reaction mixture (reverse transcriptase reaction mixture = 75 mM Tris-HCl buffer pH 8.3; 15 mM $MgCl_2$; 180 mM KCl; 15 mM dithiothreitol; 1.5 mM dATP; 1.5 mM dGTP; 1.5 mM TTP; 0.75 mM dCTP; 50 μCi/120 μl [^{32}P]dCTP; 100 μg/ml oligo dT_{12-18});
 50 units (<5 μl) reverse transcriptase.
(iii) Allow the reaction to proceed at 42°C for 60 min, then 46°C for 10 min and finally terminate by phenol extraction.
(iv) Separate the cDNA product from the unincorporated deoxynucleotide triphosphates by gel filtration on Sephadex G50, collect and then concentrate by ethanol precipitation. Starting with 10 μg of ds RNA a yield of 0.75–2 μg cDNA can be expected at this stage.

6.4.3 *Removal of RNA template and Size Fractionation of cDNA*

Resuspend the cDNA/RNA hybrids resulting from reverse transcription in 100 μl 5 mM Tris-HCl buffer pH 8.0 – 2 mM EDTA and then treat with 0.3 M NaOH at 70°C for 20 min to degrade the RNA template.

Direct molecular cloning of non size-fractionated cDNA has been found in numerous systems to result in preferential cloning of the small cDNA molecules in the population. Therefore to increase the probability of isolating full length clones, the cDNA should be size-fractionated on denaturing gels prior to cloning. The alkali-treated cDNA is loaded onto a 1.5% alkali agarose gel (29) and electrophoresed at 150 mA for 16 h to fractionate the various sizes of cDNA molecule. Individual cDNA bands are located by autoradiography, excised and the cDNA recovered by electroelution as described earlier for recovery of fractionated genomic RNA species.

6.4.4 *Generation of ds cDNA, Molecular Cloning of cDNA and Clone Identifiction*

(i) Resuspend the ethanol precipitate of recovered cDNA in 100 μl 2 x SSC (SSC = 0.15 M NaCl, 0.015 M sodium citrate) and incubate at 64°C for 2 h to allow the complementary cDNA strands to rehybridise.
(ii) After ethanol precipitation to concentrate the ds cDNA, fill in possible ragged ends on the molecules using the Klenow fragment of *E. coli* polymerase I (30).
(iii) Clone the flush-ended ds cDNA into a suitable plasmid vector (pBR322 or one of its derivatives) by use of either the G:C tailing approach (31) or following the addition of restriction enzyme site containing linkers (32).

Antibiotic-resistant *E. coli* clones containing plasmids with viral inserts can then be identified using Grunstein-Hogness filter hybridisation (33) with ^{32}P-labelled unfractionated viral mRNA as the probe. Subsequent to this the particular cloned inserts can be identified by hybridisation using ^{32}P isolated genomic RNA species as the probes. For the standard molecular cloning procedures required in this last part of the procedure, readers are recommended to consult reference 34.

7. REFERENCES

1. Joklik,W.K. (1983) *The Reoviridae,* Joklik,W.K. (ed.), Plenum Press, p. 1.
2. Moussa,A.Y. (1980) *Virology,* **106**, 173.
3. Winton,J.R., Lannan,C.N., Fryer,J.L. and Kimura,T. (1981) *Fish Pathol.,* **15**, 155.
4. Rosen,L. (1968) *Virol. Monogr.* **1**, 73.
5. Smith,R.E., Zweerink,H.J. and Joklik,W.K. (1969) *Virology,* **39**, 791.
6. Flewett,T.H. and Woode,G.N. (1978) *Arch. Virol.,* **57**, 1.
7. McNulty,M.S. (1978) *J. Gen. Virol.,* **40**, 1.
8. Wyatt,R.G., James,W.D., Bohl,E.H., Thiel,K.W., Saif,L.J., Kalica,A.R., Greenberg,H.B., Kapikian,A.Z. and Chanock,R.M. (1980) *Science (Wash.),* **207**, 189.
9. Urusawa,T., Urusawa,S. and Tanaguichi,K. (1981) *Microbiol. Immunol.,* **25**, 1025.
10. Sato,K., Inaba,Y., Shimozaki,T., Fujii,R. and Matsumoto,M. (1981) *Arch. Virol.,* **69**, 155.
11. Kutsuzawa,T., Konno,T., Suzuki,S., Kapikian,A.Z., Ebina,T. and Ishida,N. (1982) *J. Clin. Microbiol.,* **16**, 727.

12. Matsuno,S., Inouye,S. and Kono,R. (1977) *J. Clin. Microbiol.*, **5**, 1.
13. Graham,D.Y. and Estes,M.K. (1980) *Virology*, **101**, 432.
14. Malherbe,H.H. and Strickland-Cholmley,M. (1967) *Arch. Gesamte Virusforsch.*, **22**, 235.
15. Bridger,J.C. and Woode,G.N. (1975) *Br. Vet. J.*, **131**, 528.
16. Borsa,J. and Graham,A.F. (1968) *Biochem. Biophys. Res. Commun.*, **33**, 895.
17. Skehel,J.J. and Joklik,W.K. (1969) *Virology*, **39**, 822.
18. Furuichi,Y. and Shatkin,A.J. (1976) *Proc. Natl. Acad. Sci. USA*, **73**, 3448.
19. Pelham,H.R.B. and Jackson,R.J. (1976) *Eur. J. Biochem.*, **67**, 247
20. McCrae,M.A. and Joklik,W.K. (1978) *Virology*, **89**, 578.
21. Laemmli,U.K. (1970) *Nature*, **227**, 680.
22. Sugden,B., De-Troy,B., Roberts,R.J. and Sambrook,J. (1975) *Anal. Biochem.*, **68**, 36.
23. Sanger,D.V. and Mertens,P.P.C. (1983) in *ds RNA Viruses*, Compans,R.W. and Bishop,D.H., (eds.) Elsevier, p. 182.
24. Nagy,E. and Dobos,P. (1984) *Virology*, **137**, 58.
25. McCrae,M.A. and McCorquodale,J.G. (1982) *J. Virol.*, **44**, 1076
26. Both,G.W., Bellamy,A.R., Street,J.E. and Siegman,L.J. (1982) *Nucleic Acid Res.*, **10**, 7075.
27. Cashdollar,L.W., Esparza,J., Hudson,G.R., Chmelo,R., Lee,P.W.K. and Joklik,W.K. (1982) *Proc. Natl. Acad. Sci. USA*, **79**, 7644.
28. Imai,M., Richardson,M.A., Ikegami,N., Shatkin,A.J. and Furuichi,Y. (1983) *Proc. Natl. Acad. Sci. USA*, **80**, 373.
29. Emtage,J.S., Catlin,G.H. and Carey,N.H. (1979) *Nucleic Acid Res.*, **6**, 1221.
30. Jacobson,H., Klenow,H. and Ovargaard-Hansen,K. (1974) *Eur. J. Biochem.*, **45**, 623.
31. Bolivar,F., Rodriguez,R.L., Greene,P.J., Betlach,M.C., Heynecker,H.L., Boyer,H.W., Crosa,J.H. and Falkow,S. (1977) *Gene*, **2**, 95.
32. Kurtz,D.T. and Nicodemus,C.F. (1981) *Gene*, **13**, 145.
33. Grunstein,M. and Hogness,D.S. (1974) *Proc. Natl. Acad. Sci. USA*, **72**, 3961.
34. Maniatis,T., Fritsch,E.F. and Sambrook,J. (1982) *Molecular Cloning: A Laboratory Manual*, published by Cold Spring Harbor Laboratory Press, New York.

APPENDIX A

Production of Micrococcal Nuclease-treated Rabbit Reticulocyte Translation System

(i) On day 1 inject six 2.5 kg New Zealand white rabbits subcutaneously in the neck folds with 1 ml of a vitamin solution containing 0.1 mg vitamin B_{12} and 1 mg folic acid in isotonic saline and 0.6 ml of a 2.5% (v/v) phenyl hydrazine solution in saline.

(ii) On days 2–5 repeat the phenylhydrazine injection. The animals become progressively more anaemic.

(iii) On day 7 anaesthetise the rabbits and bleed by cardiac puncture. All syringes, needles and glassware used in the lysate preparation procedure should be rinsed with a solution of 1000 units/ml heparin before use. Store the collected blood on ice until all the rabbits have been bled, and carry out the remainder of the preparation procedure at 4°C.

(iv) Collect the reticulocytes by centrifugation at 2000 r.p.m. at 4°C for 15 min. Remove the serum supernatant and wash the cells four times in buffer containing 0.14 M KCl, 50 mM NaCl and 5 mM $MgCl_2$, using centrifugation as above to collect the cells after each wash.

(v) Estimate the final volume of the packed cells and lyse the cells by addition of an equal volume of ice-cold water. Centrifuge the lysed cell suspension immediately at 10 000 *g* for 10 min, collect the supernatant, dispense into 2 ml aliquots, snap freeze and store in liquid nitrogen.

(vi) Micrococcal nuclease treatment of the lysate is carried out as follows.

(a) Thaw 2 ml of lysate.

(b) Immediately add 80 μl of 1 mM haemin, 20 μl of 5 mg/ml creatine kinase and 20 μl of 100 mM $CaCl_2$.

(c) Add 30 μl of 1 mg/ml micrococcal nuclease and incubate at 20°C for 12.5 min.

(d) Add 10 μl of 0.5 M EGTA to chelate out the calcium that is essential for nuclease activity. Aliquot into volumes appropriate for use in one assay and store in liquid nitrogen. Repeated freeze-thawing of treated lysate should be avoided.

CHAPTER 8

Simian Virus 40 and Polyoma Virus: Growth, Titration, Transformation and Purification of Viral Components

HANS TÜRLER and PETER BEARD

1. INTRODUCTION

Simian virus 40 (SV40) and polyoma virus are the best known members of the papova virus group. Although neither virus normally induces serious illness in its natural host, monkeys and mice, respectively, SV40 and polyoma virus have been intensively studied. The reasons for this interest are at least 2-fold. Firstly, SV40 and polyoma induce tumours when injected into susceptible rodents, and transform rodent cells in tissue culture. Transformation results from the action of just one or two known viral proteins, so there is a good chance of identifying their mode of action and the mechanisms leading to cell transformation *in vitro* and tumour formation *in vivo*. Secondly, because of the small size of these viruses and their genomes, they rely on their host cells to provide almost all the enzymes and factors involved in replication and in transcription and maturation of viral mRNAs. The circular, double-stranded viral DNA of 5243 bp for SV40, (strain 776) and of 5292 bp for polyoma (A2 strain) is organised with cellular histones into 'minichromosomes' similar in structure to the much larger and more complex cellular chromosomes. SV40 and polyoma are thus useful tools with which fundamental processes including regulation of gene expression in eukaryotic cells can be probed. For the structure, biology, genetics and DNA nucleotide sequences the reader should consult reference 1.

In this chapter we describe the basic methods for preparation of virus stocks, purification and assays of virus, assays of virus-transformed cells, and the isolation of viral DNA and of nucleoprotein complexes from infected cells.

2. SAFETY RULES FOR WORKING WITH SV40 AND POLYOMA VIRUS

There is no evidence to suggest that working with SV40 or polyoma virus presents a health hazard. Nevertheless, the following rules should be followed for the safety of yourself and the people working around you. Also, they will help prevent accidental infection of your cell lines and primary cultures.

(i) Use, if possible, separate rooms or separate laminar flow hoods and incubators for cell culture (preparation of primary cultures and maintenance of cell lines) and for virus-infected cultures.

(ii) Do not pipette by mouth. No drinking, eating or smoking is allowed in rooms where viruses are handled.
(iii) Wear laboratory coats and gloves for manipulation of virus or virus-infected cultures; wash your hands before leaving the virus room.
(iv) All glassware that has been used for virus or virus-infected cells should be soaked in clorox or 5% hypochlorite.
(v) Disposable glassware and plastic should be wrapped in aluminium foil or in plastic bags and autoclaved before disposal.

3. CELL CULTURE, MEDIA AND BUFFERS

3.1 Cell Culture

SV40 and polyoma virus may be grown in primary cell cultures or in established cell lines. Each system presents some advantages and some disadvantages.

Primary cell cultures are prepared from tissue obtained from animals. Provided that the animals are bred under controlled conditions then the characteristics of the cells and virus-cell interactions are reproducible. Primary cultures have the disadvantage that they are laborious to prepare especially when large amounts are needed. For the preparation of primary cell cultures the reader is referred to a detailed description in a companion volume in this series (2).

Many cell lines, permissive and non-permissive, for polyoma virus and SV40 are available. The cells are easy to handle and large amounts can be prepared. However, cell lines have the disadvantage that they may change their characteristics with continued passage resulting in changes in virus-cell interactions and often in a drop in progeny virus production. Furthermore, cell lines may become infected with mycoplasma (3). For maintenance and storage of established cell lines see reference 2.

3.2 Host Cells for SV40 and Polyoma Virus

Cells and cell lines frequently used for SV40 and polyoma virus are listed in *Table 1*.

We use primary cultures of mouse kidney cells (4) for preparation of polyoma virus stocks and for studies on productive infection with polyoma virus and the transforming infection with SV40 (5). Mouse 3T6 cells are used when large amounts of polyoma DNA or RNA are needed (6). SV40 is propagated in CV-1 cells of low passage number.

The proportion of infected cells after exposure to virus varies with the different cell types. For a cell line it may change with increasing passage number and it may depend on whether cells are growing or resting. The easiest test to determine the proportion of infected cells is immunofluorescent staining of the intranuclear viral tumour antigens 1 day after infection (see Section 6.4).

3.3 Media and Sera

Pre-mixed, powdered cell culture media are commercially available. All cells listed in *Table 1* can be cultured in Dulbecco's minimal essential medium (DMEM).

Table 1. Host Cells for SV40 and Polyoma Virus.

Host cells for SV40
Productive infection (permissive cells):
primary cultures of African green monkey kidney cells
cell lines derived from monkey kidney: CV-1, BSC-1, VERO
Transforming infection (non-permissive cells):
primary cultures of baby mouse kidney cells
primary or secondary cultures of mouse, hamster, rat embryo cells
mouse, hamster and rat cell lines, e.g., 3T3, FR 3T3 (7)
Host cells for polyoma virus
Productive infection (permissive cells):
primary cultures of baby mouse kidney cells
primary or secondary cultures of mouse embryo cells
mouse cell lines, e.g., 3T6, 3T3
Transforming infection (non-permissive cells):
primary or secondary cultures of hamster and rat embryo cells
hamster and rat cell lines, e.g., BHK-21 (C13), FR 3T3

The culture medium has to be supplemented with serum. For established cell lines, use of foetal calf serum (FCS) is recommended. Although most suppliers of cell lines indicate addition of 10% FCS we now use routinely only 5%. For primary cell cultures, 10% calf serum (CS) or 10% newborn calf serum (NCS) is added to the culture medium.

3.4 Buffers and Solutions

The composition of several buffers and solutions used in cell culture and viral infections are listed in *Table 2*.

4. PREPARATION OF VIRUS STOCKS

SV40 and polyoma virus preparations obtained by high multiplicity of infection may contain variant and defective viruses (evolutionary variants). They arise by rearrangement, reiteration and deletion of viral DNA sequences and by substitution of viral DNA with host cell DNA. By maintaining the viral origin of replication, defective viruses are perfectly able to replicate in the presence of wild-type virus and to form progeny particles. They may even, by forming molecules with two or more origins, have a selective replicative advantage (8). Without a noticeable decrease in the particle number, the infective titre of such virus preparations falls after 3–5 serial infections to 10% or less of the original preparation. For some experiments, presence of variants and defective particles will not matter; for others, it is indispensable to work with a defined, pure virus preparation. In this case the virus should be checked (see Section 6.3) and if necessary new stocks should be prepared by plaque-purification and infection at low multiplicity.

The following procedure, which is suitable for both SV40 and polyoma virus, yields 150–250 ml of virus stock suspension. This is sufficient to infect 300–500

Table 2. Buffers and Solutions.

TD (Tris-Dulbecco)	
NaCl	8.0 g
KCl	0.38 g
Na_2HPO_4	0.1 g
Tris-HCl (7–9)	3.0 (Tris-base 2.4 g)
Streptomycin	0.1 g
Penicillin	5 x 10^5 units
Dissolve in 900 ml of water, adjust the pH to 7.4 and fill up to 1000 ml[a].	
PBS (Dulbecco's phosphate-buffered saline)	
NaCl	8.0 g
KCl	0.2 g
Na_2HPO_4 (anhydrous)	1.15 g
KH_2PO_4	0.2 g
$MgCl_2$ $6H_2O$	0.1 g
$CaCl_2$ (anhydrous)	0.1 g
Dissolve NaCl, KCl and the phosphates in 700 ml of water, adjust the pH to 7.4, then add the solutions of $MgCl_2$ in 100 ml H_2O and $CaCl_2$ in 100 ml H_2O. Fill up to 1000 ml[a].	
Trypsin	
Trypsin 1:250	2.0 g
NaCl	8.0 g
KCl	0.3 g
Na_2HPO_4	0.07 g
KH_2PO_4	0.02 g
Streptomycin	0.1g
Penicillin	5 x 10^5 units
Phenol red 0.5%	3 ml
Dissolve trypsin in 800 ml of water, salts and antibiotics in 150 ml H_2O. Mix the two solutions and add the phenol red. Adjust the pH to 7.0 and fill up to 1000 ml[a].	
Versene	
NaCl	8.0 g
KCl	0.2 g
Na_2HPO_4	1.15 g
KH_2PO_4	0.2 g
Na_2-EDTA	0.2 g [disodium salt or 'Versene' (BDH) or 'Titriplex III' (Merck)]
Phenol red 0.5%	3 ml
Dissolve the salts in 900 ml of water, add phenol red and adjust the pH to 7.2. Fill up to 1000 ml[a].	
Trypsin/Versene is a mixture of 1 part trypsin and 4 parts Versene.	
HBE (isolation of viral chromosomes)	
10 mM Hepes (potassium salt, pH 7.8)	
5 mM KCl	
1 mM EDTA	
0.5 mM dithiothreitol	

[a]Sterilise these solutions by filtration.

Petri dish cultures (9 cm diameter). For SV40 use confluent cultures of CV-1 cells, for polyoma virus confluent primary cultures of mouse kidney cells.

(i) Start from a well-isolated, clear plaque (for plaque assay see Section 6.1). Punch through the agar overlay with a sterile Pasteur pipette to reach the

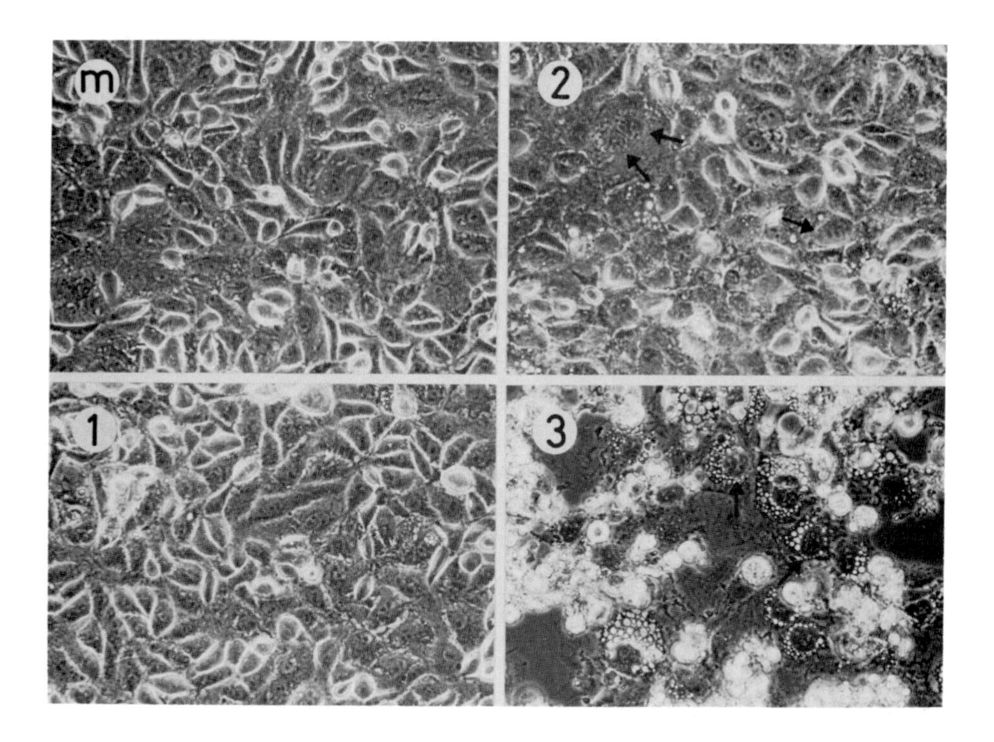

Figure 1. Infection of CV-1 cells with SV40. **(m)** Mock-infected culture 1 day after mock-infection. **(1)** Infected culture after 1 day, no signs of infection are visible. **(2)** Infected culture after 2 days, cells with small vacuoles and enlarged nuclei with dark internal structure (indicated by arrows). **(3)** Infected culture after 3 days, disintegration of cell layer, the remaining cells have many vacuoles and often enlarged, dark nuclei (indicated by arrows), detached cells appear as light round spots.

centre of the plaque and the bottom of the plate. Aspirate the agar into the pipette and transfer it to 1 ml of medium containing 5% CS.

(ii) Sonicate briefly with an ultrasonic disintegrator equipped with a 3 mm end-diameter probe or by placing the tube into an ultrasonicator bath.

(iii) Infect two cultures of confluent cells as follows:

- (a) Aspirate the medium completely and pipette 0.5 ml of the plaque suspension on each dish.
- (b) Incubate the cultures at 37°C for 90 min for adsorption of the virus to the cells. During this period rock the dishes from time to time to re-distribute the virus suspension.
- (c) After the adsorption period, cover the cultures with 10 ml of pre-warmed medium containing 5% CS.

(iv) Incubate the cultures for 6 – 8 days until clear signs of cell lysis are observed (see *Figures 1* and *2*).

(v) Scrape whatever remains of the monolayer with a sterile silicone rubber and collect the culture medium and the cell debris into a small flask.

(vi) Freeze and thaw once, then sonicate to disintegrate the cell debris. This 'plaque lysate' has a titre of 2 x $10^8 - 10^9$ plaque forming units (p.f.u.)/ml.

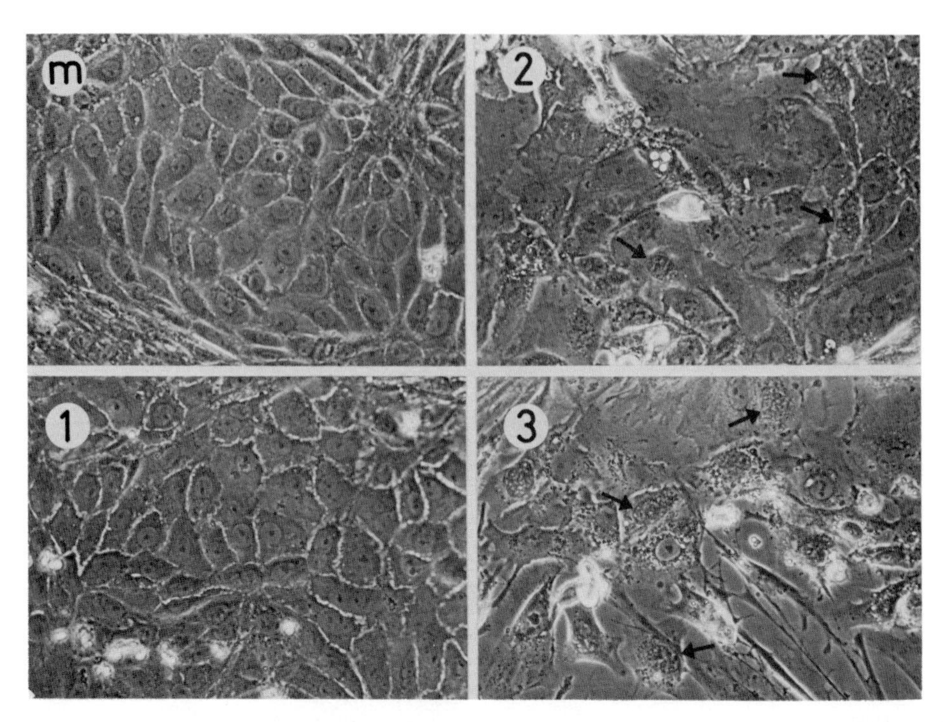

Figure 2. Infection of primary mouse kidney cell cultures with polyoma virus. **(m)** Mock-infected culture 1 day after mock-infection. **(1)** Infected culture after 1 day, nuclei are slightly enlarged and more transparent. **(2)** Infected culture after 2 days, nuclei containing virus particles have a grainy internal structure (indicated by arrows), in the centre a hole left by a disintegrated nucleus. **(3)** Infected culture after 3 days, disintegration of the cell layer, many enlarged nuclei with grainy structures. Microphotos of *Figures 1* and *2* were taken with a Zeiss ICM 405 inverted microscope with phase contrast and a magnification of 188x in the laboratory of Dr U.Laemmli.

To determine the exact titre you may perform a plaque assay (see below), but for routine work this is not necessary. Dilute an aliquot of your plaque lysate 1000-fold, e.g., 50 μl into 50 ml medium containing 5% CS, and store the rest of the plaque lysate at $-20°C$.

(vii) Infect 15–25 cultures with 0.5 ml of the 1000-fold diluted plaque lysate which corresponds to a multiplicity of infection of 0.01–0.05 p.f.u./cell (there are $5 \times 10^6 - 10^7$ cells on a 9 cm dish).

(viii) Incubate the cultures for 6–8 days, then collect the medium and the cell debris as described above. Freeze and thaw the lysate once. Sediment the cell debris by centrifugation in a sterile tube at 8000 *g* for 10 min at 4°C. Pour the supernatants into a sterile flask and resuspend the pellet using 1/20 of the volume of the supernatants. Sonicate the cell debris and centrifuge at 10 000 *g* for 20 min. Add the supernatant to the flask containing the first supernatant. This is now your virus stock suspension. Treat the cell debris with clorox or 5% hypochlorite and discard.

(ix) The virus stock suspension should be tested by one or more of the following assays:

(a) plaque assay to determine p.f.u./ml (see Section 6.1);

(b) extraction of viral DNA and analysis by electrophoresis (see Section 6.3);

(c) infection of permissive cells and determination of the proportion of T-antigen-positive cells by immunofluorescent staining (see Section 6.4).

For most experiments titres of $5 \times 10^8 - 1 \times 10^9$ p.f.u./ml are ideal. Usually virus stock suspensions prepared by this procedure have virus titres close to this and can be used directly or may even be diluted to the desired titres. If higher titres are required, concentrate the virus by ultracentrifugation for 3 h at 80 000 *g* at 4°C and resuspend the pellet in the appropriate volume of culture medium or buffer containing 5% CS. Alternatively, you may start from a purified, dialysed virus preparation (see Section 5).

(x) Distribute your stock suspension into sterile tubes (5 – 10 ml per tube) and store at −20°C. Repeated freezing and thawing leads to loss in infectivity titre. The presence of the serum helps to conserve the virus.

5. PURIFICATION OF SV40 AND POLYOMA VIRUS

SV40 and polyoma virus are not sensitive to lipid solvents such as chloroform, nor to CsCl solutions, which can therefore be used during the purification of the viruses.

The method given is suitable for both SV40 and polyoma virus. It employs 20 Petri dishes (9 cm diameter) of the appropriate cells in a confluent monolayer.

(i) Remove the medium and inoculate each culture with 0.5 ml of virus stock suspension having $10^8 - 10^9$ p.f.u./ml, which corresponds to a multiplicity of infection (m.o.i.) of about 2 – 50 p.f.u./cell. Put the dishes into the 37°C incubator and rock them from time to time to re-distribute the virus suspension. After 90 min incubation, add 10 ml of pre-warmed medium containing 5% CS and return the cultures to the incubator.

(ii) Watch the progress of the infection by observing the cells under the microscope (see *Figures 1* and *2*). Usually, cell lysis is extensive after 3 – 4 days, but this can vary with the cell cultures and with the m.o.i. used. Scrape the cells with a sterile silicone rubber and harvest the medium and cells together. The virus is in both the medium and the cellular material.

(iii) Centrifuge the lysate at 8000 *g* for 10 min to pellet the cell debris. Keep the supernatant and use 1/20 of it to resuspend the pelleted cell debris.

(iv) Release the virus from the cell debris by freezing and thawing three times or more efficiently by sonication of the suspension. Pellet the cell debris again at 8000 *g* for 10 min and add the supernatant to the original medium. Sterilise the pellet and discard it.

(v) Centrifuge the combined supernatant for 3 h at 80 000 *g* at 4°C. Resuspend the virus pellets in a small volume of 10 mM Tris-HCl pH 7.5 depending on the type of gradient purification to follow.

5.1 **Equilibrium Centrifugation in a CsCl Density Gradient**

(i) Re-suspend your virus pellets in 2.0 ml of 10 mM Tris-HCl pH 7.5. The suspension should be homogeneous and without clumps. Transfer the suspension to an ultracentrifuge tube (e.g., Beckman rotor SW60 or SW50.1) and adjust it with buffer to 2.5 g by weighing. Add 1.20 g of CsCl and make a homogeneous solution by gentle pipetting. The density of the solution should be 1.30 – 1.33 g/ml and should be checked by measuring the refractive index (see *Table 3*). Fill the centrifuge tube with paraffin oil and equilibrate your tubes and balance tubes.

(ii) Centrifuge at 35 000 r.p.m. for about 20 h at 20°C. Virus particles band at a density of 1.33 g/ml; empty capsids, which are always present in SV40 and polyoma virus preparations, band at a density of 1.29 g/ml. Cell debris is on top of the gradient. Often the two bands of virus particles and empty capsids can be seen by eye (11).

(iii) Collect the gradient by dividing it into 16 – 20 fractions, i.e., about 4 – 6 drops per fraction. There are several ways of doing this (see Chapter 2,

Table 3. Density and Refractive Index of CsCl Solutions.

wt% g CsCl/100 g solution[a]	*g CsCl/ ml H_2O*	*Density ϱ (25°C)*	*Refractive index η (25°C)*[b]
Virus purification			
29.84	0.42	1.283	1.361
		1.294	1.362
31.66	0.46	1.305	1.363
		1.315	1.364
33.33	0.50	1.326	1.365
		1.337	1.366
35.03	0.54	1.348	1.367
CsCl/ethidium bromide gradients for DNA			
47.39	0.90	1.533	1.384
	0.92	1.544	1.385
48.67	0.95	1.555	1.386
	0.97	1.565	1.387
49.85	0.99	1.576	1.388
	1.01	1.587	1.389
51.06	1.04	1.598	1.390
CsCl gradients for DNA			
55.53	1.25	1.685	1.398
	1.27	1.696	1.399
56.53	1.30	1.707	1.400
	1.33	1.717	1.401
57.56	1.36	1.728	1.402
	1.39	1.739	1.403
58.56	1.41	1.750	1.404

[a]wt% = 137.48 – 138.11 (l/ϱ) (9).
[b]ϱ25° = 10.8601η 25° – 13.4974 (10).

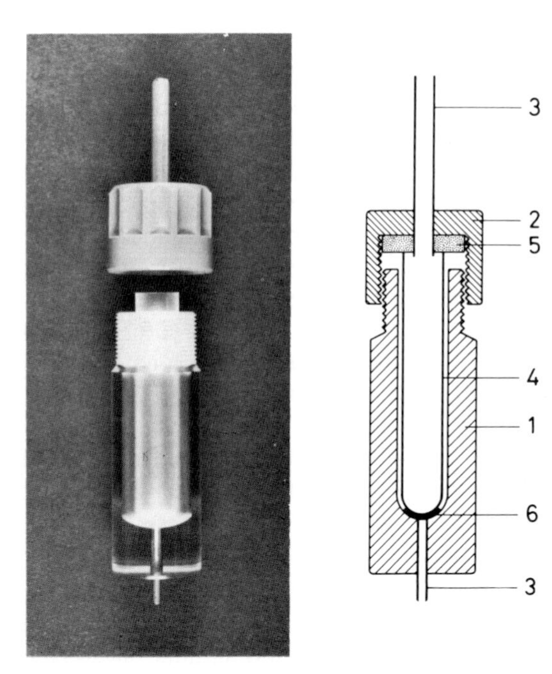

Figure 3. Gradient collector. **1:** tube holder, **2:** screw cap, **3:** metal tubing, **4:** centrifuge tube (SW60), **5:** silicone rubber disc, **6:** piece of Parafilm.

Figure 3). We use a simple tube holder with a screw cap. The bottom of the tube is pierced with a hypodermic needle and the dripping out of the gradient is controlled by placing the forefinger on the opening of the screw cap (see *Figure 3*).

(iv) Identify the virus-containing fractions by one of the following methods:
  (a) dilute each fraction to 1 ml with 10 mM Tris-HCl pH 7.5 and read the optical density at 260 nm (thereafter the virus can be concentrated again by centrifugation;
  (b) with good virus preparations the virus-containing fractions can be recognised by eye as turbidity;
  (c) with polyoma virus, the virus and the empty capsid bands can be detected by the haemagglutination assay (see Section 6.2);
  (d) if the virus has a radioactive label ([^{3}H]thymidine or [^{35}S]methionine) count radioactivity in small aliquots of each fraction.

(v) Pool the fractions containing the virus and dialyse in the cold against 10 mM Tris-HCl pH 7.5 or 1 mM sodium phosphate pH 6.8. Alternatively, you may remove CsCl by diluting the virus with buffer and by two subsequent sedimentations of the virus particles (80 000 *g*, 3 h at 4°C).

(vi) If the virus is used for infecting cell cultures do not forget to re-adjust the salt concentration to isotonic conditions. Sterilise the virus suspension by filtration through an 0.2 μm membrane filter (e.g., 'Millex', Millipore or 'Val-Filt', Van Leer Medical).

5.2 Velocity Sedimentation in Sucrose Gradients

(i) Re-suspend your virus pellets in 0.5 ml of 10 mM Tris-HCl pH 7.5.

(ii) Prepare a sucrose gradient of 10–30% w/v in 10 mM Tris-HCl pH 7.5 in an ultracentrifuge tube for the Beckman rotor SW40 (with smaller virus preparations, SW60) using a two chamber mixing device (see Chapter 1, *Figure 3*) or by layering successively equal volumes of 30%, 23%, 17% and 10% (w/v) sucrose solutions on top of each other to fill the tube. In the second case let the gradient diffuse for 2 h at room temperature.

(iii) Layer the virus suspension on top of the gradient and centrifuge at 30 000 r.p.m. for 50 min (20°C) with the SW40 rotor or at 35 000 r.p.m. for 30 min (20°C) with the SW60 rotor. Fractionate the gradient and determine which fractions contain the virus as described above [see Sections 5.1 (iii) and (iv)].

(iv) Dialyse or recover the virus by sedimentation from diluted fractions. If the presence of sucrose does not interfere with subsequent manipulations, e.g., DNA or protein analysis by gel electrophoresis, this step can be omitted.

5.3 Modifications for Use with Polyoma Virus

Polyoma virus tends to attach to cell debris present in the crude lysate at acidic pH. Therefore, adjust the lysate to 20 mM sodium phosphate pH 5.6, cool to 4°C, and collect the cellular material by centrifugation as described above. The supernatant containing little virus can be discarded (after sterilisation with clorox or hypochlorite). The virus is released from the pelleted material after re-suspending it in 10 ml Tris-Dulbecco (TD) by addition of neuraminidase (1 ml receptor destroying enzyme, Microbiological Associates) and incubation at 37°C for 12–15 h.

Centrifuge again at 8000 *g* for 10 min, keep the supernatant for further purification by equilibrium centrifugation or velocity sedimentation and discard the pellet.

6. ASSAYS OF SV40 AND POLYOMA VIRUS

We consider here different assays for testing the quality of a virus stock or a virus preparation. The plaque assay (Section 6.1) measures biological activity, i.e., the infectivity titre. This is the best way to characterise a new virus stock, but is rather time-consuming. In most cases it is sufficient to determine the proportion of infected cells by immunofluorescent T-antigen staining (Section 6.4) and compare it with the values obtained with previous stocks. The haemagglutination assay (Section 6.2), like optical density or electron microscopy, measures the number of virus particles, including empty capsids, pseudovirions (particles containing fragmented host DNA instead of circular viral DNA) (12) and defective particles. The ratio of p.f.u. to haemagglutination units (HAU) is a rough measure for the presence or absence of defective particles. For good virus stocks it should be about $5 \times 10^5 - 10^6$ (13). However, the presence of variant or defective viruses is detected more easily by extracting the viral DNA from a small volume of stock suspension or lysate and analysing it either directly or after digestion with a restriction enzyme by gel electrophoresis (Section 6.3).

6.1 Plaque Assay

A plaque is a small area where the cell monolayer has been destroyed by virus. The principle of the plaque assay is to infect the monolayer with virus diluted to contain only a few infectious units, then to cover the monolayer with agar which prevents progeny virus from diffusing all over the plate. Virus from a single lysed cell can still infect adjacent cells, giving rise after several cycles to a plaque (14,15) (see *Figure 4*).

(i) Make serial 10-fold dilutions of the virus in TD buffer containing 2% CS. Infect confluent monolayer cultures in 5 cm dishes with 0.1 ml of the appropriate dilutions (to titrate virus stocks use 10^{-5} to 10^{-8}), making duplicate dishes for each dilution. Mock-infect two cultures with 0.1 ml

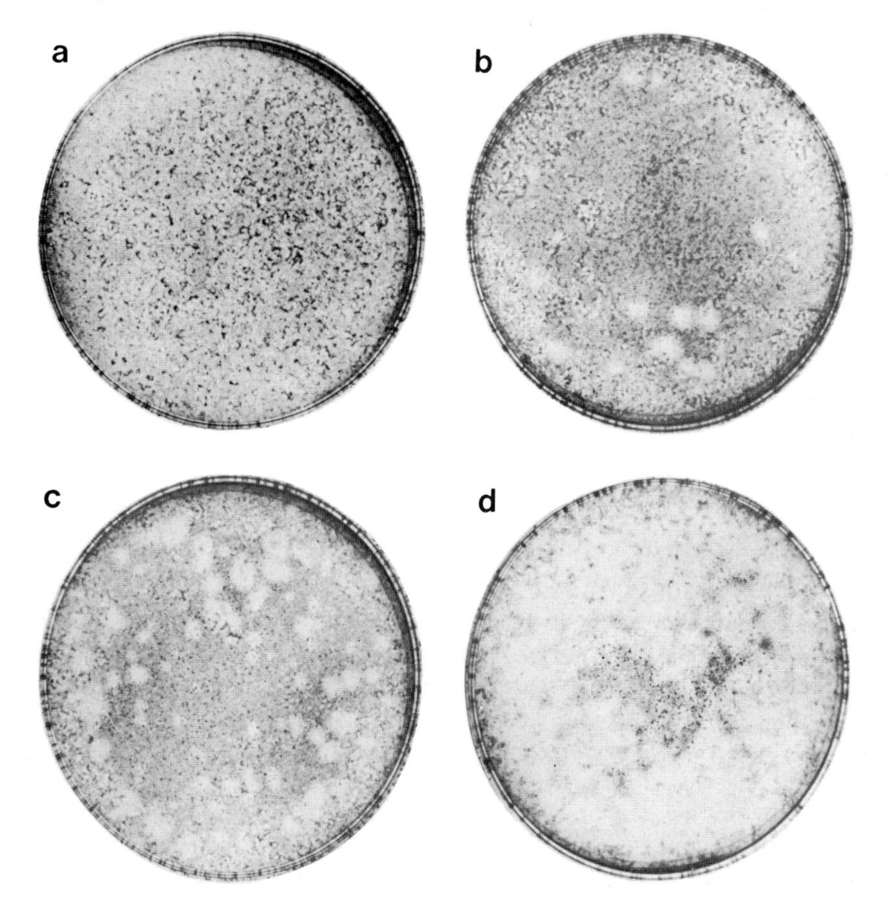

Figure 4. Plaque assay of polyoma virus with primary mouse kidney cell cultures. **(a)** Mock-infected control culture. **(b)** Infected with 10^7-fold dilution of a polyoma stock suspension showing 14 plaques. **(c)** Infected with 10^6-fold dilution, 90 – 100 plaques. **(d)** Infected with 10^5-fold dilution. The titre of the stock suspension is about 1.2×10^9 p.f.u./ml. (Reproduced with permission from reference 15).

TD containing 2% CS without virus to serve as controls. Let the virus adsorb for 90 min at 37°C in the incubator, then rinse the monolayers gently with 5 ml of pre-warmed TD and add 5 ml medium containing 2% CS. Replace the cultures in the incubator.

(ii) About 15 h after infection prepare the agar medium. Make a sterile solution of 1.8% agar (Special Agar-Noble, Difco Laboratories) by autoclaving. This can be prepared in advance and stored at 4°C; for use re-melt in a boiling water bath. Cool the agar solution to 45°C and add an equal volume of 2x concentrated medium with 4% CS pre-warmed to 45°C. Remove the medium from the cultures and replace by 5 ml of agar medium. Leave the cultures at room temperature until the agar medium is solid (~15 min) then put them back into the incubator.

(iii) Check the condition of the cells and the colour of the agar medium from time to time. Additional agar medium can be added after 5 days. Usually about 8 days after infection the cultures are ready for staining (the time may vary between 6 and 10 days).

(iv) Prepare the staining medium by adding 1/100 volume of 1% neutral red in water (sterilised by autoclaving) to medium containing 2% CS and put 5 ml on each dish. Leave overnight in the incubator.

(v) Remove the staining medium, put the dishes back into the incubator for 1–2 h then count the plaques. Since neutral red stains only living cells, the plaques appear as clear areas in a red background (see *Figure 4*). Observation of the control cultures should allow the variations in staining unrelated to virus infection to be recognised. Stained cultures can be kept for 1 or 2 more days in the incubator, before the monolayer will lyse. During this period, count the plaques twice a day and observe the dishes for appearance of new plaques and enlargement of doubtful plaques.

6.2 **Haemagglutination Assay for Polyoma Virus**

Haemagglutination is the agglutination of red blood cells (rbc) by virus particles which possess, on their surface, sites recognised by receptors on the rbc. This assay does not distinguish between infectious and non-infectious particles. Not all viruses haemagglutinate and not all species' rbc can be agglutinated. Polyoma virus agglutinates guinea pig erythrocytes (16), but SV40 does not agglutinate any of the easily available rbc. Haemagglutination can be blocked by antibodies directed against the virus particles and this is used in the haemagglutination-inhibition assay to detect anti-viral antibody.

The principle of the haemagglutination assay is to mix virus with the rbc and then to let the cells sediment. Normally, the cells settle in a round-bottomed well to form a dense ring or spot. Agglutinated rbc, on the other hand, form a diffuse layer uniformly covering the whole bottom (see *Figure 5*). By mixing the rbc with serially diluted virus, a dilution end-point for agglutination can be estimated.

(i) Prepare a suspension of guinea pig red blood cells.
   (a) Obtain the blood (1–2 ml) from an anaesthetised animal by heart puncture and mix with 2 ml of sterile 0.4% sodium citrate in TD.

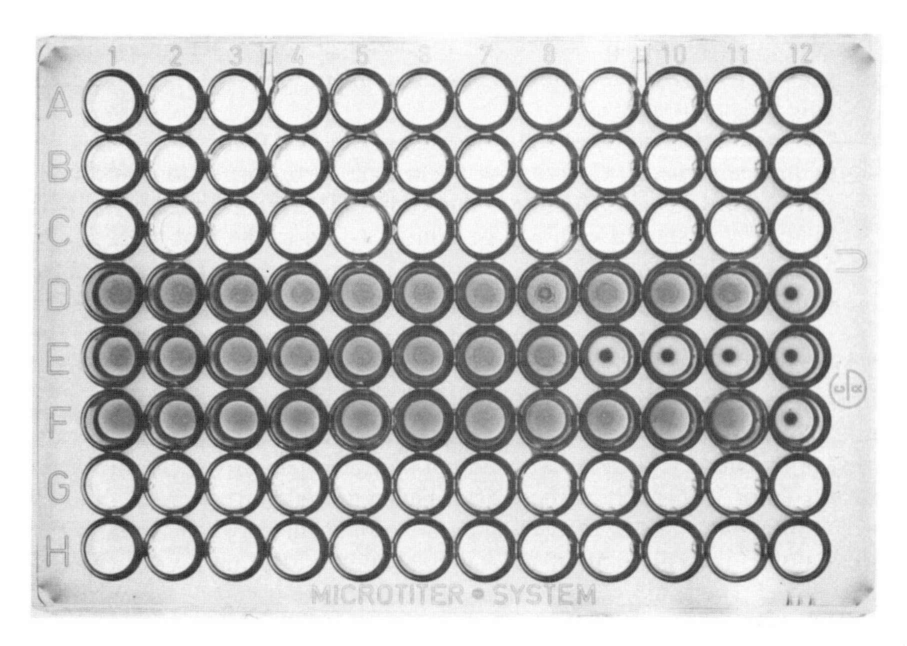

Figure 5. Haemagglutination assay of polyoma virus. Samples of three virus preparations were diluted 100-fold and titrated as described. Preparations D and F have a titre of 4 x 10^5 HAU/ml (highest positive dilution in row 11, i.e., diluted 4096-fold). Preparation E has 5 x 10^4 HAU/ml (positive in row 8 with a 512-fold dilution).

(b) Sediment the rbc by centrifugation at 1500 r.p.m. for 5 min, remove the supernatant with a pipette and re-suspend the cells in 5 ml TD.

(c) Centrifuge the cells and wash them once more with 5 ml TD.

(d) Pipette from the pellet 0.5 ml into 100 ml sterile TD. Keep the suspension at 4°C for at least 24 h before use. Being careful and using clean, sterile glassware, the suspension can be kept for about 2 weeks.

(e) When the TD becomes red, the cells have begun to lyse and a new suspension should be prepared. Guinea pig red blood cells can also be purchased (e.g., Flow).

(ii) For the test use a plastic plate with 96 wells; a series of round-bottomed plastic tubes is also suitable. Prepare dilutions of the virus to be tested in TD, dilute by a factor of 10^2, 10^3 or 10^4 depending on the expected virus concentration. If this cannot be guessed start with both a 10^1- and a 10^3-fold dilution.

(iii) Put 50 μl of TD into each well of the plastic plate. Make duplicates for each determination, so add 50 μl of the starting dilution to two wells of the first row, mix well and transfer 50 μl of this mixture to the wells of the second row. Mix and continue, thus making a series of 2-fold dilutions. Special equipment for the haemagglutination assay can be purchased: pipettes that deliver 50 μl per drop and metal samplers for mixing and transferring 50 μl from one well to the other ('Titertek' from Flow, 'Microtiter' from Cooke). If you use plastic tubes the buffer volume should be increased to 200 μl.

(iv) Add 50 μl of the rbc suspension to each well (200 μl for tubes). Agitate gently by rocking the plate and leave it undisturbed for at least 6 h, preferably overnight at 4°C.
(v) Determine the end-point, i.e., the highest dilution that still gives agglutination (see *Figure 5*). The titre of the original virus suspension expressed in HAU is the reciprocal of the highest dilution giving a positive result. The dilution you started with will be diluted 4-fold in the first well, 512-fold after eight steps and 8000-fold after 12 steps. One HAU corresponds to about 10^7 physical particles counted by electron microscopy (11).

6.3 Analysis of Viral DNA by Electrophoresis in an Agarose Minigel

One of the simplest ways to estimate roughly how much virus is in a stock suspension, in a lysate, or in a virus preparation and at the same time to check the quality of its DNA, is to analyse an aliquot by electrophoresis of the viral DNA. Gel electrophoresis has already been described in detail in this series (17) and elsewhere (18). Here we use a 'minigel' apparatus which allows a quick analysis and detection of as little as 10 ng of DNA corresponding to about 2 x 10^9 virions or 2 x 10^7 p.f.u. The equipment for minigel electrophoresis can be home-made or obtained commercially (e.g., from Bethesda Research Laboratories).

6.3.1 *Agarose Minigels*

A 10 x 7 cm 1% agarose gel is cast in the central part of a plastic box with platinum electrodes at each end. Alternatively, the gel may be cast on a 7 x 5 cm glass plate by pipetting 12 ml of agarose solution on the plate. The gel is electrophoresed after flooding it with electrophoresis buffer, so that there is about 1 cm of liquid above the gel.

6.3.2 *Extraction of Viral DNA*

(i) Take 0.1 ml of stock suspension or of a cell lysate with about 5 x $10^8 - 10^9$ p.f.u./ml, add 5 μg of DNase I and 10 μg of RNase A and incubate at 37°C for 1 h.
(ii) Add 25 μl of 5% SDS, 50 mM EDTA, pH 7.5, and keep for 10 min at room temperature.
(iii) Extract with 125 μl of phenol/chloroform 1:1 (v/v) saturated with 0.1 M Tris-HCl pH 8 by shaking for 2 min, centrifuge and recover the aqueous phase.
(iv) Precipitate the DNA with 250 μl of ethanol and keep at −20°C for 30 min.
(v) Centrifuge, remove the supernatant, dry the pellet and re-dissolve it in 15 μl of 10 mM Tris-HCl, 1 mM EDTA, pH 7.5.
(vi) To an aliquot of this solution add glycerol to 10% v/v and bromophenol blue to 0.01% w/v and load on a minigel together with a standard preparation of viral DNA, e.g., 20 and 100 ng. Part of the viral DNA can be used for digestion with a restriction enzyme (e.g., *Hind*III for SV40 and *Hpa*II for polyoma).

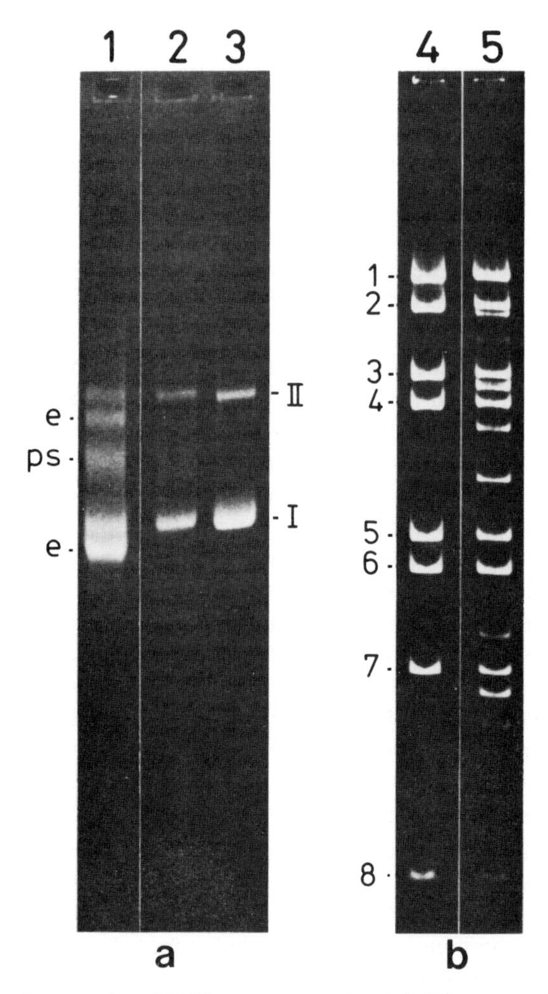

Figure 6. **(a)** Electrophoresis of viral DNA on agarose minigel (0.8% agarose, 40 mM Tris-acetate, 2 mM EDTA, pH 7.5). **(1)** Polyoma DNA containing evolutionary variants (e) extracted from 0.5 ml lysate; **(2)** polyoma DNA extracted from 0.5 ml stock suspension of plaque-purified virus; **(3)** SV40 DNA extracted from 0.1 ml stock suspension of plaque-purified virus. I: form I DNA (supercoiled), II: form II DNA (nicked circles), ps: pseudovirion DNA (fragments of mouse DNA). **(b)** Electrophoresis of *Hpa*II-digested polyoma DNA (5% acrylamide, 89 mM Tris-borate, 2 mM EDTA, pH 7.8). **(4)** Polyoma DNA extracted from cells that had been infected with plaque-purified virus, *Hpa*II fragments 1 – 8; **(5)** polyoma DNA extracted from cells that had been infected with serially passaged virus, evolutionary variants generate additional bands.

(vii) Electrophorese at 50 – 100 V for about 1 h until the bromophenol blue used as a marker reaches the bottom of the gel (voltage and time depend on the dimensions of the apparatus).

(viii) Add ethidium bromide to the electrophoresis buffer to obtain 2 μg/ml and stain the gel for 15 min.

The presence of variant viruses is revealed by additional bands usually migrating ahead of the main form I and form II bands and by additional fragments in the restriction enzyme digest (see *Figure 6*).

6.4 Proportion of Infected Cells by Immunofluorescent Staining of Intranuclear Tumour Antigens

Tumour (T)-antigens is a name given to the early gene products of SV40 and polyoma virus. The name derives from the fact that these proteins were first identified in virus-induced tumours and in virus-transformed cells. The T-antigens are now known to comprise a family of two (for SV40) or three (for polyoma virus) related proteins which are all encoded by the early region of the viral genome (for details see reference 1). The main component, the large T-antigen of 88 kd (SV40) and 100 kd (polyoma) accumulates in the nucleus of infected and transformed cells.

The presence of intranuclear T-antigen can easily be detected by immunofluorescence. This provides an assay for the proportion of infected cells in a culture, which reflects the infective titre of the virus stock. This assay has also found application in transfection experiments in which the viral early gene regions have been joined to foreign or mutated promoter sequences. The expression of T-antigens is then monitored as a test for the functioning of the inserted promoter. Fixed cells are first reacted with a virus-specific anti-tumour serum. This is normally obtained from hamsters, rats or mice carrying tumours that have been induced either by inoculation with virus or by injection of syngeneic virus-transformed cells. Instead of an anti-tumour serum, monoclonal antibodies against large T-antigens can also be used. In a second immune reaction, cells are stained with a fluorescein-conjugated antiserum directed against the immunoglobulins of the anti-tumour serum. In a fluorescence microscope T-antigen-containing nuclei are easily distinguished from nuclei of uninfected cells (see *Figure 7*).

The details of the assay are as follows:

(i) About 24 h after infection, wash the cultures grown on glass coverslips twice with cold phosphate-buffered saline (PBS). Let the monolayers dry in air for about 10 min, then fix the cells by immersing the coverslip into a mixture of methanol/acetone 3:7 (v/v) cooled to −20°C and keep for 10 min at −20°C. Let the coverslips dry in air. Coverslips with fixed cells can be kept for several weeks at −20°C and can still be used for the immunofluorescence assay.

(ii) Cut the coverslip with a diamond; a piece of 0.5−1 cm^2 is sufficient for the assay. Wet the cells by dipping the coverslip into PBS and then place it into a 3 cm Petri dish containing a piece of damp filter paper. Be sure that the cell layer is on the upper side. Add 10 μl of anti-tumour serum or monoclonal antibody. Depending on its quality this can be diluted 5- to 100-fold in PBS. Incubate for 30 min at room temperature, then wash three times with PBS, allowing 2−3 min for each wash.

(iii) Add 10 μl of fluorescein-conjugated anti-immunoglobulin serum (e.g., rabbit anti-hamster IgG serum) diluted 10- to 100-fold in PBS containing 5% FCS. Such sera are commercially available and have to be tested to find the optimal dilution for good fluorescent staining with little background fluorescence. Incubate again for 30 min at room temperature, then wash three times with PBS. Put a small drop of glycerol/PBS 9:1 (v/v) on a

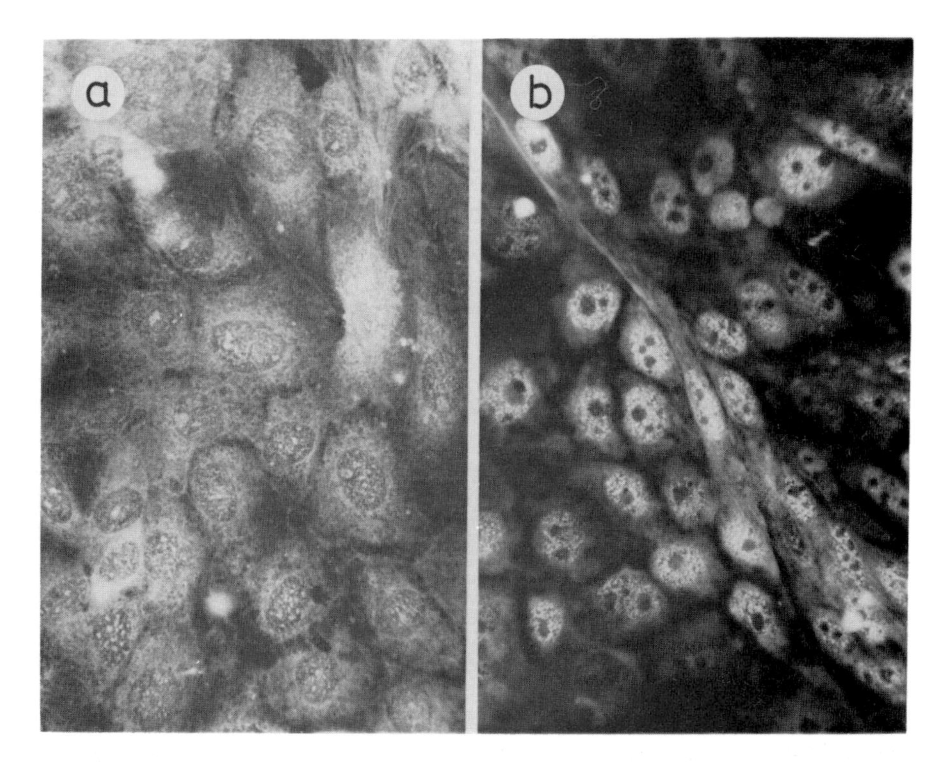

Figure 7. Immunofluorescent staining of polyoma T-antigens. **a:** Mock-infected, **b:** infected cultures of primary mouse kidney cells.

clean, alcohol-washed microscope slide. Let the coverslip drain, place it for a moment on a filter paper to dry its lower side, then mount it with the cell layer facing down (into the glycerol/PBS) on the microscope slide. In this way the preparations may be kept for a few days at 4°C if protected from light. If you are going to observe your preparation immediately under the microscope you may let it dry in the air and then mount the coverslip cell layer up on a slide.

(iv) Observe the coverslip under a dark-field u.v.-illuminated microscope. Cells containing T-antigen show yellow-green fluorescent nuclei with dark nucleoli (see *Figure 7*). Later in infection, i.e., 36–48 h after infection when progeny virus assembly takes place, the T-antigens are no longer evenly distributed in the nuclei, but are present in patches. Compare your infected culture with an equally treated uninfected control culture. Only score as positive cells where the stained nucleus is clearly seen. Rounded or dead cells usually give a higher background fluorescence.

7. TRANSFORMATION ASSAYS

Cells transformed by SV40 or polyoma virus are distinguished by a set of properties, not all of which, however, are manifest in all transformed cell lines. Transformed cell properties include growth in suspension, i.e., anchorage-independence, formation of disoriented, dense colonies on plastic, decreased serum requirement and high saturation density (19). Here we use the first two of these properties to identify and isolate clones of transformed cells.

7.1 Infection of Non-permissive Cells

(i) Prepare three cultures of non-permissive cells, one of them containing one or two glass coverslips.

(ii) Infect the coverslip-containing culture and one of the other cultures with your virus stock suspension. The other culture is mock-infected with the same volume of medium containing 5% FCS. After the adsorption period (90 min at 37°C) the cultures are covered with 10 ml of medium containing 5% FCS. (The transformation efficiency is usually higher for growing than for resting cells.)

(iii) On day 1 after infection, fix the coverslip cultures which will be used to determine the proportion of infected cells by immunofluorescent staining of intranuclear T-antigen (see Section 6.4). Trypsinise the infected and mock-infected cultures separately as follows.

- (a) Wash the culture with TD and rinse it with 2.5 ml trypsin/versene (see *Table 2*) leaving 0.1 – 0.2 ml of this solution on the culture.
- (b) Leave the dish at room temperature and wait until the cells become rounded, which normally takes 2 – 5 min.
- (c) Detach the cells by a sharp tap on the side of the dish.
- (d) Add 5 ml of medium containing 5% FCS and re-suspend the cells by pipetting.
- (e) Determine the number of cells per ml in the two suspensions.

7.2 Colony Formation on Plastic in Liquid Medium

(i) Make serial dilutions 10^{-1}, 10^{-2}, 10^{-3}, 10^{-4} by pipetting 2.5 ml of the cell suspension into 22.5 ml of medium containing 5% FCS. With each dilution seed four dishes (5 cm diameter) using 5 ml of your cell suspensions and put the dishes into the incubator. If your starting culture contained 5×10^6 cells you have plated 5×10^5, 5×10^4, 5×10^3 and 5×10^2 cells. With a higher starting cell density include an additional dilution step, so that your highest dilution has not more than 50 – 100 cells/ml.

(ii) Change the medium twice per week using separate pipettes for your mock-infected and infected cells. Observe the dishes for formation of dense colonies which will appear after about 2 weeks.

(iii) Two to four weeks after plating, reserve one or two cultures for isolation of individual transformed colonies. The remaining dishes are used for colony counting.

- (a) Aspirate the medium, wash the cells with PBS, then fix with 5 ml of 10% formol in PBS for 30 min at room temperature.
- (b) Aspirate the solution and let the dishes dry.
- (c) Colonies can be counted either directly or after staining with Giemsa or haematoxylin.

(d) Determine the transformation efficiency which is the number of transformed colonies per 100 seeded, infected cells (use the value for the proportion of infected cells determined by T-antigen staining). No dense colonies should be formed in the mock-infected cultures.

(iv) To isolate individual colonies of transformed cells use cloning cylinders, prepared from stainless steel or glass tubing having an inner diameter of 5 mm. The cylinders should be 10 mm high. Put some vacuum grease on both ends of the cylinders then autoclave them in a glass Petri dish. Aspirate the medium from the cultures reserved for this purpose and wash the cells with 5 ml of TD pre-warmed to 37°C. Remove the buffer as completely as possible. Using flamed forceps, place one of your cylinders over a well-isolated colony. The cylinder will stick to the plate by means of the vacuum grease. Put one drop of trypsin/versene into the cylinder, cover the dish and wait a few minutes; you can observe the detachment of the cells in the microscope. Then add three drops of medium containing 5% FCS, re-suspend the cells by pipetting with a Pasteur pipette and transfer the cell suspension into a 3 cm dish containing 2 ml medium with 5% FCS. Proceed in the same way with two or three other colonies.

Once the cells are growing, test the transformed cell lines for the presence of viral T-antigens by immunofluorescent staining.

7.3 **Colony Formation in Semi-solid Agar** (20)

Dissolve 1.2% (w/v) agar (Difco Bacto-agar or agarose) in water by autoclaving and store in 50 ml batches at 4°C.

7.3.1 *Preparation of Agar Underlay*

In order to prepare the agar underlay (0.6% agar) the following steps should be carried out.

(i) Melt two batches of 50 ml of 1.2% agar, cool to 45°C, add to each 40 ml of 2x concentrated medium pre-warmed to 45°C and 10 ml of FCS. Keep the mixes at 45°C.

(ii) Pipette 5 ml into each of 20 dishes (5 cm diameter). Try to avoid formation of bubbles; if there are any destroy them immediately by touching with a flamed, hot Pasteur pipette.

(iii) Mark eight dishes 10^{-1} and four dishes each 10^{-2}, 10^{-3}, 10^{-4}. Let the agar solidify at room temperature.

7.3.2 *Preparation of Semi-solid Agar Overlay*

To prepare the semi-solid agar overlay (0.3% agar or agarose), carry out the following procedure.

(i) Warm five sterile tubes containing 4 ml of medium with 10% FCS in the 45°C water bath.

(ii) Add 5 ml of your 0.6% bottom agar mix. Mark two tubes 10^{-1} and the others 10^{-2}, 10^{-3} and 10^{-4}.

(iii) From the undiluted suspension of infected cells (see above) and from the 10^{-1}, 10^{-2} and 10^{-3} dilutions, add 1 ml into the corresponding tubes.
(iv) As soon as you have added the cells mix well and pipette 2 ml on each of four agar bottom dishes for each dilution.
(v) Proceed in the same way with the suspension of cells from the mock-infected culture. However, since no colonies should grow in the semi-solid agar, plate only the 10^{-1} dilution. Let them solidify at room temperature for 15 min, then place them in the incubator.
(vi) After 5 days incubation, add to each culture 2 ml of semi-solid agar (0.3% agar in medium containing 10% FCS, prepared as described above).
(vii) After 10 days add to each culture 1 ml medium containing 10% FCS; replace this medium every 5 days. Transformed colonies should become visible by eye after incubation for 2–3 weeks, but sometimes prolonged incubation exceeding 4 weeks is necessary.
(viii) Let the colonies grow to 0.5–1 mm diameter. Count the colonies and determine the transformation efficiency.

7.3.3 *Isolation of Clones of Transformed Cells*

(i) Using a sterile Pasteur pipette, aspirate a well-isolated clone with the surrounding agar.
(ii) Transfer it into a 3 cm dish containing 2 ml TD (37°C).
(iii) Remove most of the agar from the colony with the Pasteur pipette or a sterile pin.
(iv) Wash the colony once more with 2 ml TD.
(v) Add 0.1 ml of trypsin/versene and disrupt the colony mechanically with the Pasteur pipette and the pin. Complete disruption is not necessary since the small clumps will adhere to the plate and transformed cells will grow out.
(vi) Add 0.5 ml of medium containing 10% FCS, re-suspend the cells and the clumps by pipetting.
(vii) Add 1.5 ml of medium containing 10% FCS and put the dish into the incubator. Once the cell line is established test it for intranuclear viral T-antigens.

8. ISOLATION OF VIRAL DNA AND OF VIRAL 'MINICHROMOSOMES' FROM INFECTED CELL CULTURES

8.1 Extraction of Viral DNA

Most methods for extracting the DNA of SV40 and polyoma virus begin with the selective extraction of small DNAs described by Hirt (21). This procedure makes use of the fact that when an infected cell lysate made with the detergent SDS is adjusted to 1 M NaCl and cooled, the SDS co-precipitates high molecular weight cellular DNA, but leaves the small viral DNA in the supernatant. The supernatant still contains contaminating low molecular weight cell DNA, RNA and proteins which are removed in subsequent purification steps. Covalently-closed circular DNA can be isolated by equilibrium centrifugation in a CsCl gradient containing ethidium bromide (22), a technique that is widely known and used (6,23). Here we describe an alternative which is quicker, cheaper and works as well for most purposes. It is based on the fact that covalently-closed circular DNA under appropriate conditions re-anneals spontaneously after heat denaturation, whereas nicked circular and linear DNA (e.g., cell DNA fragments) do not. Single-stranded DNA can then be separated from double-stranded DNA by extraction with phenol in the presence of 0.5 M NaCl (24). Under these conditions single-stranded DNA is sequestered at the interphase. To obtain maximal amounts of viral DNA, infected cultures are lysed shortly before extensive disintegration of the monolayer occurs. In our experience this is at around 40 h after infection for polyoma-infected mouse kidney cells, around 50 h after infection for polyoma-infected 3T6 cells and around 55 – 60 h after infection for SV40-infected CV-1 cells. The procedure given here is suitable for 10 cultures (9 cm dishes).

(i) Remove the medium, wash the cultures with 5 ml PBS, remove the buffer as completely as possible and add to each dish 1 ml of 0.6% SDS, 10 mM Tris-HCl, 10 mM EDTA pH 7.5. Keep at room temperature for 15 min, then scrape the lysate with a silicone rubber and pour it into a centrifuge tube. It is important not to shear the DNA, therefore do not use pipettes.

(ii) Add to the centrifuge tube 1/4 volume of 5 M NaCl, mix gently by inverting the stoppered tube 10 times and put it in the cold (4°C) for at least 6 h, preferably overnight.

(iii) Centrifuge at 25 000 *g* for 15 min. Recover the supernatant containing the viral DNA. If the gelatinous pellet is large (1/4 volume or more) it may be advantageous to re-extract it with 1 M NaCl overnight and to carry out a second centrifugation. If this is the case, use a larger volume of SDS to lyse the cultures (e.g., 1.6 ml per Petri dish).

(iv) Deproteinise the extracts with an equal volume of phenol/chloroform 1:1 (v/v) saturated with 0.1 M Tris-HCl pH 8. Mix well for 3 min, centrifuge at 10 000 *g* for 10 min and recover the upper, aqueous phase. Extract once more with a mixture of chloroform/iso-amyl alcohol 24:1 (v/v) and centrifuge as above.

(v) Add 2.5 volumes of ethanol to the aqueous phase and leave at −20°C overnight to precipitate the nucleic acids. Centrifuge at 10 000 *g* for 10 min and dry the pellet under vacuum. (At this stage the viral DNA can be further purified by CsCl-ethidium bromide density gradient centrifugation.)

(iv) Dissolve the DNA in 0.5 ml 50 mM Tris-HCl, 5 mM EDTA, pH 8. Add RNase A (which has been heated in 10 mM sodium-acetate pH 5.2 to 80°C for 10 min to destroy contaminating DNase activity) to a final concentration of 50 μg/ml and incubate at 37°C for 1 h. Heat the tube in a boiling water bath for 5 min, cool immediately in ice and add 5 M NaCl to obtain a final concentration of 0.5 M. Extract the solution with an equal volume of phenol saturated with 0.5 M NaCl for 3 min. Centrifuge at 10 000 *g* for 10 min, remove the upper, aqueous phase, carefully avoiding the interphase. Extract the aqueous phase once more with an equal volume of phenol saturated with 0.5 M NaCl. At this step the denatured linear and nicked circular DNA, as well as the RNase, are removed.

(vii) Precipitate the DNA with 2.5 volumes of ethanol and leave overnight at −20°C. Centrifuge at 10 000 *g* for 10 min, rinse the pellet with 70% ethanol in water kept at −20°C and re-centrifuge. Dry the pellet under vacuum and dissolve it in an appropriate volume of 10 mM Tris-HCl, pH 7.8, 1 mM EDTA. Determine the optical density at 260 nm as a measure of DNA concentration (1 OD_{260} = 50 μg/ml). From SV40-infected CV-1 cultures you can expect 10−25 μg viral DNA per 9 cm Petri dish. From polyoma-infected mouse kidney or 3T6 cultures the yield is about 5 μg per culture.

8.2 Isolation of SV40 Chromosomes

The DNA of SV40 and polyoma virus is associated with cellular histones, both in virus particles and in infected cells. The viral chromosomes have been used as simpler models of the much more complex cellular chromosomes in structural and mechanistic studies on replication and transcription.

Many methods have been described for the isolation of SV40 chromosomes from infected cells. Here we describe two, which work well for us. One of them, using a buffer of low ionic strength (HBE, see *Table 2*) has been employed for studies on *in vitro* transcription of SV40 chromosomes (25). The other uses isotonic buffers and the non-ionic detergent Nonidet P-40 (NP-40) (26). It has been employed to study the structure of the SV40 chromosome and of viral assembly intermediates. A potential problem in isolating SV40 chromosomes from infected cells is that under some extraction conditions the viral assembly intermediates and virions can dissociate to reform viral chromosomes. This does not seem to occur during the isotonic extraction, which is fairly rapid. With the 'HBE' method, where dissociation and uncoating can occur, the problem is avoided by extracting viral chromosomes at a time before large amounts of assembly intermediates are formed, i.e., at about 30 h after infection.

8.2.1 *Isolation of SV40 Chromosomes with Isotonic Buffer*

The method described here is for two 9 cm Petri dish cultures, but can be scaled up.

(i) Remove the medium, wash the cells with TD, then scrape them with a silicone rubber into 2 ml TD. Centrifuge at 1000 *g* for 5 min. Re-suspend the pellet in 2 ml TD.

(ii) Add 100 μl of 10% NP-40 (Sigma Chemical Co.) while mixing gently on a vortex. This lyses the cell membrane. Centrifuge at 1200 *g* for 5 min to pellet the nuclei, remove the supernatant and re-suspend the pellet in 0.5 ml TD. Homogenise fairly vigorously with a Dounce homogeniser using 30−40 strokes. Much of the viral chromosomal material leaves the nuclei at this step. Pellet the nuclei at 1200 *g* for 5 min and recover the supernatant, containing the viral chromosomes.

(iii) Purify the viral chromosomes by velocity centrifugation through a 5−20% (w/v) sucrose gradient in TD at 4°C. Using the Beckman SW60 rotor, which gives good resolution and uses small volumes, the centrifugation time is about 60–80 min at 40 000 r.p.m. For larger volumes of sample the Beckman SW40 can be used with correspondingly adjusted centrifugation times. The fractions containing the SV40 chromosomes can be detected by u.v. light absorption at 260 nm, by agarose gel electrophoresis of small, SDS-treated aliquots or by isotopic labelling.

8.2.2 *Isolation of SV40 Chromosomes at Low Ionic Strength*

The method described is for 10 Petri dish cultures (9 cm) of SV40 infected CV-1 cells.

(i) At 30 h after infection remove the medium and rinse the monolayers with 5 ml of ice-cold HBE (see *Table 2*). Repeat the rinsing twice while keeping all the dishes on ice. Remove the HBE by suction, then scrape the cells with a silicone rubber and transfer them to a Dounce homogeniser kept in ice.

(ii) Break the cells with 10 strokes of the homogeniser. Transfer the suspension to a plastic tube and centrifuge at 1000 *g* for 2 min to pellet nuclei. Remove the supernatant carefully with a pipette as the pellet is not firm, and resuspend the nuclei in HBE. The volume of HBE can vary, but for transcription studies the nuclei should remain concentrated so the pellet is resuspended in 150 μl HBE.

(iii) Transfer the suspension to an Eppendorf tube using a wide bore pipette and place the tube on ice. The nuclei swell and the viral chromosomes diffuse out. The time required for efficient recovery of the SV40 chromosomes must be determined empirically. We find that at least 4 h are needed, sometimes we leave overnight. However, others have found with different cell lines that 1 h may be sufficient.

(iv) Centrifuge at 25 000 *g* for 15 min to pellet the lysed nuclear material. Recover the supernatant which contains the SV40 chromosomes and transfer it into another Eppendorf tube. This crude preparation of SV40 chromosomes can be stored for a few days on ice, or may be further purified by velocity centrifugation through a 5–20% sucrose gradient as described above.

9. ACKNOWLEDGEMENTS

We acknowledge the expert help of Béatrice Bentele and Consuelo Salomon for establishing and testing these methods in our laboratories. We thank R.Weil for his permission to reproduce *Figures 4* and *7* that had been made earlier in his laboratory, M.Bensemmane for preparing cell cultures, Monique Visini for typing the manuscript and O.Jenni for drawings and photographs.

10. REFERENCES

1. Tooze,J. (ed.), (1980) *DNA Tumour Viruses,* published by Cold Spring Harbor Laboratory Press, New York.
2. Freshney,R.I., (ed.), (1985) *Animal Cell Culture. A Practical Approach,* IRL Press, Oxford and Washington, D.C., in press.
3. McGarrity,G.J., Murphy,D.G. and Nichols,W.W. (1978) *Mycoplasma Infection of Cell Cultures,* published by Plenum Press, New York and London.

4. Winocour,E. (1963) *Virology,* **19**, 158.
5. Weil,R. (1978) *Biochim. Biophys. Acta,* **516**, 301.
6. Favaloro,J., Treisman,R. and Kamen,R. (1980) in *Methods in Enzymology,* Vol. **65**, Grossman,L. and Moldave,K. (eds.), Academic Press, New York, p. 718.
7. Seif,R. and Cuzin,F. (1977) *J. Virol.,* **24**, 721.
8. Griffin,B.E. (1980) in *DNA Tumour Viruses,* Tooze,J. (ed.), Cold Spring Harbor Laboratory Press, New York, p. 61.
9. Vinograd,J. and Hearst,J. (1964) in *Progress in the Chemistry of Organic Natural Products,* Vol. **20**, Springer, Vienna, p. 372.
10. Thomas,C.A., Jr. and Berns,K.I. (1961) *J. Mol. Biol.,* **3**, 277.
11. Crawford,L. (1969) in *Fundamental Techniques in Virology,* Habel,K. and Salzman,N.P. (eds.), Academic Press, New York and London, p. 75.
12. Aposhian,H.V. (1975) in *Comprehensive Virology,* Vol. **5**, Fraenkel-Conrat,H. and Wagner,R.R. (eds.), Plenum Press, New York and London, p. 155.
13. Fried,M. (1974) *J. Virol.,* **13**, 939.
14. Dulbecco,R. (1966) in *Phage and the Origins of Molecular Biology,* Cairns,J., Stent,G.S. and Watson,J.D. (eds.), Cold Spring Harbor Laboratory Press, New York, p. 287.
15. Consigli,R.A. Zabielski,J. and Weil,R. (1973) *Appl. Microbiol.,* **26**, 627.
16. Eddy,B.E., Rowe,W.P., Hartley,J.W., Stewart,S.E. and Huebner,R.J. (1958) *Virology,* **6**, 290.
17. Sealey,P.G. and Southern,E.M. (1982) in *Gel Electrophoresis of Nucleic Acids. A Practical Approach,* Rickwood,D. and Hames,B.D. (eds.), IRL Press, Oxford and Washington, D.C., p. 39.
18. Southern,E.M. (1979) in *Methods in Enzymology,* Vol. **68**, Wu,R. (ed.), Academic Press, New York, p. 152.
19. Topp,W.C., Lane,D. and Pollack,R. (1980) in *DNA Tumour Viruses,* Tooze,J. (ed.), Cold Spring Harbor Laboratory Press, New York, p. 205.
20. MacPherson,I. and Montagnier,L. (1964) *Virology,* **23**, 291.
21. Hirt,B. (1967) *J. Mol. Biol.,* **26**, 365.
22. Radloff,R., Bauer,W. and Vinograd,J. (1967) *Proc. Natl. Acad. Sci. USA,* **57**, 1514.
23. Pagano,J.S. and Hutchison,C.A. (1971) in *Methods in Virology,* Vol. **5**, Maramorosch,K. and Koprowski,H. (eds.), Academic Press, New York and London, p. 79.
24. McMaster,G., Beard,P., Engers,H.D. and Hirt,B. (1981) *J. Virol.,* **38**, 317.
25. Beard,P. and Nyfeler,K. (1982) *EMBO J.,* **1**, 9.
26. Fernandez-Munoz,R., Coca-Prados,M. and Hsu,M.T. (1979) *J. Virol.,* **29**, 612.

CHAPTER 9

Growth, Purification and Titration of Adenoviruses

BERNARD PRECIOUS and WILLIAM C. RUSSELL

1. INTRODUCTION

Adenoviruses are medium sized non-enveloped viruses containing a linear double-stranded DNA of 20 – 30 x 10^6 molecular weight. They have characteristic particles with icosahedral symmetry, 70 – 90 nm diameter with 252 capsomeres, 12 of these, at the vertices, carrying filamentous projections (*Figure 1*). Currently there are two genera in the family *Adenoviridae*, termed *Mastadenovirus* and *Aviadenovirus*. All the members within each genus are related *via* a common group-specific antigen. Adenoviruses have been isolated from man and a variety of animals and have been further subdivided into subgroups and types (species) based on quantitative neutralisation with appropriate animal antisera. *Table 1* lists the adenoviruses currently recognised in the second report of the adenovirus study group (1). Adenoviruses in humans when associated with disease are mainly associated with respiratory and ocular conditions. The majority, however, have

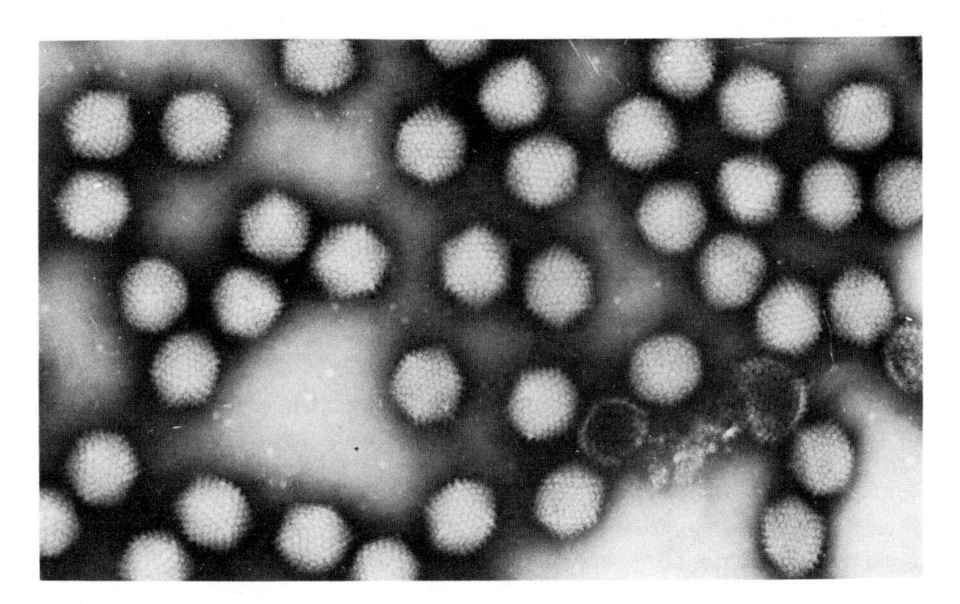

Figure 1. Electron micrograph of purified human adenovirus type 5. Freeze dried; negative staining with sodium silicotungstate (4%, pH 6.5). Magnification x150 000. Kindly supplied by Dr. M.V. Nermut, N.I.M.R., Mill Hill, London.

Table 1. Adenovirus Species Found in Man and Animals

Genus	*Host*	*Number of members*[a]
Mastadenovirus	human	37
	simian	24
	cattle	9
	pig	4
	sheep	5
	horse	1
	dog	1
	mouse	1
	tree shrew	1
Aviadenovirus	fowl	9
	turkey	2

[a]The numbers only refer to definite members. About 20 other viruses covering a wide range of hosts are candidates for inclusion in both genera.

not been implicated directly in disease. Some adenoviruses have also been found in association with diarrhoea in fecal samples.

This group of viruses has served as a useful model system for studying many aspects of eukaryotic biochemistry and genetics and there are many studies in the literature which utilise human adenovirus types 2 and 5 for this reason. These types grow readily in tissue culture yielding good crops of virus and related excess structural antigens. Human types 7 and 12 have also been extensively studied since these types possess oncogenic properties, producing tumours with varying degrees of efficiency in baby hamsters.

A number of review articles have been published which deal with various aspects of the biochemistry and virology of adenovirus (2,3).

2. CELL SYSTEMS FOR GROWTH OF ADENOVIRUSES

In general, adenoviruses grow best in epithelial cells of the host species or closely related species. The nature of the cells does seem to play an important role in determining the yield of virus. Thus, human adenoviruses types 2 and 5 replicate very efficiently in human epithelial cell lines such as HeLa or KB and relatively poorly in human fibroblast cells such as WI-38 or MRC-5. The oncogenic human adenoviruses are somewhat more fastidious in their requirements and the best virus yields require the use of primary or secondary human embryonic kidney cells. The differences in cell susceptibility to infection do not appear to reside in the availability of cell receptors since most cells will take up the viruses, but these are processed to varying degrees giving either the fully lytic or an abortive infection. In the latter situation virus DNA and structural components may be produced in the absence of significant assembly but in some cases only a limited number of early viral genes may be expressed and the cells' morphology and function modulated giving rise in the appropriate conditions to the phenomenon of transformation. In the case of lytic infection and many abortive infections the cells eventually die. The cellular functions which participate in different stages of replication and assembly of virus have not yet been defined in such a way as to allow the cells to be manipulated to give increased growth of virus (except perhaps in the case of 293 cells – see below).

2.1 Non-fastidious Human Adenoviruses

As indicated above, adenovirus types 2 and 5 are the most widely studied viruses of the group and they will grow readily in HeLa or KB cells cultivated as monolayers or in suspension in plastic or glass containers. The latter method is the more convenient, especially for the production of large amounts of virus, and the technique used is therefore described in more detail below.

2.1.1 *Suspension Culture Apparatus*

A magnetic stirring apparatus similar to that shown in *Figure 2* can be readily constructed to suit a variety of sizes and shapes of flask. The motor, preferably with an adjustable rheostat, should give a magnet speed of approximately 150 r.p.m. and since it could be operating for long periods at 37°C it should be of the 'tropicalised' variety. (An alternative to placing the apparatus in a warm room or incubator could be to utilise a stirring platform which can be immersed in a water bath – these are now commercially available.) The cells can be propagated in a variety of containers. Initially 500 ml 'blood bottles' can be used,

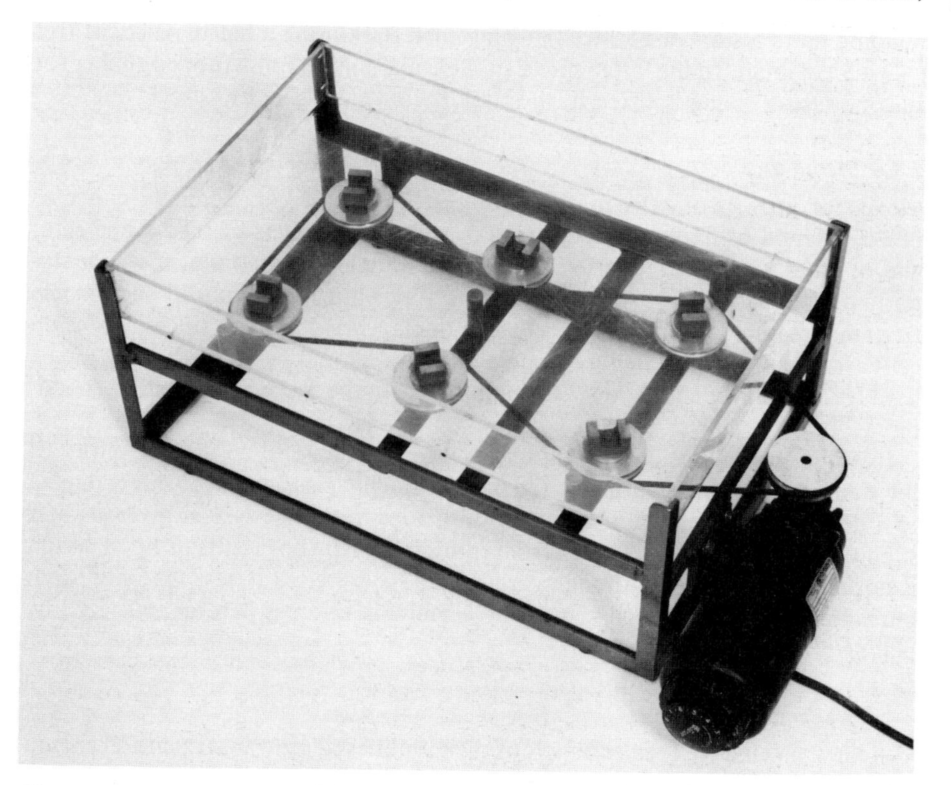

Figure 2. Simple apparatus for stirring six suspension cultures. Constructed from mild steel and perspex (dimension ~600 x 400 x 60 cm).

Table 2. Some Suppliers of More Specialised Items.

(i)	Media and Sera	
	Gibco (Europe) Ltd.	P.O. Box 35, Renfrew Road, Paisley PA3 4EF, UK
	Flow Laboratories Ltd.	P.O. Box 17, 2nd Avenue, Irvine KA12 8NB, UK
	Imperial Laboratories (Serum)	Ashley Road, Salisbury, Wiltshire SR2 7DD, UK
(ii)	Mycoplasma Screening	
	'Mycospec', B.R.L. (U.K.) Ltd.	P.O. Box 145, Science Park, Cambridge CB4 3BE, UK
(iii)	Water Purification	
	'Fistream' Fisons plc	Bishops Meadow Road, Loughborough LE11 0RG, UK
	'Vaponics'	20 Park Street, Princes Risborough NP17 9AN, UK
	The Elga Group	Lane End, High Wycombe NP14 3JN, UK
	Millipore Ltd.	Millipore House, 11-15 Peterborough Road, Harrow HA1 2YH, UK
(iv)	Peristaltic Pumps	
	LKB Instruments Ltd.	LKB House, 232 Addington Road, S. Croydon CR2 9APX, UK

building up to a standard 5 litre flat-bottomed flask using a bar magnetic stirrer (~3–5 cm long). In the latter case a suitable aluminium foil-wrapped rubber (or silicone rubber) bung is fitted to the flask before sterilising by autoclaving.

2.1.2 *Media and Sera*

Media for growth of cells in suspension are available commercially either as powder, liquid or liquid concentrate. For a list of suppliers of media and sera, and of other specialised items, see *Table 2*. It should be noted that there are important differences (e.g., in Ca^{2+} concentration) in composition of such media from that required for growth of cells as monolayers. Moreover, since suspension cells are initially propagated as monolayers, care should be taken that the media used for monolayer culture are as compatible with the suspension media as possible.

It is important, particularly with suspension cultures, to ensure the purity of the water used for preparation of the media. Usually glass-distilled water coupled to a deioniser or a filtration/reverse osmosis ion-exchange system gives a water supply of adequate purity. Calf serum (preferably new born calf) or horse serum is normally satisfactory although care should be exercised to test that any batch purchased will promote cell growth by examining samples beforehand, e.g., by plating efficiency.

2.1.3 *Cells*

Cell lines which have been selected for their abilities to grow in suspension should be used, e.g., HeLa S3 (ATCC No. CCL 2.2) or KB (ATCC No. CCL 17). The cells are normally obtained initially as monolayer cultures and should be divided and cultured in growth medium (for monolayers) containing 7.5% newborn calf serum (NCS) using standard procedures until there are sufficient cells to suspend in 200 ml of suspension medium containing 7.5% NCS at a concentration of 3 x 10^5/ml. (Note that these cells do not readily form good monolayers.) Place the suspended cells in a 500 ml serum bottle or equivalent and stir on the apparatus at 37°C for about 2 days and then count the cells in a haemocytometer. When the cell density reaches a concentration of 6 x 10^5/ml dilute 1:1 with fresh medium and calf serum. A healthy culture will have a doubling time of approximately 42–48 h. Once the culture is established with 1–2 litres of cell suspension, transfer to larger flasks can be carried out. Optimal conditions for growth of cells are best obtained by leaving about half of the volume of the container for the gas phase and ensuring adequate pH control by loosening the cap or bung while retaining sterility.

Suspension cells appear to be relatively sensitive to temperature fluctuations and particular care should be taken to ensure that the temperature does not rise above 38.5°C. Since suspension cultures take time to be established, precautions should be taken to guard against mechanical breakdown of the stirring apparatus possibly by having duplicate cultures on a separate apparatus and certainly by having an adequate supply of frozen cell stock.

2.1.4 *Screening for Mycoplasmas*

Cells should be checked on a regular basis for mycoplasmas as virus production is significantly impaired in contaminated cultures (4). A quick and effective procedure is to utilise the fluorescent staining technique (5).

(i) Grow the cells as monolayers on coverslips, wash once with a small volume of phosphate-buffered saline (PBS).

(ii) Incubate at 37°C in a humidified atmosphere (e.g., in a plastic sandwich box with damp paper tissue) for 30 min in a small volume of PBS containing 4′,6-diamidino-2-phenylindole (DAPI) at a concentration of 0.1 μg/ml. (The benzimidole derivative Hoechst 33258 is also effective.)

(iii) Wash the coverslip once and mount (unfixed) on a microscope slide and examine in a fluorescence microscope. (The pale blue fluorescence has an excitation frequency of ~365 mm and emission principally at ~450 nm).

The dye is taken up by DNA and stains the nucleus, whereas unbound dye does not fluoresce. Mycoplasmas can be visualised as characteristic fluorescent clusters in the cytoplasm and on the plasma membranes of contaminated cells. Standard PPLO agar cultures can also be used:

PPLO agar (Difco) 35 g/litre
2% horse serum (mycoplasma screened)
10% yeast extract (freshly extracted – see reference 6)

Stab the surface of the agar with a Pasteur pipette transferring a drop of cell suspension to the agar and incubate anaerobically at 37°C for 3–5 days. Characteristic 'fried-egg' colonies indicative of mycoplasmas can be seen in the vicinity of the stabs. In some cases incubation for a further 5 days may be necessary to produce recognisable colonies.

Other techniques employing monoclonal antibodies are now available commercially; these can also be used to type the mycoplasmas.

2.1.5 *Virus Infection*

(i) Pellet the suspension cells by centrifugation (5 min, 500 *g*), remove the medium and suspend the cells in 1/20th of the original volume of serum-free medium.
(ii) Add virus seed to an added multiplicity of 1–10 p.f.u./cell and incubate the re-suspended cells with gentle stirring at 37°C for 1 h.
(iii) Further dilute the cells to 3–4 x 10^5 cells/ml with warm medium containing 2% NCS and incubate for 48–60 h at 37°C.
(iv) For monolayers of cells, infect in a small volume of serum-free medium after removal of supernatant at an added multiplicity of 1–10 p.f.u./cell.
(v) After 1 h at 37°C add further medium containing 0.5% NCS (20 ml in the case of a 20 oz medical flat) and incubate the infected monolayers at 37°C until a cytopathic effect (cpe) can be seen.

In the case of adenoviruses types 2 and 5 incubation for 2–3 days will cause most cells to round up and detach from the glass. Other virus types may require longer incubation for maximum virus production.

2.1.6 *Virus Seeds*

When generating adenovirus, care should be taken to minimise defective virus production by multiple sequential tissue culture passage. To maintain reasonable homogeneity of virus populations it is preferable to produce virus stocks from a single large stock of 'seed' virus which can be prepared from a 'seed-seed' stock. This latter virus preparation should be propagated directly from plaque-purified virus and is conveniently stored in small aliquots at –70°C. Seed virus stocks are normally stored as fluorocarbon extracts (see below) with titres (in the case of adenovirus type 2 and 5) in the range 2–5 x 10^9 p.f.u./ml.

2.2 **Fastidious Human Viruses**

Not all human adenoviruses are readily propagated in HeLa or KB cells and in some cases (e.g., adenovirus type 12) much better yields can be obtained by infection of primary or secondary human embryo kidney cells. Some of the human enteric adenoviruses cannot be propagated in any of these tissue cells and use must be made of 293 cells – a human cell line transformed by adenovirus type 5 (7). Techniques for preparing these two cell cultures are described below.

2.2.1 *Human Embryo Kidney Cell Cultures*

The kidneys of human embryos obtained from therapeutic abortions and removed asceptically, are required.

(i) Decapsulate the tissue in a sterile Petri dish, finely chop with a scalpel and add to 10 ml of medium containing 1.25 units/ml of dispase (a neutral protease) in a small conical flask.

(ii) Place a small Teflon magnet in the flask and gently stir the tissue for about 20 min at 37°C.
(iii) Remove any large pieces of tissue that remain by passing through a sterile muslin filter and centrifuge the filtrate which contains mainly single cells (500 *g*, 5 min).
(iv) Gently wash the cell pellet in 10 ml of growth medium [containing 5% foetal calf serum (FCS)] to remove excess protease. Disperse the cells either on glass or plastic bottles or into 80 oz roller bottles at approximately 1×10^5 cells/ml of growth medium (200 ml in the roller bottles).
(v) At confluence, the cells are either subcultured to provide secondary cultures or are infected with a small volume of virus at a multiplicity of approximately 0.1 – 1 p.f.u./cell.

Cytopathic effects should be apparent after 2 – 3 days incubation at 37°C and the cells harvested and extracted as described below.

2.2.2 *Propagation of 293 Cells*

These cells, derived originally from human embryo kidney cells transfected with sonicated adenovirus type 5 DNA, have been shown to allow good propagation of enteric adenoviruses (which do not replicate in other human cell lines) (8). The cells are grown as monolayers using standard growth medium containing 10% NCS and are readily divided using standard trypsin/versene procedures, splitting 1→3. They do not form good monolayers and generally are very heterogeneous in morphology with many characteristic features of transformed cell lines. If the cell monolayers are required for plaque production (see below) improved monolayers can often be obtained by using FCS. (Plastic surfaces are normally preferable for good growth of these cells.)

2.3 **Viruses of Other Species**

Primary or secondary cultures of embryonic or kidney cells from the appropriate species provide the general means of cultivating the virus (for references see reference 2).

2.3.1 *Simian Adenoviruses*

Primary or secondary cultures of kidney cells of the green monkey, *Cercopithecus aethiops*, appear to be the most satisfactory cells although there have been reports of satisfactory yields after propagation in human cell lines (e.g., Hep-2).

2.3.2 *Murine Adenoviruses*

Mouse embryo tissue culture has been used generally but there have been accounts of the successful propagation of mouse adenovirus in 3T3 cells (a continuous line of mouse epithelial cells) (9).

2.3.3 *Bovine Adenoviruses*

Bovine kidney and testis cell lines have been successfully used.

2.3.4 *Avian Adenoviruses*

Tissue cultures of chick liver, kidney, spleen have proved useful in propagation of these viruses (10).

3. PURIFICATION

Adenoviruses mature in the nucleus of the infected cells and do not have any specific mechanism (other than cell disintegration) for the release of virus into the cytoplasm and the tissue culture medium. Thus most of the virus (90%) is normally cell associated and efficient release of the virus from nuclear and cellular material depends on disruption of the nuclear membrane. The latter is normally very fragile in the infected cell and virus release into the cytoplasm is quite common at later stages of infection. Procedures for the disruption of cells are many and the best technique depends to a great extent on the number and nature of the cells to be extracted. The following procedure has been used successfully for the purification of adenovirus type 5 (11).

3.1 **Extraction of Infected Cells**

(i) Pellet the infected cells by centrifugation (5 min; 500 *g*) and re-disperse the cells in a small volume of phosphate buffer (10 mM phosphate pH 6.8, 1 ml of buffer per 2 x 10^7 cells – infected cells can be conveniently stored at –70°C at this stage).

(ii) Homogenise the cell suspension for about 1 min with an equal volume of fluorocarbon (Arklone P ICI Ltd.) using an efficient blender (e.g., Waring Blender), the total mixture being immersed in an ice bath. The procedure disrupts the cells and denatures many of the cellular proteins while retaining virus integrity. Other methods of disrupting, e.g., freezing and thawing, sonication using probe or bath sonicators, can also be utilised, the choice being governed by a number of factors such as the volume of cell suspension.

(iii) Centrifuge the extract (500 *g*, 5 min) and remove and retain the aqueous layer which should be opalescent if the virus has propagated well. The fluorocarbon phase and the cellular material at the interphase can be re-extracted with an equal volume of fresh buffer and the aqueous extracts combined. (Generally it is not worth extracting more than twice.)

3.2 **Purification of Virus from Infected Cell Extracts**

Carefully pipette the aqueous phase after fluorocarbon extraction on top of two layers of caesium chloride solutions (0.8 ml of density 1.45 g/ml and 1.5 ml of density 1.33 in 5 mM Tris-HCl, 1 mM EDTA pH 7.8) in 5 ml centrifuge tubes and centrifuge at 90 000 *g* for 90 min. Alternatively, for larger volumes, two layers of 1.5 ml of density 1.45 and 2.5 ml of density 1.33 in 15 ml tubes, centrifuging at 90 000 *g* for 2 h gives similar results. (Densities of CsCl solutions can be readily determined *via* their refractive indices using an Abbe refractor and the following relationship: $\varrho 25 = 10.8661\, n_D - 13.4974$ (ϱ = density at 25° and n_D = refractive index).

Figure 3. Velocity density gradient centrifugation of extracts of adenovirus-infected cells. Virus band indicated by arrow.

Under these conditions of velocity centrifugation the virus bands as an opalescent layer at the interface between the higher and lower density solutions. Other bands higher up the tube corresponding to 'soluble' antigens – excess virus structural components – will also be evident (*Figure 3*). With some virus types an upper band (or bands) of defective or 'empty' virus particles can be detected just above the opalescent virus band. The virus band can be harvested in a variety of ways, e.g., puncturing the tube either from the bottom or the side. The most convenient and economical, however, is to pump out the gradient using a peristaltic pump (see Chapter 2, Figure 3). Fractions containing the virus can be readily distinguished by their opalescence.

Virus obtained in this fashion is still contaminated to some extent with cellular components (12) and a further equilibrium gradient centrifugation is necessary.

Dilute the virus-containing fractions from the velocity gradient with Tris-EDTA buffer (2:1) and then layer on top of a pre-formed double layer of 1.5 ml solution of CsCl of density 1.33 and 1.0 ml of density 1.45 in a 5 ml tube and centrifuge at 100 000 *g* for 18 h. The band of opalescent virus forms at a buoyant density of 1.34–1.35 g/ml ('antigens' equilibrate at lower density of ~1.30).

3.3 Properties of Purified Virus

The purified virus in suspension should be opalescent and appears to be metastable but can be stored at 4°C for about 4–5 days in the hypertonic CsCl solution. Dialysis against hypotonic buffers is not advisable since virus aggregation and disruption occurs and changes of buffer environment can normally be achieved satisfactorily by gel filtration of freshly purified virus using G-25 or G-50 coarse Sephadex and an appropriate buffer. Virus in isotonic or hypotonic buffer is not very stable although some enhancement of stability can be obtained by addition of sucrose (to 1%) or glycerol (10%).

4. TITRATION OF VIRUS

A number of methods are available for titration of adenoviruses and three of these will be described here, i.e., standard plaquing, a cytopathic effect (cpe) assay, and a fluorescent focus assay.

4.1 Adenovirus Plaque Assay

(i) Prepare monolayers of HeLa cells in 50 mm Petri dishes by seeding approximately 2 x 10^6 cells into each dish on the day previous to the assay.

(ii) Prepare serial log dilutions of the virus preparation to be titrated in serum-free medium and add aliquots to the dishes (at least three per dilution) after removal of the medium.

(iii) Incubate virus with the cell monolayers for approximately 90 min at 37°C, remove the supernatant and overlay the cells with an agar-containing medium prepared as follows:

For 100 ml medium:
20 ml 2.5% Noble's agar (in water)
20 ml 5 x medium (–glutamine)
10 ml tryptose phosphate broth (2.95 g/100 ml)
9 ml sodium bicarbonate (2.25 g/100 ml)
2 ml NCS
2 ml 1 M magnesium chloride
100 units each of penicillin, streptomycin
2 ml glutamine (2.9 g/100 ml)
and water to 100 ml

Melt agar and cool to 50°C. Mix remaining ingredients and warm to 42°C and then add to the agar, adjusting the temperature to 45°C. Add 5 ml of the overlay to each plate and then cool until set and incubate at 37°C for 4 days in an atmosphere of 5% CO_2 in air.

(iv) After 4 days add a further 5 ml of overlay and incubate for another 4–6 days. Plaques are normally evident after this time and can be visualised directly by fixing and staining (see *Figure 4*). It is evident that since the cell monolayers will need to remain intact for 8–10 days only cells which are in a healthy condition will be suitable for this assay, e.g., mycoplasma-infected cells will usually not provide stable monolayers. Ideally each Petri

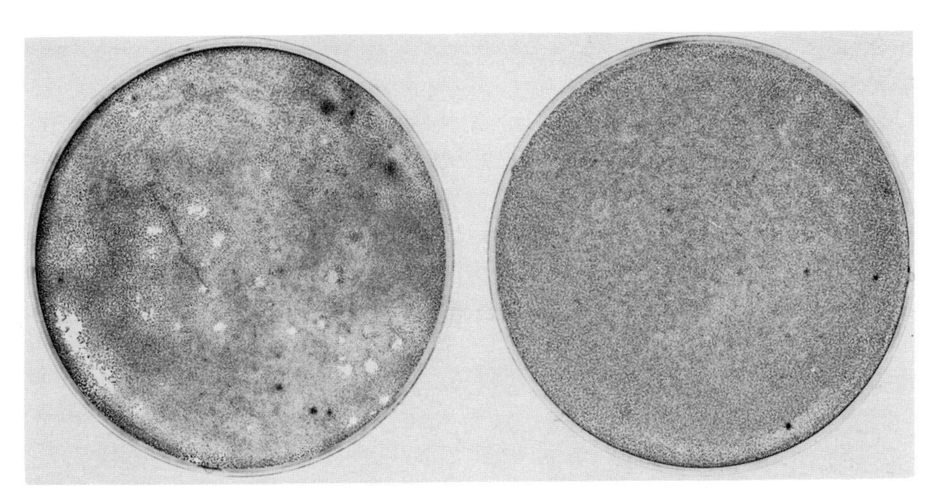

Figure 4. Adenovirus plaques. Twenty plaques can be visualised on the left-hand dish (some deterioration of the cell sheet at the edges is evident). The other dish shows the uninfected control.

dish should have 20 – 50 plaques although smaller numbers per unit would be acceptable when smaller dishes or multiwell plates are used.

(v) Fix the cell sheets with formal-saline (4% formalin in 0.15 M NaCl) for 10 min and stain with crystal-violet (0.1% in 2% ethanol in water) for 5 min. If required, vital staining with neutral red can be carried out by adding a 2 ml overlay containing 0.01% neutral red at the 6th or 7th day. In this case only 3 ml of overlay should be added at the fourth day (13).

4.2 End-point cpe Assay

(i) Dispense 100 μl aliquots of medium supplemented with 0.5% calf serum into the wells of a 96-well tissue culture microtitre plate and make appropriate dilutions of virus allowing for triplicate titrations of each of the samples.

(ii) Add 100 μl of HeLa cells from monolayers in the same medium with calf serum at a concentration of 3 x 10^5 cells/ml to each well and then incubate the plate at 37°C in an atmosphere of 5% CO_2 in air for 3 – 4 days.

(iii) Direct microscopical examination will reveal those wells which show cytopathic effects and, by having made appropriate dilutions, e.g., 10-fold and subsequent 2-fold dilutions, a good assessment of the virus titre can be made in terms of that dilution of virus which shows approximately 50% of the cell sheet destroyed.

Increased sensitivity can be achieved by prolonging the incubation time. Adding extra medium and, if necessary, replacing medium can allow a titration to proceed for weeks if maximum sensitivity is required. In all situations the viability and integrity of the uninfected cells should be monitored by direct microscopy. The cells can be fixed and stained as described above and the titre estimated by visual examination (*Figure 5*). A method for automating the cpe assay technique has been described (14).

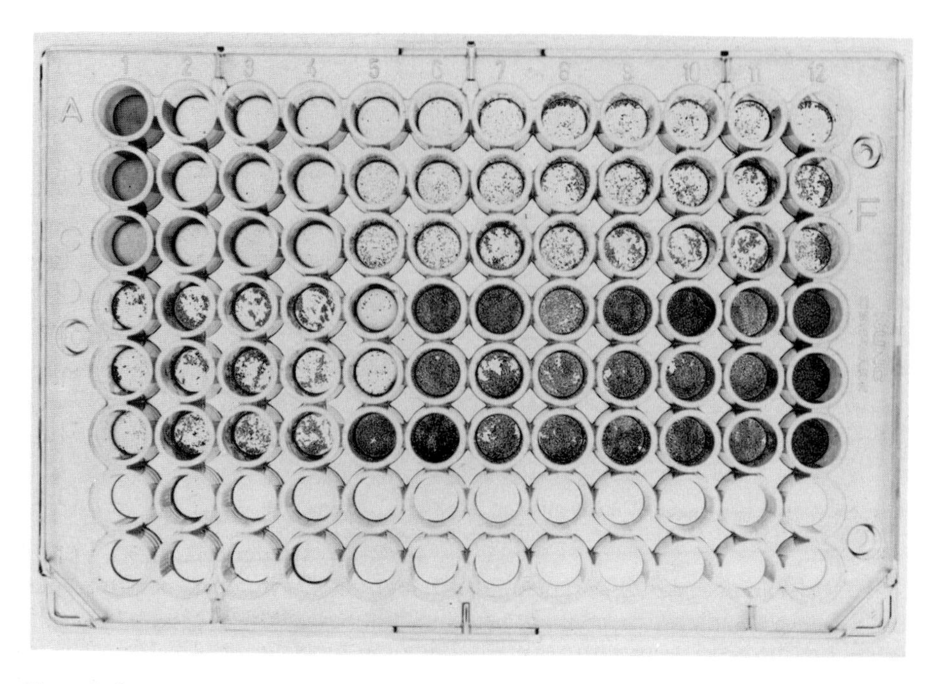

Figure 5. Cpe assay. Titration in triplicate of adenovirus seed. Wells D12, E12, F12: uninfected control cells. Wells in rows A, B and C contain 10-fold dilutions in triplicate until the 6th well (at 10^{-6}), then doubling dilutions continuing on rows D, E and F. The end-point in this case is taken at lane 4/5 and is equivalent to a titre of 2 x 10^9 cpe units/ml.

4.3 **Fluorescent Focus Assay**

(i) Prepare 100 μl dilutions of virus (allowing for triplicate titration) using medium supplemented with 0.5% NCS in the wells of a 96-well tissue culture tray.

(ii) To each of the wells add an equal volume of HeLa cells (from monolayers) in the same medium (with calf serum) at a concentration of 3 x 10^5 cells/ml. Incubate at 37°C in an atmosphere of 5% CO_2 in air for approximately 30 h. (The appropriate time will depend on the particular virus-cell system employed.)

(iii) Remove the medium from the wells by gentle suction and fix the cells in ice-cold 1% formalin – 2% sucrose for 5 min.

(iv) Carefully wash the cell sheets in PBS and then add 10 μl of adenovirus-specific antibody diluted appropriately in PBS. [A mouse monoclonal antibody (15) and a polyclonal antibody against a purified component (e.g., hexon) are equally suitable although, in the latter case, adsorption of the serum with HeLa cells may be necessary (see below).]

(v) Incubate the tray at 37°C for 30 min in a humidified atmosphere and then remove the antibody from the wells by gentle suction and wash carefully three times with about 250 μl aliquots of PBS before adding 10 μl of fluorescein-conjugated anti-mouse (or rabbit etc.) immunoglobulin at a pre-determined dilution and previously adsorbed with HeLa cells. [The adsorption can be accomplished by adding a washed suspension of HeLa cells to the conjugate and incubating for about 18 h in the cold in the presence of a protease inhibitor. Centrifugation (500 *g*; 5 min), removal of the supernatant and filtration through a nitrocellulose (300 μm pore size) membrane should produce a conjugate with minimal background fluorescence on HeLa cells.]

(vi) Again incubate at 37°C for 30 min in a humidified atmosphere, remove the conjugate and wash the wells three times in PBS.

(vii) Examine each well carefully for specific fluorescence using an inverted fluorescence microscope. If the time of fixation is suitable specific fluorescent foci should be observed in the absence of significant secondary infection and an estimate of the virus titre made. This method is based on that originally described by Philipson (16).

4.4 General Comments

The most accurate method of titration is undoubtedly the plaque assay. However, this does require lengthy incubation with the consequent reliance on very stable cell monolayers. The end-point cpe assay is probably the most convenient and can be very sensitive and is accurate enough for most needs. The fluorescent focus assay is certainly the quickest to perform but does require considerably more reagents and moreover needs some preliminary procedures to establish the optimum times for easy detection of the primary foci.

5. REFERENCES

1. Wigand,R., Bartha,A., Dreijin,R.S., Esche,H., Ginsberg,H.S., Green,M., Hierholzer,J.C., Kalter,S.S., McFerran,J.B., Pettersson,U., Russell,W.C. and Wadell,G. (1982) *Intervirology,* **18**, 169.
2. Andrewes,C.H., Pereira,H.G. and Wildy,P. (1978) in *Viruses of Vertebrates,* Balliere and Tindall, 4th Edition.
3. Tooze,J., ed. (1980) *DNA Tumor Viruses*, published by Cold Spring Harbor Laboratory Press, New York.
4. Rouse,H.C., Bonifas,V.H. and Schlesinger,R.W. (1963) *Virology,* **20**, 357.
5. Russell,W.C., Newman,C., Williamson,D.H. (1975) *Nature,* **253**, 461.
6. Lemcke,R.M. (1965) *Lab. Pract.,* **14**, 712.
7. Graham,F.L., Smiley,J., Russell,W.C. and Nairn,R.C. (1977) *J. Gen. Virol.,* **36**, 59.
8. Takiff,H.E., Straus,S.E. and Garon,C.F. (1981) *Lancet,* 832.
9. Larsen,S.H. and Nathans,D. (1977) *Virology,* **82**, 182.
10. Gelderblom,H. and Maichle-Laupe,I. (1982) *Arch. Virol.,* **72**, 289.
11. Winters,W.D. and Russell,W.C. (1971) *J. Gen. Virol.,* **10**, 181.
12. McIntosh,K., Payne,S. and Russell,W.C. (1971) *J. Gen. Virol.,* **10**, 251.
13. Williams,J.F. (1970) *J. Gen. Virol.,* **9**, 251.
14. McAleer,W.J. Miller,W.J., Hurni,W.M., Machlowitz,R.A. and Hilleman,M.R. (1983) *J. Biol. Stand.,* **11**, 241.
15. Russell,W.C., Patel,G., Precious,B., Sharp,I. and Gardner,P.S. (1981) *J. Gen. Virol.,* **56**, 393.
16. Philipson,L. (1961) *Virology,* **15**, 263.

CHAPTER 10

Growth, Assay and Purification of Herpesviruses

R.A. KILLINGTON and K.L. POWELL

1. INTRODUCTION

The herpes group of viruses comprises some 70 or more members which infect a diverse range of hosts (e.g., oysters, fish, fungi and man). Definition and classification of the group is detailed by Matthews (1). The herpesviridae are divided into three sub-families, α-herpesvirinae (to include herpes simplex virus types 1 and 2, pseudorabies virus, equine abortion virus, infectious bovine rhinotracheitis virus and probably feline herpesvirus), β-herpesvirinae (to include human and murine cytomegalovirus) and γ-herpesvirinae (to include Epstein-Barr virus and herpesvirus saimiri). The division into sub-families is based upon a number of criteria, including virus replication, host range, polypeptide composition and genome structure.

In this chapter we discuss the growth, assay and purification of one virus selected from each of the above sub-families. The major effort, however, will be directed toward herpes simplex virus, the most extensively studied herpesvirus, with sub-sections on human cytomegalovirus and herpesvirus saimiri.

2. HERPES SIMPLEX VIRUS TYPES 1 AND 2

These two viruses are members of the α subgroup of the herpetoviridae and are responsible for a broad range of mild to severely debilitating conditions in man, including gingivostomatitis, stomatitis, meningitis, encephalitis, and venereally transmitted genital lesions. There is some evidence which links herpes simplex virus type 2 (HSV-2) with cancer of the cervix (2), and herpes simplex type 1 (HSV-1) has been implicated as being associated with other malignant conditions (3). The virus has thus been the subject of much attention, particularly over the last 10 years, and highly efficient methods of virus growth, assay and purification have evolved. Most of the methods used to study the two types of virus are similar and our discussion will assume this unless stated otherwise.

2.1 **Preparation of Virus Stocks**

2.1.1 *Cells*

As the virus has a wide host range a number of different cells can be used for the preparation of virus stocks. The choice of cells to be used is not trivial, however,

Figure 1. Apparatus for the large scale culture of tissue-culture cells. This machine can be made easily in any competent workshop.

since the yield of virus varies very widely from cell to cell. Hep-2 cells (4) have been used widely to make HSV stocks as have baby hamster kidney (BHK) c13 cells (5), but in our laboratory the yield of different strains may vary by up to 10-fold between the cell lines. Of the two, BHK cells are probably more robust, but both grow to a high cell density on glass Winchester or plastic roller bottles. Conditions of optimum pH must be maintained in such large vessels. If the bicarbonate-CO_2 buffer system is used, it is necessary to gas the bicarbonate prior to sterilisation for use in tissue culture medium, or alternatively the roller bottle itself can be gassed. It is possible to either purchase or make apparatus suitable for turning large roller cultures; the apparatus shown in *Figure 1* was made from a Dexion frame, Meccano wheels and Parvalux tropical motors.

As a rule the cells outlined above can be grown in any tissue culture medium but we conventionally use the autoclave formulation of the Glasgow modification of Eagle's medium with 10% calf serum. For BHK cells this can be supplemented with a solution of 10% tryptose phosphate broth. Other factors may also affect the choice of cell line to be used, these relating mainly to the experiment and assay method required (see below). Finally, it is clear that the cell line to be used for virus stock preparation should be tested for mycoplasma contamination prior to its use.

2.1.2 *Methods of Infection*

The method is essentially that described by Watson *et al.* (5). It is, however, good practice initially to derive master and sub-master (one passage from master) stocks of virus which if possible are spread between more than one −70°C cabinet. The working stocks of virus are routinely grown from the sub-master stock, thus avoiding the continued passage of virus. This avoids the risk of creating stocks of virus so far apart in passage number as to make direct comparison between experiments invalid. For the majority of virus stocks we routinely use roller bottle cultures which contain healthy almost confluent cell monolayers. Such bottles frequently contain as many as 5×10^8 BHK cells or 3×10^8 Hep-2 cells and thus provide a good basis from which to grow a high titre stock of virus. The monolayers are washed to remove loosely adhering cells (a particular problem if using Hep-2 cells) and virus is added by dilution from the sub-master stock. We normally choose to infect cells at a multiplicity of infection (m.o.i.) of between 0.001 and 0.01 (p.f.u./cell) to avoid the generation of defective virus (6,7).

2.1.3 *Growth of Stock Virus*

Any standard tissue culture medium can be used in the preparation of virus stocks. As for cell growth, we conventionally use medium with 10% calf serum. Reducing the concentration of serum brings about some reduction in virus yield but this may not be significant compared with the cost. The virus stocks should be prepared according to the following procedure.

(i) Adsorb virus (m.o.i. 0.001 − 0.01 p.f.u./cell) for 1 h in a small volume of medium (just sufficient to cover the monolayer; ~20 ml in the case of a roller bottle).

(ii) Add the required amount of medium for cell incubation (~200 ml for a roller bottle).

(iii) Incubate the cells for between 2 and 3 days at a suitable incubation temperature. Both 32°C and 37°C may be used for growth of virus stocks. It is possible however, that continued growth of virus at the lower temperature of 32°C might lead to a virus stock which contains a high proportion of temperature-sensitive virus, or at least virus which grows better at 32°C than 37°C. For this reason we normally use an incubation temperature of 37°C.

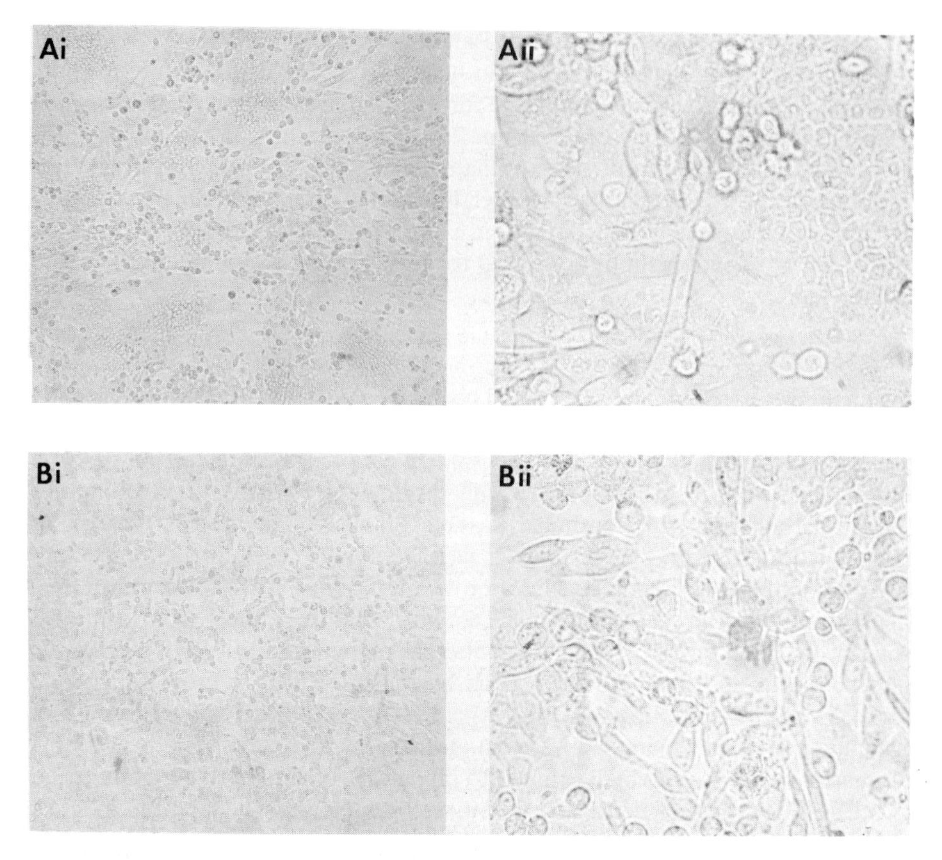

Figure 2. Demonstration of cytopathic effect in a cell monolayer previously infected with a syncytial strain (**A**) or non-syncytial strain (**B**) of HSV-1 (i and ii in each case represent 100 and 400 x magnification images).

(iv) After 2 days at 37°C the virus will have produced considerable cytopathic effect (c.p.e.) in the cells, although the time required to do this will vary depending primarily on the virus strain and cell line used. In general, HSV-2 strains grow faster than their HSV-1 counterparts. Virus c.p.e. is usually designated as being syncytial (e.g., strain HFEM of HSV-1) or non-syncytial (e.g., strain MP17 of HSV-1). The morphological appearance of such c.p.e. is shown in *Figure 2*.

(v) Harvest the cells as soon as they have undergone complete c.p.e. The method used for collection of the cells will depend on the strain of virus and the cells used. With strain HFEM in BHK cells very little cell lysis or release from substrate occurs. The bulk of the medium can be poured from the culture vessel without significant loss of the virus. Then the cells are scraped into the remainder of the medium using a sterile rubber policeman and collected by low-speed centrifugation. With other strains of virus, e.g., strain 186 (HSV-2), infected cells regularly detach from the culture vessel and can only be efficiently collected by centrifuging the cells from the medium. In cell/virus combinations where a considerable amount of lysis occurs it may be necessary to use the medium as part of the virus stock. The cells collected by centrifugation are then re-suspended in a medium of choice, e.g., sterile distilled water, or normal tissue culture medium. Stocks may be re-suspended in medium with an additional cryoprotectant such as glycerol, although this is not strictly necessary.

(vi) The majority of the virus is tightly bound to cells and a significant problem with the preparation of herpes virus stocks is the breakage of cells to obtain a cell-free preparation of virus. This can be done by cycles of freezing and thawing which break the cells but may cause a loss of virus titre. The use of sonicator probes is generally accepted as being unsafe due to the probable production of aerosols of infectious virus, so the best technique is to use a water bath sonicator for cell breakage in a closed vessel. Unfortunately, no one brand of sonicator can be recommended and it is essential that each bath be checked for its efficiency. Sonicate the virus stock for the period required for efficient cell breakage in ice-cold water, which should not be allowed to get hot!

(vii) Remove the cell debris by sedimentation, dispense the stock in small volumes, and store as detailed below.

For long-term storage of herpes virus stocks it is best to freeze the virus at −70°C. On no account should herpesvirus be stored at −20°C as this results in a considerable drop in infectivity titre. For short-term storage HFEM (HSV-1) is stable for a few days at 4°C, but this may vary between strains. It is important to thaw virus rapidly, and wherever possible avoid repeated freezing and thawing of virus. However, we have demonstrated that strain HFEM (HSV-1) virus stored at high titre in growth medium loses only 0.5 log infectivity titre when frozen and thawed on five successive occasions. We have also found that stocks of virus when frozen at −70°C and thawed rapidly do not lose significant titre over a 5-year period. Before use the stock should be tested for sterility, both on solid media and in enrichment culture. Ideally the stock should be checked by electron microscopy to avoid the use of stocks with high particle/infectivity ratios. The particle counting methods to be used are described in Section 2.2. In practice, however, we have found that the methods outlined above should give stocks with titres of approximately 10^9-10^{10} p.f.u./ml and low particle/infectivity ratios.

2.2 Electron Microscopy (Particle Counts)

Particle counts are performed by the 'loop drop' method of Watson (8). A virus dilution in water (or in 0.2 mg/ml bovine serum albumin (BSA) for virus preparations of below this protein concentration) to give 10^9-10^{10} particles/ml is mixed with an equal volume (0.1 ml) of calibrating material. (The reference particles used are 'DOW' (264) polystyrene latex calibrated to contain $3.0-3.5 \times 10^9$ latex particles/ml). Add 0.12 ml of a neutral 2% (w/v) solution of sodium phosphotungstate (PTA) to the mixture and dry one drop of the resulting material on to a 400 mesh 'formvar' coated specimen grid. The grids are viewed in a suitable electron microscope at a screen magnification of x 40 000. Particles are classified as 'naked', 'enveloped' (penetrated) or 'enveloped' (unpenetrated), following the nomenclature used in *Figure 5*. The unpenetrated enveloped particles appear as solid white spheres, but can be shown to be capable of quantitative conversion to penetrated enveloped particles. Normally five groups of 20 latex particles are counted for each virus sample.

2.3 **Infectivity Assays**

Herpes simplex viruses are relatively easy to titrate by a variety of methods. We routinely use either a suspension assay (9) or a monolayer assay to titrate virus preparations.

2.3.1 *Suspension Assay*

(i) Prepare dilutions of virus in tissue culture medium, and add 8 x 10^6 BHK cells to 2 ml of each virus dilution.
(ii) After the virus/cell mixture has been incubated, with agitation, at 37°C for 30 min, add 8 ml of 2% carboxymethyl cellulose (CMC) in tissue culture medium and mix with the virus and cells (for virus suspension assays we find 5% calf serum satisfactory).
(iii) Divide the mixture between two 60 mm tissue culture Petri dishes and incubate for 2–3 days at 37°C in an incubator with an atmosphere of 5% CO_2 in air.
(iv) Fix the cell monolayers using conventional fixatives and stain with Gentian Violet or Methylene Blue (1% aqueous solutions).

Plaques can be seen without the aid of magnification, but for counting purposes, view them with the aid of a stereomicroscope. Herpes simplex plaques vary in morphology. Some strains are capable of causing cell fusion, which results in the formation of large syncytial plaques (e.g., strain HFEM) whereas others (e.g., strain MP17) induce cell rounding. The comparative plaque morphologies are seen in *Figure 3*. However, in many cases continued incubation of cell sheets previously infected with syncytial strains often results in the loss of syncytia and the plaque takes on the non-syncytial appearance.

2.3.2 *Monolayer assay*

If a monolayer assay is preferred, i.e., infection of a pre-existing cell sheet, then the choice of cell line is important. As a rule, BHK cells infected as monolayers result in the formation of smaller plaques and our experience suggests that the plaquing efficiency is lower in monolayers of BHK cells. We do, however, routinely use Vero cells (10) for monolayer assays, where the plaquing efficiency is equally as good as, or in some strains much better than the BHK suspension assay.

(i) Seed the Vero cells at 2.0 x 10^6 cells/dish 18–24 h prior to infection.
(ii) Drain the cells of medium and infect with 0.1 ml of virus dilution (diluted in tissue culture medium), and leave for 1 h at 37°C for adsorption.
(iii) Add 5 ml of 2% CMC in tissue culture medium (5% calf serum is satisfactory) and incubate at 37°C. The time needed for incubation varies between

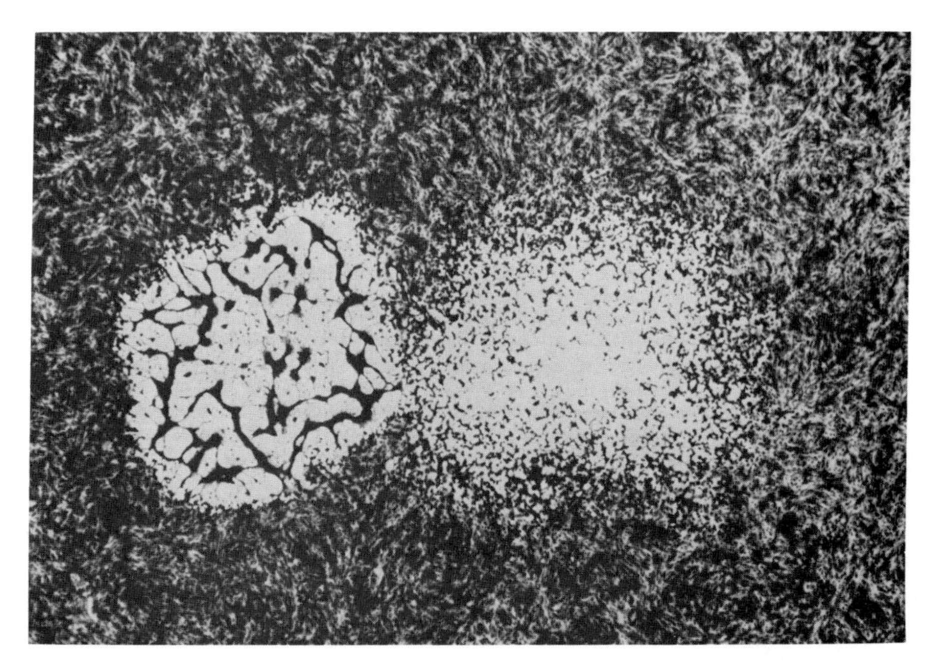

Figure 3. Two plaque morphologies of HSV-1. The cells in the plaque on the left are fused (syncytial) whereas those on the right are rounded (non-syncytial).

strains, some fast-growing HSV-2 strains taking only 2 days, others 4 days, to produce readily discernible plaques.

(iv) Fix and stain the plates as described above.

2.3.3 *Overlay Media*

Whilst a number of different overlay media have been used in both suspension and monolayer titrations (e.g., heparin, pooled human sera and various grades of agar) we consider CMC to be the most practical, as it is easy to use and does not inhibit plaque size. One important feature of CMC, however, is that gross disturbance during incubation may lead to secondary plaques or 'rocketing' plaques, and hence this should be avoided.

2.3.4 *Pock assay*

Both HSV-1 and -2 form pocks on the choriallantoic membrane of fertile hens eggs, and it has been suggested that this technique might usefully distinguish between the two types of virus (11).

2.4 Preparation of Virus-infected Cell Extracts

Whole virus-infected cell homogenates or detergent extracts of these are routinely used as a source of antigen in a number of serological tests, including enzyme-linked immunosorbent assay (ELISA) and radioimmunoassays. They are also used for immunisation purposes, either to raise hyperimmune antisera or to develop monoclonal antibodies against HSV proteins. The cell line of choice to infect would depend on the purpose of growing antigen.

(i) For immunoassays, use Vero or BHK cells, infected at an m.o.i. of at least 10 (p.f.u./cell) in a similar manner to that described for virus stocks.
(ii) 18 – 24 h post-infection, scrape the cells into the medium and pellet by low-speed centrifugation.
(iii) Wash the cell pellet twice in PBS and re-suspend in distilled water, if possible at a concentration of 1 x 10^8 cells/ml.
(iv) Disrupt the infected cell suspension in an ultrasonic bath and store at – 70°C. The polypeptides present in such a fraction are shown in *Figure 6*.

For some procedures, e.g., immunoprecipitation, a soluble antigen fraction is required. This supernatant fraction is acquired by centrifugation (38 000 r.p.m. for 1 h in a Sorvall AH627 rotor) of the disrupted infected cell suspension. Detergents may be used to increase the yield of membrane proteins (12). Test the potency of the antigen by either the ELISA or Ouchterlony gel diffusion test, using rabbit hyperimmune serum raised against herpes virus-infected cells. For immunisation purposes the procedure is similar but the choice of cell different. For preparation of monoclonal antibodies we routinely infect mouse L-cell monolayers (13) whereas for derivation of rabbit sera we infect rabbit kidney cells, RK13, (5,14) grown in 5% rabbit serum.

2.5 Preparation of Purified Virus

Herpes simplex virus provides a more difficult challenge than most viruses with regard to the preparation of pure virus particles. Techniques which work well with one virus strain in a particular cell line do not work for other strains of virus in the same cells or for the same strain of virus in different cells. For this reason we give two methods here. The first we have found to work very well for strain HFEM and certain other strains of virus in Hep-2 cells only. It is simple and gives excellent preparations of virus. The second gives good samples of virus with all strains used and with many different cell types. It is therefore advisable to examine the one-step growth curve pattern of virus strains in different cell lines, in order to determine the optimum conditions for virus yield and starting material for subsequent purification. *Figure 4* shows the growth curve for Hep-2 cells infected with strain HFEM at a high m.o.i. following incubation at 32°C. These optimum conditions form the basis for the starting material outlined in Section 2.5.1 below.

2.5.1 *Method 1 (15)*

(i) Infect confluent monolayers of Hep-2 cells in roller cultures at a m.o.i. of 20 – 25 (p.f.u./cell) in a fashion similar to that described for virus stock preparation.
(ii) Wash the infected cells after virus adsorption, add fresh medium and incubate the cells at 32°C for 2 – 3 days.
(iii) If virus-specific proteins are to be radioactively labelled (i.e., with [^{35}S]-methionine or [^{14}C]amino acids), add the isotope after 4 h incubation in

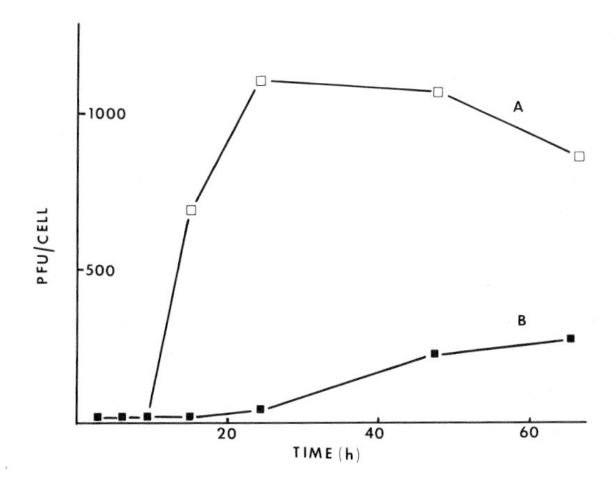

Figure 4. Growth of HSV-1 (strain HFEM) in Hep-2 monolayers at 32°C. Virus was adsorbed at an m.o.i. of 20 (p.f.u./cell) and cellular (**A**) or extracellular (**B**) samples assayed by plaque assay at the indicated times. The curve of HSV-2 under these conditions is not shown, but has significantly higher titres of extracellular virus due to cell lysis.

growth medium with 1/10th normal methionine concentration and 2% calf serum.

(iv) At the end of the incubation, centrifuge the medium at low speed to remove cell debris.

(v) Harvest the virus either by precipitation from the medium with polyethylene glycol (PEG, molecular weight 6000, 8% w/v in the presence of 0.5 M NaCl) or by centrifugation (12 000 r.p.m., 2 h in a GSA rotor, Sorvall RCS-5B centrifuge).

(vi) Re-suspend the virus in a low molarity Tris buffer pH 7.8 containing 50 mM NaCl. It is preferable to allow re-suspension overnight if this is convenient.

(vii) Layer the suspension of virus over a 30 ml gradient of 5–45% sucrose (prepared as described in Chapter 1) in the same buffer and centrifuge for 1 h at 12 500 r.p.m. (Sorvall AH627 rotor). At the end of this period a fluffy white band of purified virus should be clearly visible at the centre of the gradient. This visible band contains the single peak of infectious virus in the gradient.

At this point the virus is of adequate purity (~50 μg of protein/10^{10} virus particles) for many purposes including the preparation of virus DNA. The virus may be recovered by simple sedimentation. To produce high quality preparations of virus a wide variety of other techniques can be used, including a second sucrose gradient or caesium chloride gradients. Both these techniques yield virus with a protein/particle ratio better than 20 μg/10^{10} particles.

2.5.2 *Method 2*

This method is derived from Spear and Roizman (16) and Heine *et al.* (17).

(i) Infect the cells exactly as described for method 1.

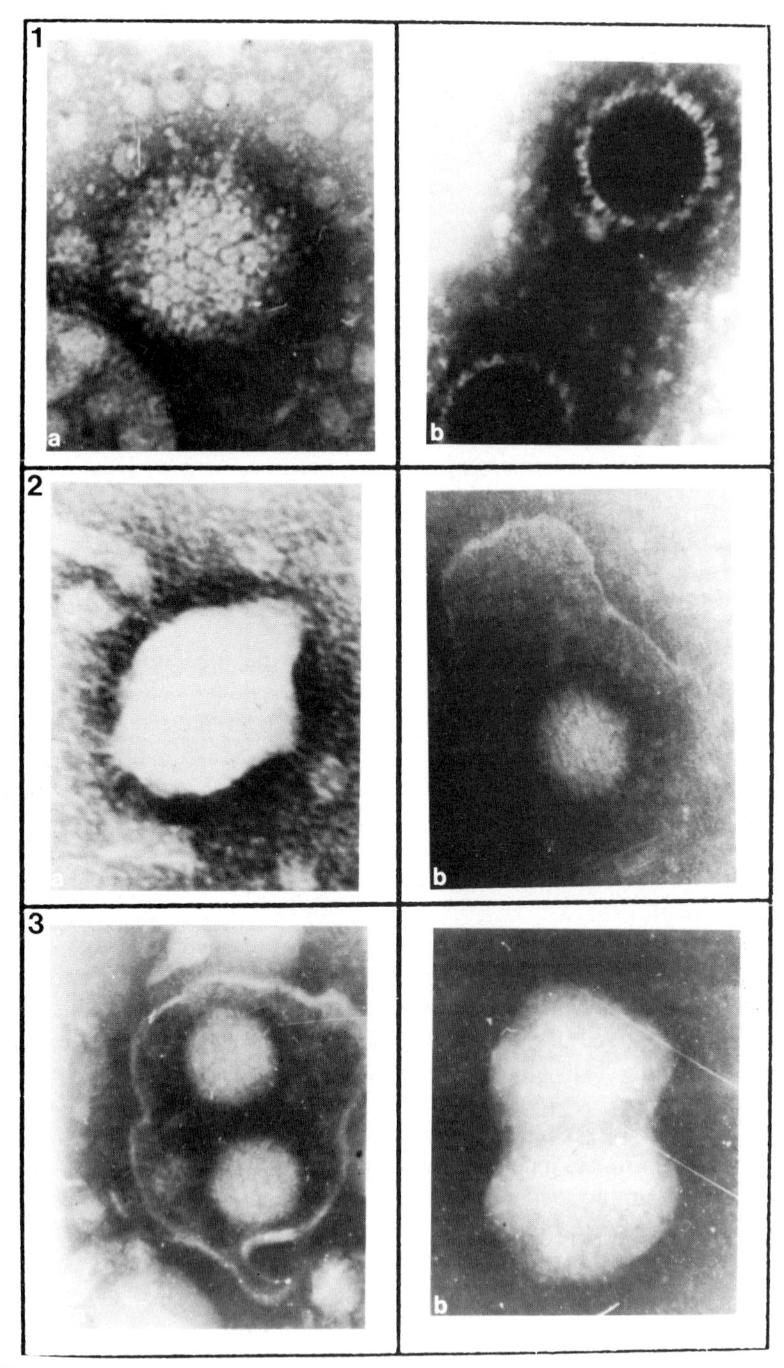

Figure 5. Morphology of the types of particles found in purified preparations of HSV. Using the methods of purification described, more than 90% of the particles are of type **2a** or **2b**. **(1)** Naked particles, **a** unpenetrated, **b** penetrated. **(2)** Enveloped particles, **a** unpenetrated, **b** penetrated. **(3)** Two particles in one envelope, **a** penetrated, **b** possible unpenetrated equivalent of **a**. No attempt has been made to preserve uniform magnification.

(ii) Incubate the cells at 37°C for 18 – 24 h and harvest by scraping from the glass and low-speed centrifugation.
(iii) Re-suspend the cell pellet so obtained in Reticulocyte Standard Buffer (RSB) and allow the cells to swell for 10 min. The cytoplasm may then be obtained from the cells by Dounce homogenisation.
(iv) Centrifuge the Dounce-homogenised cells at low speed to remove nuclei and cell debris.
(v) Layer the supernatant (containing the majority of the infectious virus) onto a 5 – 40% dextran gradient in Tris buffer (17), and centrifuge at 12 500 r.p.m. for 1 h (Sorvall AH 627 rotor); the virus is obtained by removing the visible band in the centre of the gradient.
(vi) Collect the virus from this band by sedimentation (20 000 r.p.m. for 1 h, Sorvall AH 627 rotor).

At this point the virus is of adequate purity for many purposes but can easily be improved by the methods mentioned above. Purified virus derived by either method may be re-suspended in distilled water or a suitable buffer and aliquots taken for infectivity assay, total particle count and protein estimation. Virus can be frozen at – 70°C, but on thawing this leads to partial disintegration of the viral envelope. *Figure 5* shows examples of virus particles found in preparations of pure virus.

2.5.3 *Analysis of Virus Structural Polypeptides*

It is inappropriate in this chapter to discuss in detail the methods available for analysis of the structural polypeptides of the purified virus. We routinely use SDS electrophoresis in polyacrylamide slab gels as originally described by Dimmock and Watson (18) and modified for the use of polyacrylamide slabs (19). Gels are stained with Coomassie brilliant blue and exposed to X-ray film for autoradiographic analysis. The profile of structural polypeptides examined in this fashion for both HSV-1 and HSV-2 is shown in *Figure 6*. Further characteristics of these polypeptides are summarised in *Table 1*.

2.6 **Other Alphaherpesvirinae**

We have grown and purified in our laboratory a number of other α-herpesvirinae, including bovine mammillitis virus (BMV), pseudorabies (Aujeszky's disease) virus (PRV), equine herpes virus type 1 (EHV-1), feline rhinotracheitis virus (Fetr) and channel catfish virus (CCV). Whilst the methods are grossly similar to those described above we highlight selected details of these viruses below.

2.6.1 *Bovine Mammillitis Virus and Pseudorabies Virus*

Both of these viruses grow to high titre stocks in BHK cells and can be readily assayed by the suspension assay method of Russell (9). For purification of either virus, infect BHK cells at a high m.o.i. and incubate at 37°C for either 24 h (PRV) or 48 h (BMV) at which point virus may be purified using the methods described in Sections 2.5.1 and 2.5.2. above. This yields good preparations of enveloped particles, the structural profiles of which can be seen in *Figure 7*.

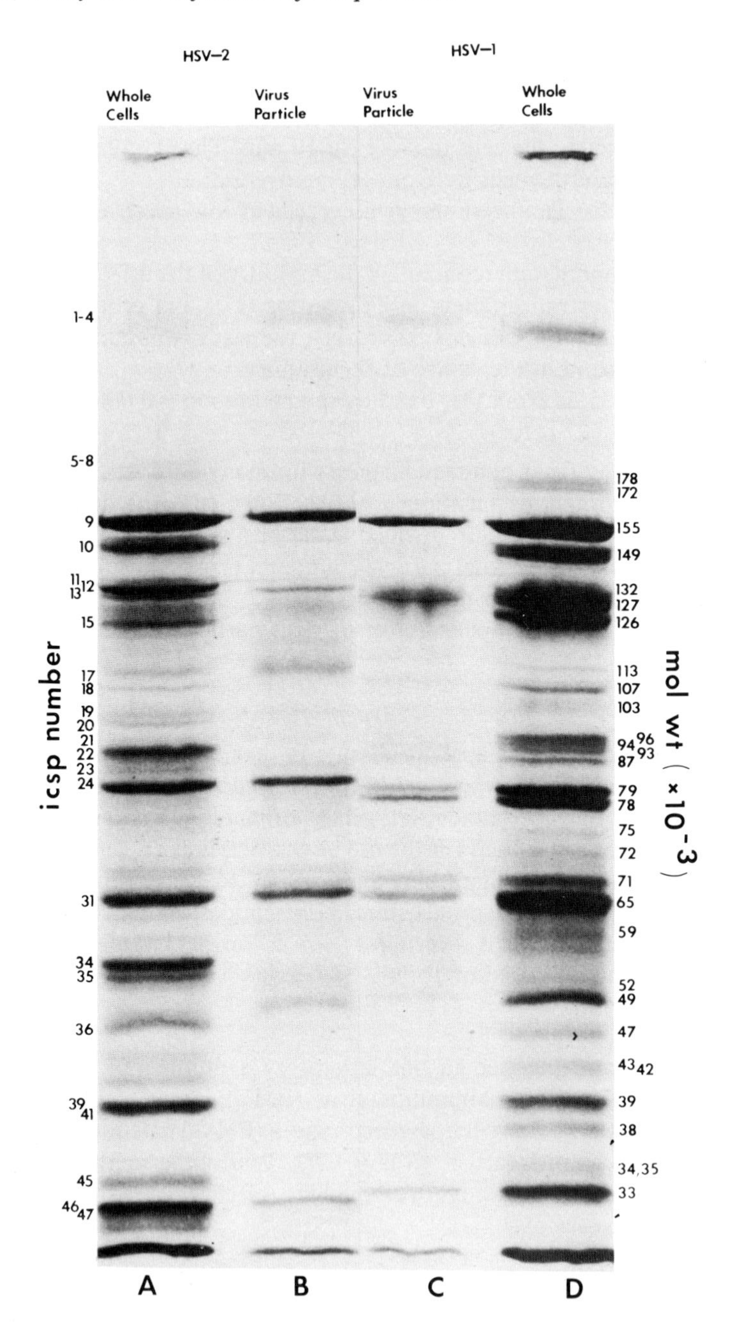

Figure 6. Profiles of HSV-2 and HSV-1 virus specific polypeptides as revealed by polyacrylamide gel electrophoresis. Tracks (**A**) and (**D**) are extracts of cells previously infected with either HSV-2 (**A**) or HSV-1 (**D**). Tracks (**B**) and (**C**) represent the profiles of purified virions, either HSV-2 (**B**) or HSV-1 (**C**). The figures on the left refer to infected cell specific polypeptide number (HSV-2), those on the right to molecular weight x 10^{-3}.

Table 1. Properties and Nomenclature of Selected HSV-1 and HSV-2 Polypeptides.

HSV-2				HSV-1		
Polypeptide[a] number	*Structural[b] virus protein*	*Molecular[c] weight ($x\ 10^{-3}$)*	*Characteristics*	*Molecular[c] weight ($x\ 10^{-3}$)*	*Structural[b] virus protein*	*Polypeptide[d] number*
ICSP 5 – 8	NS	182 – 186	Immediate early phosphoprotein	175	NS	ICP 4
ICSP 9	VP5	157	Major capsid protein	155	VP5	ICP 5
ICSP 10	NS	153	Ribonucleotide reductase	149	NS	ICP 6
ICSP 11 ICSP 12	NS	146 143	Major DNA-binding protein	132	NS	ICP 8
ICSP 13 – 15	VP7 – 8.5	126 – 140	Glycoproteins gA, gB (gC for HSV-1)	112 – 126	VP7 – 8.5	ICP9 – 14
ICSP 22	NS	85 – 90	Alkaline nuclease (phosphoprotein)	85 – 90	NS	ICP 19
NN	NN	92	Glycoprotein gE (FC-binding protein)	80	NN	NN
ISP 34 – 35	NS	54	DNA polymerase-associated protein	48	NS	ICP 29
NN	NS	42.4	Thymidine kinase	44	NS	NN
NN	NN	53	Glycoprotein gD responsible for cross-neutralisation between HSV-1 and HSV-2	59	VP18	ICP 31
ICSP 46 – 47	VP22 – 23	28 – 30	Capsid protein	30 – 33	VP22 – 23	ICP 39 – 40

[a]Nomenclature of Powell and Courtney (19).
[b]Nomenclature of Spear and Roizman (16).
[c]Molecular weight may show strain variability.
[d]Honess and Roizman (20).
NS = Non-structural.
NN = Not-named; in most cases not known or disputed.

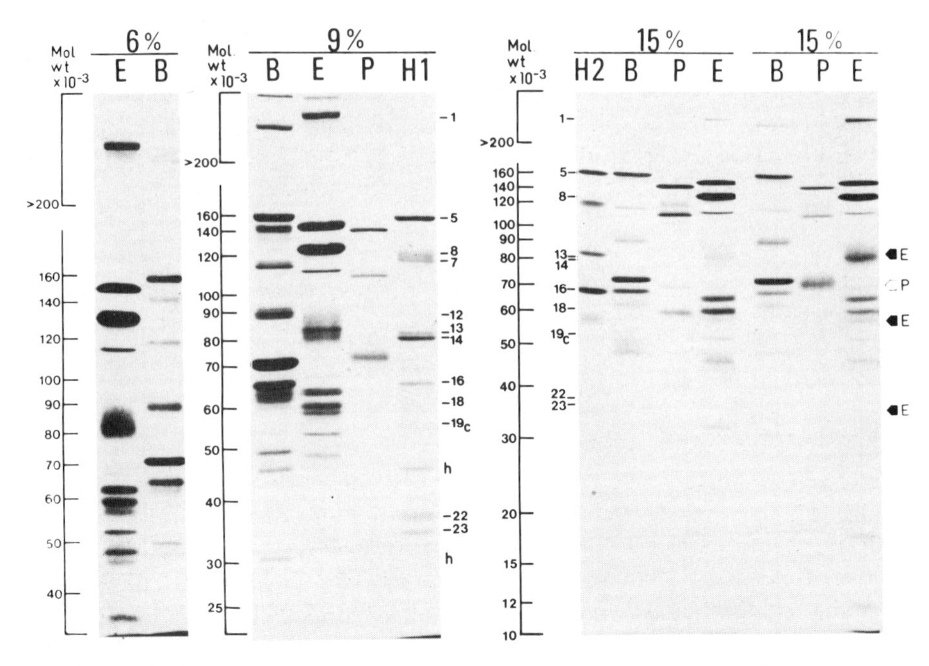

Figure 7. Polypeptides from purified enveloped particles of EHV-1 (**E**), BMV (**B**), PrV (**P**), HSV-1 (**H1**) and HSV-2 (**H2**) separated by electrophoresis on 6%, 9% and 15% polyacrylamide slab gels in the presence of SDS. The 6% and 9% gels and the right-hand portion of the 15% gel are photographs of Coomassie brilliant blue-stained gel films. The left-hand portion of the 15% gel is a corresponding autoradiogram of [^{35}S]methionine radioactivity. Filled and open arrows indicate examples of polypeptides of equine abortion virus (**E**) and pseudorabies virus (**P**) which have unusually low ratios of [^{35}S]-methionine radioactivity to total protein estimated by Coomassie brilliant blue staining. Positions of selected HSV-1 polypeptides are shown to the right of the 9% gel and the left of the 15% gel.

2.6.2 *Equine Herpes Virus Type 1*

Restriction endonuclease digestion suggests that strains of EHV-1 fall into either sub-type 1 or sub-type 2. The former have been associated with abortion in pregnant mares as well as being isolated from the respiratory tract, whereas the latter have been referred to as 'respiratory-type' viruses and are not associated with abortion (21–23). Sub-type 1 viruses grow in a variety of tissue culture lines. In this laboratory we routinely use RK13 cells, whereas in other laboratories the choice may be a cell of equine origin (e.g., Kentucky equine foetal dermis fibroblasts). Sub-type 1 virus has also been successfully grown in L-M cell cultures (24). It has also been possible to adapt some sub-type 1 strains to growth in hamsters (25). Sub-type 2 virus isolates grow only in cells of equine origin. Virus of both sub-types has been successfully purified using the method 1 described in Section 2.5.1 above. The comparative polypeptides of EHV-1 (sub-type 1), BMV, PRV, HSV-1 and HSV-2 are shown in *Figure 7*.

2.6.3 *Feline Rhinotracheitis Virus*

This virus appears to be specific to cells of feline origin. Growth conditions are as for HSV-1, and we have purified the virus using the method described in Section

2.5.1 above. We routinely use a feline cell line provided by Dr. P. Talbot (Wellcome Laboratories, Beckenham) although equally suitable are the feline embryo lung cells derived by Dr. O.S. Jarrett and available from Flow Laboratories, Irvine, Scotland. Dr. R. Maes (Department of Microbiology, Michigan State University) has also successfully purified the virus from culture media of infected Crandell-Rees feline kidney cells, using potassium tartrate gradients.

2.6.4 *Channel Catfish Virus*

This virus has a limited host range and grows best in Brown Bullhead (BB) cells (26). Methods of growth and purification are essentially as for HSV-1 with the exception that BB cells and virus are grown at 28°C.

3. HUMAN CYTOMEGALOVIRUS

The various animal cytomegaloviruses comprise the β-sub-group of the herpetoviridae. Human cytomegaloviruses are important human pathogens particularly in the immunocompromised host and during pregnancy. As such they are responsible for a number of fatalities and a high degree of debilitation in patients with naturally- or artificially-induced immune suppression (e.g., in leukaemia and during tissue transplantation). During pregnancy the virus is responsible for malformation of the foetus, abortion and a mild to severe infection of the neonate. For this reason the intensity of research into cytomegalovirus (CMV) has increased in the last 10 years but our knowledge of the virus is not as advanced as that of herpes simplex.

3.1 **Preparation of Virus Stocks**

This virus is much slower than HSV in its replication cycle, is more strongly cell-associated and is extremely cell-specific. Whilst many laboratories use MRC-5, Flow 5000 or Hep-2 cells, W. Gibson (personal communication) recommends early passages of either human foreskin fibroblasts (HFF) or human embryonic lung (HEL) cells. Gibson cultivates cells in DMEM containing 4500 mg/litre glucose and 10% foetal calf serum (FCS). In our own laboratory we use MRC 5 cells grown in the Glasgow modification of Eagle's medium supplemented with 10% FCS (growth medium) and prepare virus stocks as follows.

(i) Just prior to confluence, add CMV at a low m.o.i. (<0.1 p.f.u./cell) in a minimal volume of medium for 1 – 2 h at 37°C.
(ii) After adsorption, add the required volume of growth medium and re-incubate the cells at 37°C.
(iii) Seven to ten days post-infection replace the growth medium with fresh medium (with 2% FCS) and re-incubate at 37°C, until 2 – 3 weeks post-infection when the cytopathic effect (c.p.e.) is evident.
(iv) Watch the cells daily until the c.p.e. is extensive, at which point the cells are scraped into the medium and pelleted for 10 min at 2000 r.p.m.
(v) Re-suspend the pellet in a small volume of growth medium, disrupt in an ultrasonic bath, and store at – 70°C.

(vi) Much of the virus is in the extracellular medium; recover this by centrifuging the supernatant from the low-speed spin at 12 000 r.p.m. for 2 h (GSA rotor, Sorvall RCS-5B centrifuge).

(vii) Re-suspend the resulting pellet in a small volume of growth medium, disrupt slightly in an ultrasonic bath and store at −70°C. The virus is not as stable at −70°C as is HSV and should not be subjected to frequent rounds of freezing and thawing. As with HSV the virus should be re-thawed quickly in warm water.

3.2 Growth Cycle

Smith and DeHarven (27) did comparative experiments for HSV and CMV in relation to the growth curve and times of appearance of certain events within the infected cell. Their data are shown in *Figure 8* and *Table 2*. One has to sound a note of caution, however, as to whether such a growth curve is an overall picture of cell asynchrony or truly represents the timing of events within an infected cell (see herpesvirus saimiri later in this chapter). As mentioned above, the growth cycle is much slower for CMV than for HSV.

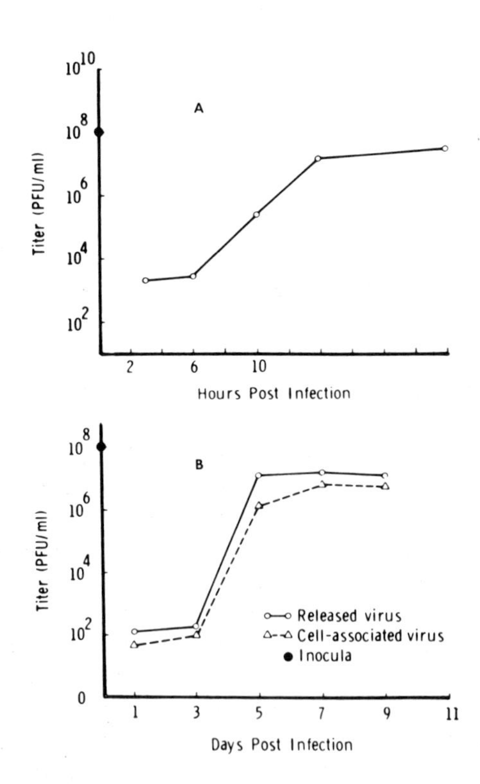

Figure 8. Comparison of HSV and CMV one-step growth curves in human fibroblast cells. (**A**) Titres of released infectious HSV (○——○). (**B**)Titres of released (○——○) and cell-associated (△——△) infectious CMV. The input m.o.i. for both virus infections was 10 (p.f.u./cell). Data from Smith and DeHarven (27).

Table 2. Comparision of Timing of Major Events in CMV- and HSV-infected Cells as Determined by Electron Microscopy[a].

Event	*Onset*	
	CMV (days)	*HSV (h)*
Cell fusion and rounding	0.5	6
Golgi alteration	1	8
Viral eclipse	2	2
Appearance of increased number of perichromatin-like granules	2	3
Condensation of the chromatin	Not observed	3
Assembly of the first capsids in the nucleus	3	4
Appearance of the first capsids with dense cores	3.5	5
Envelopment at the nuclear membrane	3.5	6
Appearance of the naked capsids in the cytoplasm	4	6
Appearance of dense cytoplasmic aggregates	4	Not observed
First released particles	4	8
Nuclear membrane reduplication	4	8
Envelopment at cytoplasmic membranes	4	8
Cell lysis	7 – 8	24 – 48

[a]Data from Smith and DeHarven (27).

3.3 Virus Assay

Many attempts have been made to devise suitable plaque assays for this slow-growing virus, and many laboratories use end-point dilution $TCID_{50}$ methods. The problem lies in trying to maintain a healthy cell monolayer over the time span required for plaque appearance. In our laboratory we use the following modification of the plaque assay method devised by Wentworth and French (28).

(i) Seed MRC-5 cells into either 24-well Linbro panels or 30 mm Petri dishes of tissue culture quality. When confluent, infect the cells with 200 μl samples of virus suitably diluted in growth medium.

(ii) After a 1 h adsorption period remove the eluate, replace with 1% agarose (in medium containing 2% FCS) and incubate at 37°C.

(iii) One week post-infection add further agarose (in medium containing 2% FCS) and re-incubate the dishes at 37°C.

(iv) On microscopic examination 2 – 3 weeks post-infection small foci of infected cells can be seen. When optimum (before deterioration of non-infected cells), fix the cell monolayers in formol-saline solution, carefully remove the agar, and stain the cell sheet with 1% Gentian Violet solution. The small foci of infection (or plaques) are countable with the aid of a stereomicroscope. Alternatively, stain the dishes in 1% Neutral Red. We have also successfully derived CMV plaques using CMC (0.8% in medium containing 2% FCS) instead of agarose.

3.4 **Black Plaque ELISA**

This method can be used for assay of any virus but is particularly useful with CMV where the plaques are particularly small, as it allows earlier quantification of plaques.

(i) Infect monolayers as for the normal assay.
(ii) Six to 7 days post-infection decant the CMC and wash the cell sheet three times in PBS before adding 0.25% glutaraldehyde in PBS for 5 min to fix the cells.
(iii) After a further wash in PBS, add a suitable dilution of antiserum or monoclonal antibody directed against CMV and leave for 2 h at room temperature.
(iv) Remove the antibody and wash the cell sheet in PBS, then add a second conjugated antibody, directed against the first (i.e., anti-rabbit or anti-mouse) and leave for 3 h at room temperature. The conjugate is normally horseradish peroxidase.
(v) After washing three times in PBS, add Hanker-Yates substrate [75 mg Hanker-Yates reagent (Polysciences (U.K.) Ltd., Northampton), 50 ml 0.1 M Tris-HCl pH 7.5 and 0.5 ml 1% H_2O_2]; the colour forms in 5 – 30 min.
(vi) Wash the plates with distilled water, invert and allow to dry in the dark. The darkly stained foci are easily quantified.

3.5 **Conditions of Growth used for CMV Purification, and Purification Procedures**

(i) Just prior to confluence, infect cell monolayers with a virus at an m.o.i. of 5 – 20 (p.f.u./cell).
(ii) Following a 60 min adsorption period, remove the medium and replace with fresh medium containing 2% FCS, then incubate the infected cell cultures at 37°C.
(iii) If required, replace the medium 48 h post-infection with radioisotope labelling medium. This can be either 1 μCi/ml [^{35}S]methionine in medium (1/10th methionine concentration) containing 2% FCS or 2 μCi/ml [^{14}C]-amino acids in medium (1/10th amino acid concentration) containing 2% FCS.
(iv) By 3 days post-infection, c.p.e. (cell swelling and rounding) should be evident and this progresses until approximately 7 – 10 days post-infection when infected cells are detaching themselves from the plastic. For virus purification, allow the c.p.e. to continue until maximum effect is achieved. For the purification of nucleocapsids, cells should be harvested earlier.

For use in immunological tests (e.g., ELISA) and for the purpose of immunisation of animals we routinely purify CMV from the extracellular medium, using the procedures outlined above for HSV. However, for the comparative examination of protein profiles etc. the connoisseur is advised to use the method outlined below, suggested by Dr. Wade Gibson. For maximum infectivity and low contamination, the starting material should be the extracellular medium, but in order

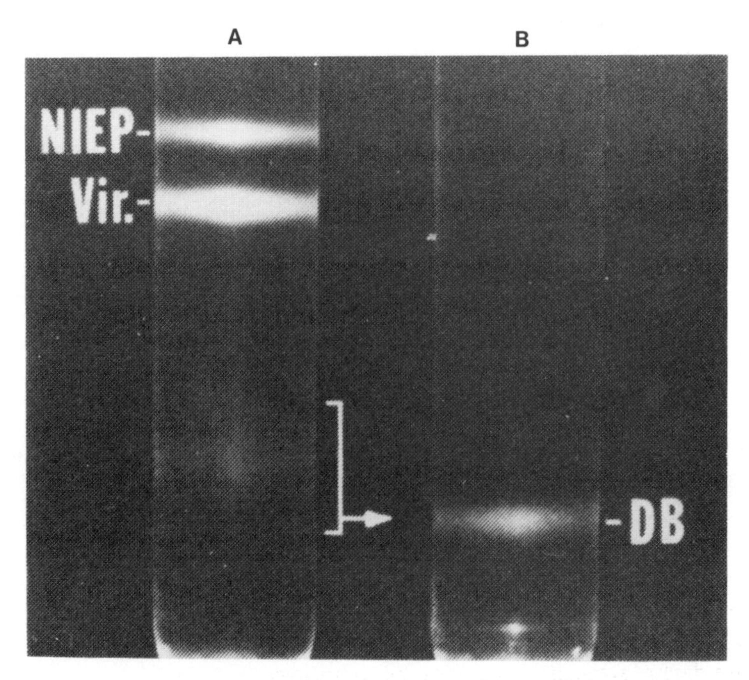

Figure. 9. Separation of extracellular HCMV particles by centrifugation. (**A**) Clarified medium for AD169-infected HFF cells was layered above a glycerol-tartrate gradient and centrifuged (40 000 r.p.m., 4°C, 15 min) in a Beckman SW41 rotor. Illumination from the top of the gradient revealed two sharp light-scattering bands designated as non-infectious enveloped particles (NIEP) and virions (Vir.) and a broad area containing dense bodies (DB). (**B**) Broad dense body zone was collected from the first gradient, diluted with TN buffer and banded to equilibrium in a similar gradient (40 000 r.p.m., 4°C, 18 h). Illumination from the top reveals a narrower band, varying in tinge from reddish at the top to brownish at the bottom. (Data from Irmiere and Gibson, 31.)

to avoid loss of infectivity and to maintain particle integrity the virus should not be recovered by pelleting. The negative viscosity/positive density gradient system of Barzilai *et al.* (29) first applied to CMV purification by Talbot and Almeida (30) has proved to be an efficient and reliable procedure to recover the particles from the medium in the following way.

(i) Scrape infected cells into the medium, and after a low-speed centrifugation (4°C, 1500 g, 10 min) layer 3 ml of the medium (without pelleting) onto 9 ml gradients of 30% glycerol/35% tartrate (prepared as described in ref. 30).

(ii) Centrifuge at 4°C, 40 000 r.p.m., 20 min in a Beckman SW41 rotor. Three bands are seen by light scattering in the resultant gradient (*Figure 9*). The bands are non-infectious enveloped particles (NIEP), virions and dense bodies in order of increasing density (31). Respective particle bands can be diluted in buffer and re-banded if desired.

Gibson recommends the following gradient variations which have proved useful in his laboratory.

(i) To process greater volumes, pour 18 ml gradients into 38 ml tubes for the

Beckman SW27 rotor and overlay with 20 ml of clarified medium. Centrifuge at 25 000 r.p.m. for 40 min at 4°C in the SW27 rotor.

(ii) Substitute 0.04 M phosphate buffer, pH 7.4, for the Tris-HCl buffer normally used, if an alternative buffer is desired.

(iii) Substitute sodium tartrate for potassium tartrate if the potassium salt interfers with assays (e.g., direct combination of gradient aliquots with SDS-containing buffer for polyacrylamide gel analysis). The sodium salt is slightly less soluble at 4°C than the potassium salt.

(iv) Use gradients of sucrose (15 – 50% w/v prepared in 0.04 M phosphate buffer, 0.15 M NaCl and adjusted to pH 7.4) as an alternative when the tartrate ion is not needed. Conditions of centrifugation are the same.

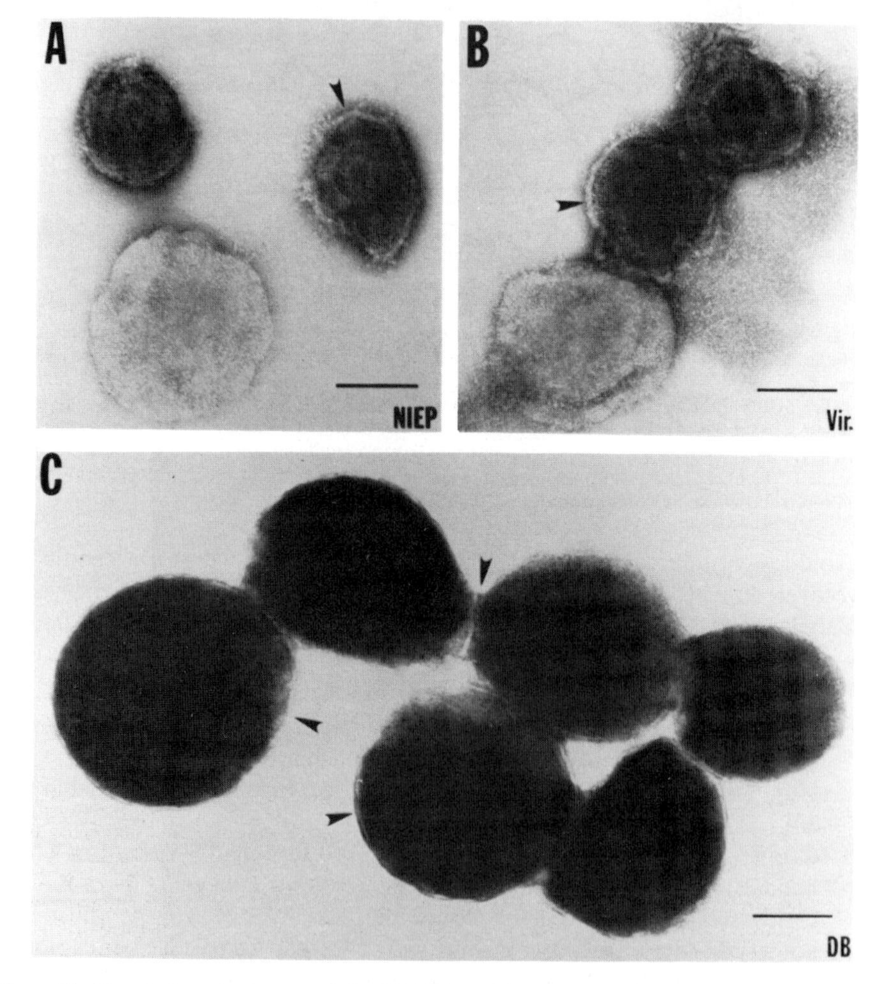

Figure 10. Negatively stained extracellular HCMV particles recovered from a glycerol-tartrate gradient. NIEP (**A**), virions (**B**) and dense bodies (**C**) were collected from a gradient, pelleted, resuspended in TN buffer (0.05 M Tris-HCl, pH 7.4, 0.10 M NaCl) adsorbed to grids and stained with 1% uranyl formate. Particles in lower left-hand corners of **A** and **B** are non-penetrated. Arrows indicate limiting membrane structure; bars represent 100 nm. (Data from Irmiere and Gibson, 31.)

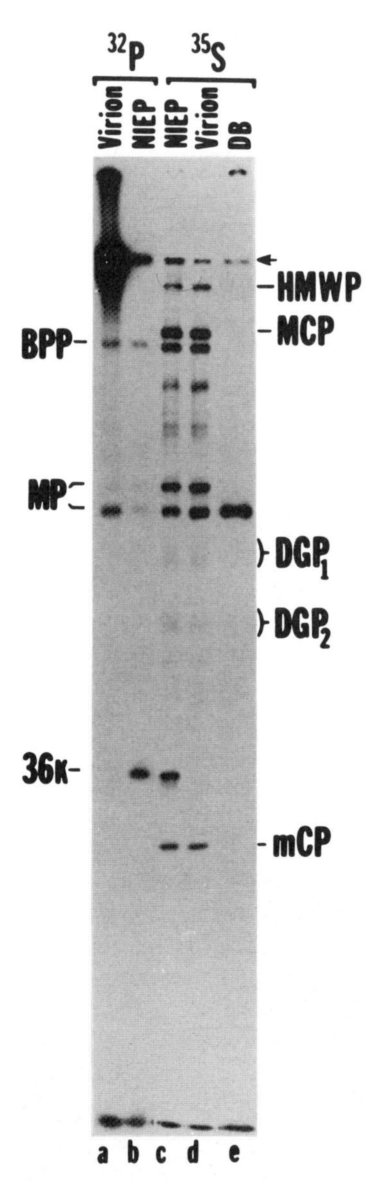

Figure 11. Fluorograms showing radiolabelled HCMV proteins. CMV-infected cells were radiolabelled with either [^{32}P]orthophosphate or [^{35}S]methionine. Extracellular particles, recovered using glycerol-tartrate gradients, were solubilised and the proteins separated by electrophoresis in a 10% 'high bis' polyacrylamide gel. Shown here are fluorographic images of the resulting gel. Different exposure times were required for channels **a, b** and **e**. Shown here are NIEP and virion proteins labelled with [^{32}P]phosphate (**a,b**) or [^{35}S]methionine (**c,d**) and dense body proteins labelled with [^{35}S]-methionine (**e**). Protein designations are as follows: HMWP (212 kd, high molecular weight protein), MCP (153 kd, major capsid protein), BPP (149 kd, basic phosphoprotein), MP (74 kd and 69 kd, matrix proteins), DGP_1 and DGP_2 (62 kd and 54 kd, discrete glycoproteins), 36 kd (36 000 dalton NIEP-specific 'assembly' protein) and mCP (34 kd, minor capsid protein) (see Gibson, 35). Arrow at top of right-hand margin indicates top of resolving gel. Intense exposure in stacking gel and at top of resolving gel in the ^{32}P-labelled virion preparation is due to viral DNA which is not present in NIEP. (Data from Irmiere and Gibson, 31.)

3.6 **Types of CMV Particle**

The morphological appearance of the negative-stained CMV particle types is shown in *Figure 10.* A comparison of the structural polypeptide profile following polyacrylamide gel electrophoresis is shown in *Figure 11.*

3.6.1 *NIEP*

Irmiere and Gibson (31) give a well described account of the characteristics of NIEP. Although similar in appearance and protein composition to virions, NIEP lack DNA and are therefore non-infectious.

3.6.2 *Dense Bodies*

These are simpler than NIEP and virions in structure and composition. They appear as large solid spheres filled with homogeneous material and bounded by an outer membrane (31 – 34). They have no DNA and 90% of their protein mass is represented by the 69-kd matrix protein.

3.6.3 *Virions*

Gibson (35) has presented excellent data on the structural profiles of various strains of human CMV and has compared these with simian CMV. His analysis is shown in *Figure 12* and his interpretation of the data is presented in *Table 3.*

4. HERPESVIRUS SAIMIRI

Herpesvirus saimiri (HVS) can be isolated from the blood and tissue culture cells of most healthy squirrel monkeys (*Saimiri sciureus*) but appears to be non-pathogenic in this, its natural host. The virus, however, has become of interest since it was shown to be extremely oncogenic in other primate tissues, in particular marmoset monkeys of the genus *Saguinus* (36). The latter hosts develop malignant tumours of the lymphatic system within 2 months of infection. The virus has been designated as belonging to the γ-sub-group of the herpetoviridae.

4.1 **Preparation of Virus Stocks**

Whilst a number of tissue culture cells have been tested, the most successful cells for lytic infection and growth of all virus strains are Owl Monkey kidney (OMK-210) cells (37,38). Vero cells are also permissive, although some strains of virus do not grow well in these cells. The OMK cells appear to grow well in most growth media supplemented with calf serum.

For virus stocks infect roller bottle cultures of OMK cells (containing $\sim 2 \times 10^8$ cells) at a low m.o.i. (<0.1 p.f.u./cell) and incubate at 37°C for 4 – 5 days. Incubation at 34°C may be optimum but, as with HVS, this may lead to problems with regard to temperature sensitivity. The virus shows cell to cell spread and the c.p.e. (at least for strain HVS 11) is characterised by the appearance of large balloon-like structures, which are not true syncytia.

The appearance of c.p.e., although normally taking 4 – 5 days, may vary from strain to strain. The progeny virus is released into the extracellular medium which

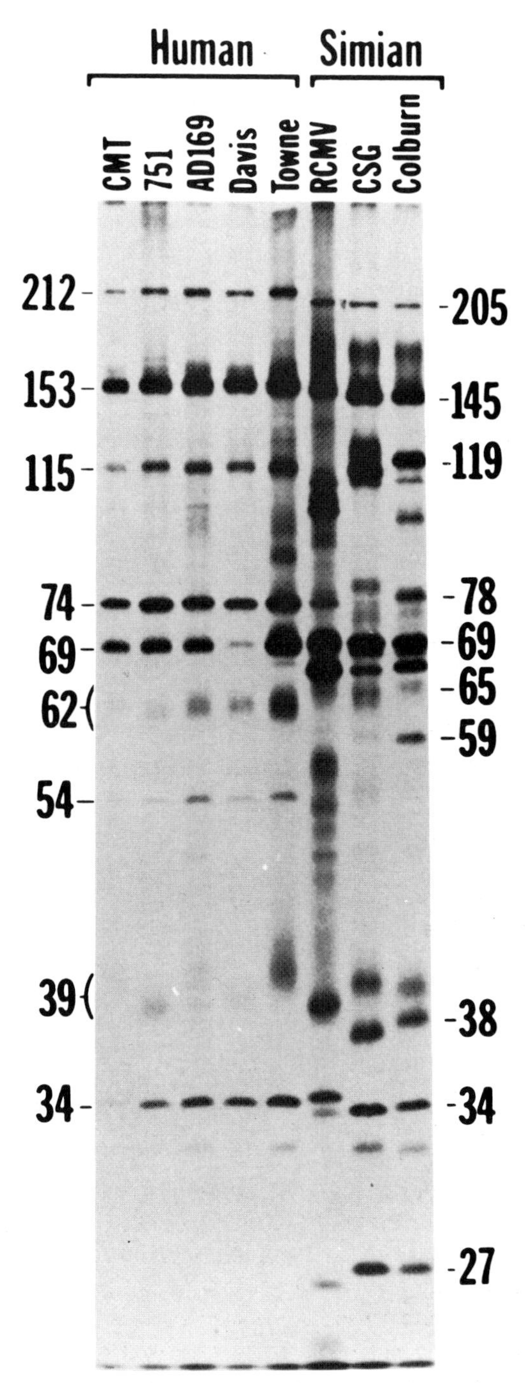

Figure 12. A comparison of the virion proteins of HCMV and SCMV isolates. Radiolabelled virions were recovered, solubilised and subjected to electrophoresis in a 10% plyacrylamide gel. Numbers in the right and left hand margins indicated the estimated molecular weights (x 10^{-3}) of the corresponding protein bands. (Data from Gibson, 35.)

Table 3. Characteristic Properties of CMV Virion Proteins.

Protein counterparts of different CMV strains (mol. wt. x 10^{-3})				*Protein designation*	*Characteristics*
Colburn[a]	*CSG*[a]	*RCMV*[a]	*HCMV 751*[b]		
205	205	207	212	High molecular weight (HMWP)	Largest virion protein
163	263		145	Acidic glycoprotein (AGP)	Glycosylated, acidic
145	143	150	153	Major capsid protein (MCP)	Principal capsid constituent, abundant
129	129	129	140		Nonvirion
119	114	100	149	Basic phosphoprotein (BPP)	Phosphorylated, basic, abundant
119	n.t.	n.t.	–	GP119	Glycosylated, acidic
112	112	112	115		Virion constituent
100	n.t.	n.t.	–	GP100	Glycosylated, acidic
98	–	–	–		
94	90	92	79	Immediate-early (IE)	Phosphorylated, acidic, nonvirion
78	82	75	80		Phosphorylated, virion
69	69	69	74	Upper matrix (MP)	Phosphorylated, abundant
66	66	66	69	Lower matrix (MP)	Phosphorylated, in dense body, abundant
65	64	65	62	Discrete glycoprotein (DGP_1)	Glycosylated, abundant
61	n.t.	n.t.	54	Discrete glycoprotein (DGP_2)	Glycoslyated
59	59	–	57	GP59, 57	Glycosylated
51	50	51	52	DNA-binding protein (DBP)	Phosphorylated, abundant, nonvirion, nuclear
40	41	39	42		Virion constituent
38	37	39	39		Virion constituent
37	37	38	35	Assembly protein	Phosphorylated, in B-capsids and NIEP, abundant
34	34	34	34	Minor capsid protein (mCP)	Capsid protein, slightly basic
32	32	–	32		Minor virion constituent
27	27	26	–		Major virion constituent

[a]Colburn, CS6 and RCMV are simian isolates.
[b]HCMV 751 a human isolate (data from Gibson, 35).

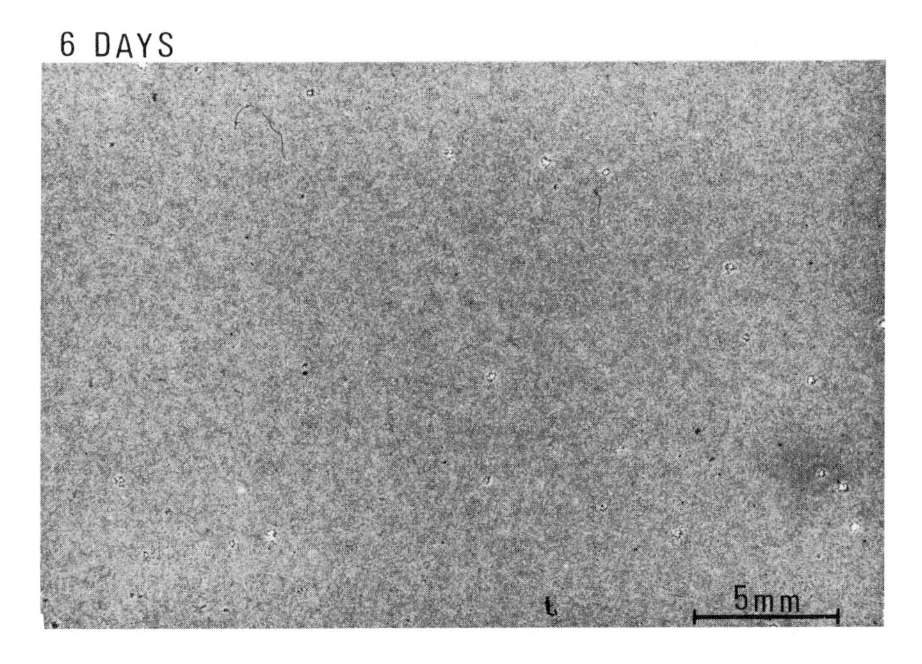

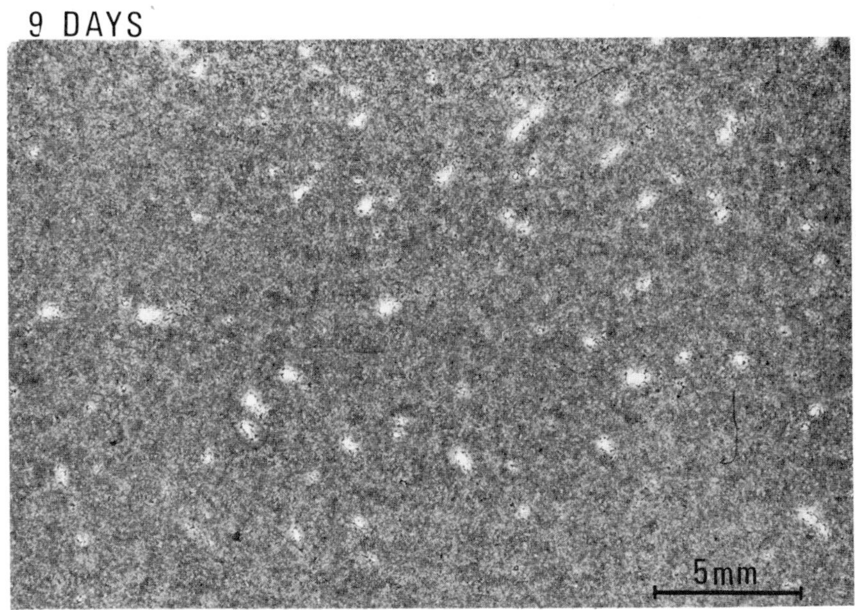

Figure 13. Morphological appearance of HVS plaques at either 6 or 9 days post-infection (kindly provided by R.E. Randall and R.W. Honess).

contains approximately 2 x 10^6 – 2 x 10^7 p.f.u./ml of 50 ml medium per roller bottle. Either store the virus at –70°C at such a titre, or first concentrate by pelleting. Pelleted virus routinely has a particle:infectivity ratio of 200 – 500.

HVS strain 11 is surprisingly stable at 4°C, losing only 0.5 $\log_{10}$ infectivity after 1 month's storage in growth medium.

4.2 Virus Assay

Suitable dilutions of virus are routinely assayed on either Vero or OMK cell monolayers grown on 25 cm² plastic medical flat bottles, using standard plaque assay techniques. Some batches of calf serum inhibit plaque formation in OMK and therefore serum should routinely be heated at 56°C for 30 min prior to use.

(i) Adsorb 1 ml aliquots of virus dilution for 1–2 h at 37°C.

(ii) Decant the 1 ml and replace by medium containing 2% calf serum and CMC (0.5%).

(iii) After incubation at 37°C plaques can be seen as small foci at 7–10 days post-infection. After approximately 14 days plaques are 2–3 mm in diameter, although this is of course strain-specific.

Plaques usually appear slightly earlier in monolayers of primary marmoset cells. *Figure 13* shows the morphology of HVS plaques at six and nine days post-infection.

4.3 Growth Cycle

Infection at high m.o.i. and harvesting of virus for assay at frequent intervals is the normal method for plotting virus growth curves. A growth curve under such

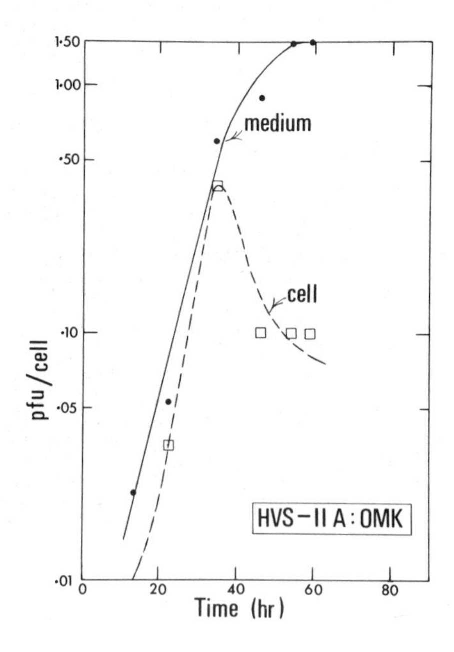

Figure 14. Representative growth curve for HVS in OMK cells at 37°C. The solid line represents appearance of released virus. The dashed line represents cell-associated virus. The m.o.i. was approximately 5 p.f.u./cell. (Kindly provided by R.E. Randall, R.W. Honess and P. O'Hare.)

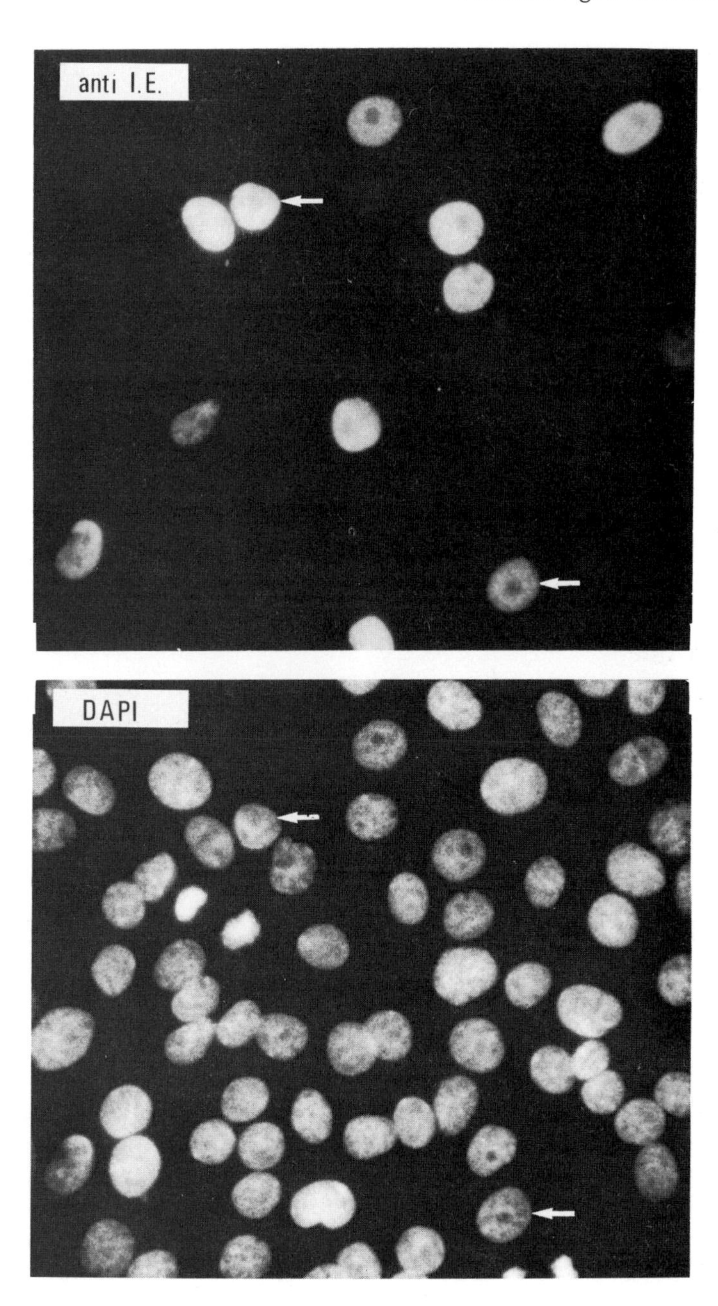

Figure 15. Demonstration of asynchrony in OMK cells previously infected with HVS at an m.o.i. of 5 p.f.u./cell. Cells were fixed at 15 h post-infection and stained by indirect immunofluorescence (rhodamine) with a monoclonal antibody to the major immediate early protein of HVS (Randall *et al.* (31)). The same cells were also counterstained with 2 mM 4′,6-diamidine-2-phenylindole (DAPI) which binds to DNA and thus stains nuclei (40). DAPI responds to u.v. excitation and gives blue fluorescence whereas rhodamine responds to green excitation and gives red fluorescence. Only a small proportion of the cells are showing synthesis of virus proteins. (Kindly provided by R.E. Randall and R.W. Honess.)

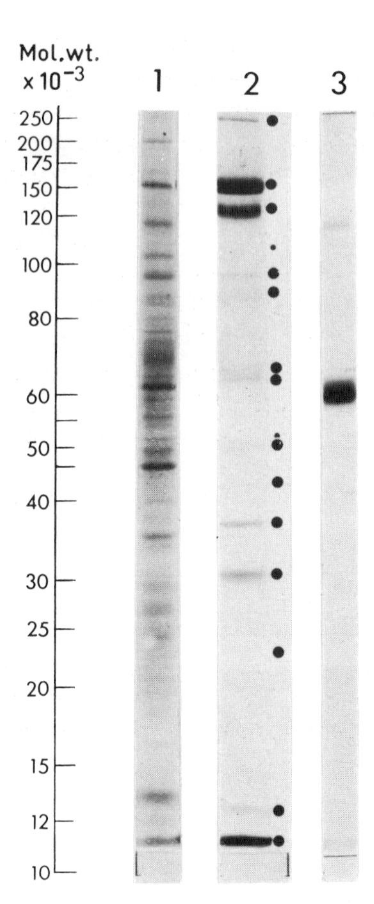

Figure 16. Polyacrylamide gel electrophoresis of (**1**) HVS-infected OMK cells (stained with Coomassie brilliant blue). (**2**) A purified preparation of HVS (stained with Coomassie brilliant blue) and (**3**) a purified preparation of HVS (labelled with ^{32}P). (Kindly provided by R.E. Randall and R.W. Honess.)

conditions can be devised for HVS and appears to be long (*Figure 14*). However, one has to be wary as to what such a growth curve represents. Randall and Honess (personal communication) have demonstrated that the infection of cells used under such conditions is totally asynchronous and many cells show a delay in the initiation of virus replication. Indeed it appears that, once initiated, the cycle of growth within one infected cell is over within 18 h. The virus replication cycle also seems to be dependent on the cell cycle. These intricate studies have been carried out by immunofluorescence using monoclonal antibodies directed against infected cell antigens (*Figure 15*).

4.4 Preparation of Purified Virus

(i) Infect monolayers on roller bottles as for virus stocks, using a low m.o.i., and at 4 – 5 days post-infection pellet the virus in the extracellular medium at 25 000 r.p.m. in a Beckman SW27 rotor for 30 min. If required, add radioactive label at 2 days post-infection.

(ii) Re-suspend the pellet in a suitable buffer (Tris-HCl with 50 mM EDTA and 50 mM salt) and whenever possible leave overnight at 4°C.

(iii) Layer the disrupted pellet onto either a glycerol (20 – 40%) or sucrose (20 – 40%) gradient and centrifuge for 20 – 30 min at 19 000 r.p.m. in a Beckman SW27 rotor. Observe the tube under light scattering conditions, when one major central band should be seen.

(iv) Remove this band, which contains more than 90% enveloped particles, and pellet for 1 h at 25 000 r.p.m. before re-suspending in a suitable buffer.

The initial gradient run often results in a pellet containing high concentrations of virus which can be resuspended and purified in a new gradient. Fractions of pure virus may be taken for protein, particle and infectivity determinations. For use in polyacrylamide gel electrophoresis disrupt the virus using the methods described above for HSV. A structural polypeptide pattern for HVS strain 11 is shown in *Figure 16*.

5. ACKNOWLEDGEMENTS

We would like to thank colleagues who provided data for this chapter whether they were used or not, namely Gretchen Caughman, Bernard Fleckenstein, Ian Halliburton, Elliot Kieff, Roger Maes, George Miller, Dorothy Purifoy and Norman Ross. We are particularly indebted to the extensive data supplied by Wade Gibson, Rick Randall and Bob Honess. Last (but by no means least) we thank our patient typist Chris Moorhouse.

6. REFERENCES

1. Matthews,R.E.F. (1982) *Intervirology,* **17**, Nos. 1-3.
2. Rawls,W.E., Bacchetti,S. and Graham,F.L. (1977) *Curr. Top. Microbiol. Immunol.,* **77**, 71.
3. Scully,C. (1983) *Oral Surg.,* **56**, 285.
4. Spear,P.G. and Roizman,B. (1968) *Virology,* **36**, 545.
5. Watson,D.H., Shedden,W.I.H., Elliot,A., Tetsuka,T., Wildy,P., Bourgaux-Ramoisy,D. and Gold,E. (1966) *Immunology,* **11**, 399.
6. Bronson,D.L., Dreesman,G.R., Biswal,N. and Benyesh-Melnick,M. (1973) *Intervirology,* **1**, 141.
7. Frenkel,N., Jacob,R.J., Honess,R.W., Hayward,G.S., Locker,H. and Roizman,B. (1975) *J. Virol.,* **16**, 153.
8. Watson,D.H. (1962) *Biochim. Biophys. Acta,* **61**, 321.
9. Russell,W.C. (1962) *Nature,* **195**, 1028.
10. Yasamura,T. and Kawakita,Y. (1963) *Nippon Rinsho,* **21**, 1201.
11. Plummer,G., Goodheart,C.R., Miyagi,M., Skinner,G.R.B., Thouless,M.E. and Wildy,P. (1976) *Virology,* **60**, 206.
12. Spear,P.G. (1975) in *Oncogenesis and Herpesviruses,* Vol. **II**, part 1, Epstein,M.A. and Zur Hausen,H. (eds.), IARC, Lyon, p. 49.
13. Killington,R.A., Newhook,L., Balachandran,N., Rawls,W.E. and Bachetti,S. (1981) *J. Virol. Methods,* **2**, 223.
14. Beale,A.J., Christofinis,E.C. and Furminger,I.G.S. (1963) *Lancet,* **11**, 640.
15. Powell,K.L. and Watson,D.H. (1975) *J. Gen. Virol.,* **29**, 167.
16. Spear,P.G. and Roizman,B. (1972) *J. Virol.,* **9**, 143.
17. Heine,J.W., Honess,R.W., Cassai,E. and Roizman,B. (1974) *J. Virol.,* **14**, 640.
18. Dimmock,N.J. and Watson,D.H. (1969) *J. Gen. Virol.,* **5**, 499.
19. Powell,K.L. and Courtney,R.J. (1975) *Virology,* **66**, 217.

20. Honess,R.W. and Roizman,B. (1973) *J. Virol.,* **12**, 1347.
21. Sabine,M., Robertson,G., Whalley,J. (1981) *Aust. Vet. J.,* **57**, 148.
22. Studdert,M., Simpson,T., Roizman,B. (1981) *Science (Wash.),* **214**, 562.
23. Allen,G.P., Yeargen,M.R., Turtinen,L.W., Bryans,J.T. and McCollum,W.H. (1983) *Aust. J. Vet. Res.,* **44**, 263.
24. O'Callaghan,D.J., Cheevers,W.P., Gentry,G.A. and Randall,C.C. (1968) *Virology,* **36**, 104.
25. Darlington,R.W. and Randall,C.C. (1963) *Virology,* **19**, 332.
26. Wolf,K. and Darlington,R.W. (1971) *J. Virol.,* **8**, 525.
27. Smith,J.D. and DeHarven,E. (1973) *J. Virol.,* **12**, 919.
28. Wentworth,B.B. and French,L. (1970) *Proc. Soc. Exp. Biol. Med.,* **135**, 253.
29. Barzilai,R., Lazarus,L.H. and Goldblum,N. (1972) *Arch. gesamte Virusforsch.,* **36**, 141.
30. Talbot,P. and Almeida,J.D. (1977) *J. Gen. Virol.,* **36**, 345.
31. Irmiere,A. and Gibson,W. (1983) *Virology,* **130**, 118.
32. Craighead,J.E., Kanich,R. and Almeida,J.D. (1972) *J. Virol.,* **10**, 766.
33. Sarov,I. and Abady,I. (1975) *Virology,* **66**, 464.
34. Stinski,M.F. (1976) *J. Virol.,* **19**, 594.
35. Gibson,W. (1983) *Virology,* **128**, 391.
36. Melendez,L.V., Hunt,R.D., Daniel,M.D., Garcia,F.G. and Fraser,C.E.O. (1969) *Lab. Anim. Care,* **19**, 378.
37. Daniel,M.D., Silva,D. and Ma,N. (1976) *In Vitro,* **12**, 290.
38. Todaro,G.J., Scherr,C.J., Sen,A., King,N., Daniel,M.D. and Fleckenstein,B. (1978) *Proc. Natl. Acad. Sci. USA,* **75**, 1004.
39. Randall,R.E., Newman,C. and Honess,R.W. (1984) *J. Gen. Virol.,* in press.
40. Russell,W.C., Newman,C. and Williamson,D.M. (1975) *Nature,* **253**, 461.

FOR FURTHER READING

The Herpesviruses, Roizman,B. (ed.), Plenum Press, New York.
Roizman,B. and Furlong,D. (1974) in *Comprehensive Virology*, Vol. **3**, Fraenkel-Conrat,H. and Wagner,R.R. (eds.), Plenum, New York, pp. 219.

CHAPTER 11

Techniques in Clinical Virology

P. MORGAN-CAPNER and J.R. PATTISON

1. INTRODUCTION

The laboratory diagnosis of virus infection in clinical practice is made using three basic techniques – microscopy, cell culture and serology. In general one or other of these techniques is most appropriate for resolving a given diagnostic situation and this often depends upon the stage at which a patient is investigated in relation to the onset of the infection. For example, all three approaches may be useful in the diagnosis of varicella-zoster virus infection but the success of microscopy and cell culture depends upon the availability of satisfactory specimens taken relatively early in the illness.

The success of diagnostic tests depends greatly upon the quality of the specimens received. In this respect there is no substitute for close links between the laboratory and the patient and this may involve laboratory staff in obtaining the specimens themselves. Details of the nature of the specimens required and their transport to the laboratory are given by Lennette and Schmidt (1) and by Grist *et al.* (2).

Many of the reagents required for diagnostic virology tests are commercially available, and indeed an ever increasing number of kits is being marketed. Often there is more than one satisfactory commercial source for a reagent. Therefore we have tried to avoid favouring a particular commercial firm and in this chapter we mention a specific supplier only when it is the sole source of the reagent in question. Otherwise reference has to be made to a general list of suppliers (*Table 1*).

A comprehensive account of all the techniques available for the diagnosis of

Table 1. Examples of Commercial Sources of Reagents and Equipment.

Flow Laboratories:	Woodcock Hill, Harefield Road, Rickmansworth, Hertfordshire WD3 1PQ, UK
Gibco Europe:	Unit 4, Cowley Mill Trading Estate, Longbridge Way, Uxbridge, Middlesex UB8 2YG, UK
Tissue Culture Services:	10 Henry Road, Slough, Berkshire SL1 2QL, UK
Wellcome Diagnostics:	Temple Hill, Dartford, Kent DA1 5BR, UK
Northumbria Biologicals:	South Nelson Industrial Estate, Cramlington, Northumberland NE23 9HL, UK
Oxoid:	Wade Road, Basingstoke, Hampshire RG24 0PW, UK
Dynatech Laboratories Ltd.:	Daux Road, Billingshurst, Sussex RH14 9SJ, UK
Sterilin Ltd.:	43/45 Broad Street, Teddington, Middlesex TW11 8QZ, UK
Abbott Laboratories Ltd.:	Brighton Hill Parade, Basingstoke, Hampshire RG22 4EH, UK

human virus infections would be outside the scope of this chapter. Rather we have attempted to describe examples of the basic approaches, which, granted specimens of reasonable quality, will provide a well attested introduction for workers entering the field. Thereafter, with increased experience, the same basic techniques can be applied to more difficult problems.

2. ELECTRON MICROSCOPY

For virus diagnosis electron microscopy may be used for examination of thin sections of tissue or, more frequently, for examination of material such as faeces or fluid obtained from the vesicles present in certain rashes such as chickenpox. When examining such material, viruses are visualised by using negative staining techniques which delineate the virion by outlining its components with electron-dense material. This technique can only be applied directly to specimens which contain very large numbers of virions (at least $\sim 10^6$/ml) as in faeces or vesicle fluids. Alternatively, in specimens containing small numbers of virus particles, the chances of seeing the virus can be increased by concentration using ultracentrifugation or by aggregating particles with specific antibody (immuno-electron microscopy). This latter technique can also be useful in identifying viruses. We describe here a technique for the demonstration of rotaviruses in faeces by direct electron microscopy and an immuno-electron microscopy technique which has proved useful in the detection of parvoviruses in human serum. Current electron microscopy techniques are discussed in more detail by Field (3).

2.1 Direct Electron Microscopic Examination of Faeces

(i) Dip a Pasteur pipette tip in faeces and transfer sufficient to produce a 1 cm smear on a glass slide.
(ii) Re-suspend the faecal smear in a volume of electron microscopy negative stain such that the resulting suspension is just translucent. An appropriate negative stain is 2% phosphotungstic acid (PTA) in distilled water adjusted to pH 7.2 with concentrated potassium hydroxide.
(iii) Place a small drop of the suspension onto a carbon-formvar coated electron microscope grid (Agar Aids, 66a Cambridge Road, Stansted, Essex) which is held in a pair of fine forceps.
(iv) Leave for 30 sec.
(v) Remove excess fluid by touching the side of the grid with the edge of a torn piece of filter paper.
(vi) Leave to air dry.
(vii) If considered necessary, any viable virus can be inactivated by irradiating each side of the grid with u.v. light of an intensity of 440 000 μW sec/cm^2 using an R-52 Mineralight short wave u.v. lamp with filter (Ultra-Violet Products Ltd., Science Park, Milton Road, Cambridge, UK), a distance from light to grid of 15 cm and an irradiation period of 5 min per grid side.
(viii) Examine by transmission electron microscopy at a screen magnification of 30 000 – 50 000 for the characteristic virions of rotavirus (*Figure 1*).

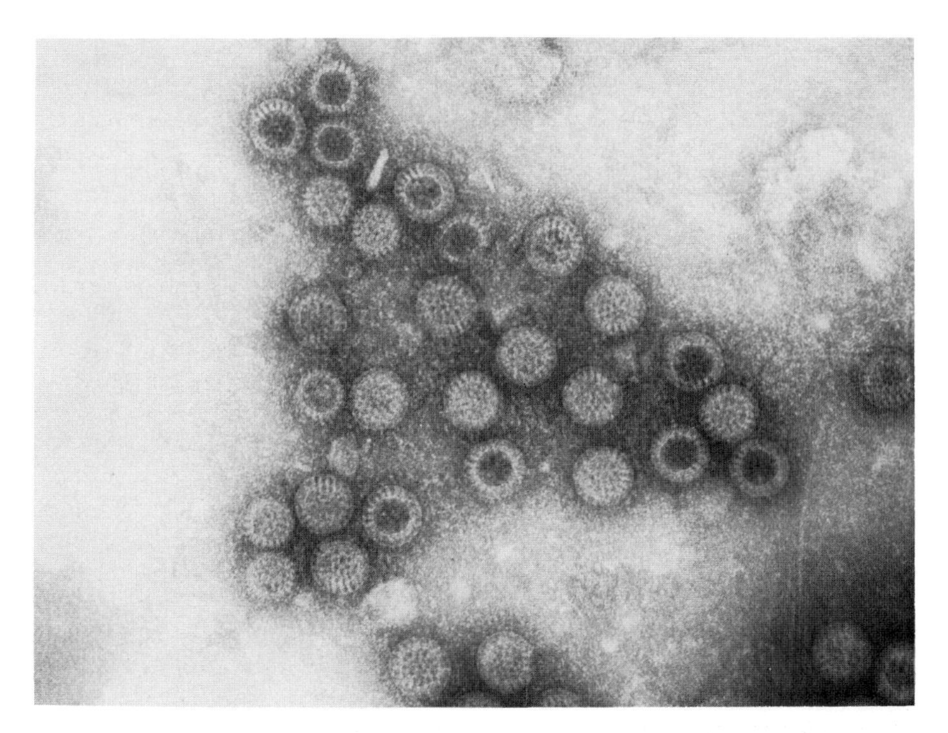

Figure 1. Electron micrograph of rotavirus.

2.2 Immuno-electron Microscopy

The following technique for immuno-electron microscopy is only one example of a number of such methods that are available. A further suitable technique uses protein A bound to the grid to retain the virus-specific antiserum (3).

The concentration of anti-parvovirus antiserum used may have to be established by trial and error in the range of 1 in 10 to 1 in 1000 but the concentration we quote has been found satisfactory for routine use. Similarly, various concentrations from 1 in 10 to 1 in 100 of the serum containing parvovirus may have to be tested in order to obtain the optimum virus/antibody interaction.

(i) Dilute 10 μl human parvovirus antiserum 1 in 100 in phosphate buffered saline (PBS) and warm to 56°C in a water bath.
(ii) Melt a 10 ml volume of 2% agarose in PBS in a universal and cool to 56°C in a water bath.
(iii) Mix 1 ml of 1 in 100 antiserum with 1 ml 2% agarose in a universal at 56°C.
(iv) Pipette 200 μl of the mixture into two wells of a microtitre plate.
(v) Allow to set at room temperature. Such plates may be stored at 4°C for several weeks when sealed with adhesive tape.
(vi) Add 10 μl of a 1 in 10 dilution in PBS of parvovirus-containing serum to the well containing agarose/antiserum.

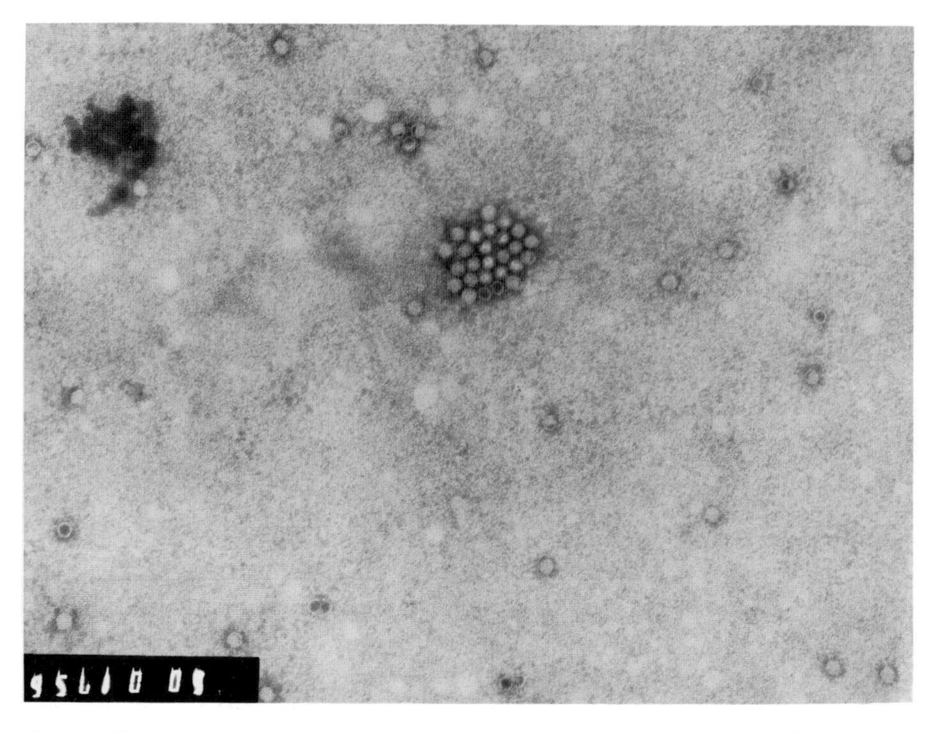

Figure 2. Electron micrograph of parvovirus showing aggregation by parvovirus-specific antibody.

(vii) Place an electron microscope grid with the carbon formvar-coated surface (the 'less shiny' side) face down onto the surface of the drop of serum.
(viii) Leave for 2 h at 37°C in a humid container.
(ix) Remove the grid with fine forceps and add a drop of 2% PTA [see Section 2.1 (ii)] to the surface of the grid which was in contact with the serum.
(x) Leave for 30 sec, remove excess stain, air dry and inactivate as before [Section 2.1 (iv – vii)].
(xi) Examine by transmission electron microscopy at a screen magnification of 30 000 – 50 000 for the characteristic 'clumps' of virus particles (*Figure 2*).

3. VIRUS ANTIGEN DETECTION

Detection of virus in tissue or body fluid may be achieved by demonstrating virus-specified protein using the specificity of an antigen-antibody reaction. The antigen detected is usually a protein of the virus capsid. Detection of the antigen-antibody reaction is achieved either by labelling the anti-viral antibody itself (the direct technique) or by labelling an antibody directed against the virus-specific antibody (the indirect technique). Antibodies may be labelled with fluorescein, radioactive iodine or an enzyme which will induce a colour change in an appropriate substrate. Alternatively, red blood cells may be coated with virus-specific antibody, the presence of antigen being detected by haemagglutination.

In current practice the main uses for these techniques are in the detection of hepatitis B antigens in the blood or the detection of antigens of a number of viruses which cause respiratory tract infection.

A number of satisfactory commercial kits are available (e.g., by Abbott Laboratories Ltd., UK) for the detection of hepatitis B antigens and red blood cells, radioisotope and enzyme markers are incorporated into the various test systems. These techniques are not described further since it is sufficient in general to follow the manufacturers' instruction. Rather we describe an immuno-fluorescent technique for the detection of respiratory syncytial virus antigen in nasopharyngeal aspirates.

3.1 **Immunofluorescence Detection of Respiratory Syncytial Virus Antigen in a Nasopharyngeal Aspirate**

The technique for obtaining nasopharyngeal aspirates is described by Gardner and McQuillin (4). The laboratory procedure is performed in two stages. First is the processing of the nasopharyngeal mucus so that cells are deposited on small areas of a glass slide. The fixed slide preparations obtained may be stored at −20°C for many months. The second stage of the procedure is the staining for respiratory syncytial virus (RS virus) antigen using an indirect technique with RS virus-specific antibody and fluorescein-conjugated anti-globulin prepared against the globulin of the species of animal in which the RS virus antiserum was prepared.

3.1.1 *Preparation of Slides of Nasopharyngeal Aspirate*

(i) Wash out the mucus from the mucus extractor with 1−2 ml PBS and transfer to a centrifuge tube.
(ii) Centrifuge for 10 min at 1500 r.p.m. (~500 *g*) in a bench centrifuge.
(iii) Remove the supernatant (which may be used for virus isolation).
(iv) Gently re-suspend the cell deposit with 2−3 ml of PBS using a wide-bore Pasteur pipette until the deposit is broken up into a fairly smooth suspension.
(v) Transfer the suspension to a test tube.
(vi) Add a further 2−4 ml of PBS and mix again with a Pasteur pipette. Discard any large fragments of mucus.
(vii) Centrifuge for 10 min at 1500 r.p.m. (~500 *g*) in a bench centrifuge.
(viii) Discard the supernatant and re-suspend the deposit in sufficient PBS such that the suspension runs freely down the side of the test tube.
(ix) Place drops of suspension onto discrete areas on a glass slide.
(x) Allow to air dry.
(xi) Fix the slides in acetone for 10 min at 4°C.
(xii) Remove the slides from the acetone and allow to air dry.
(xiii) Use the slides immediately or store at −20°C.

3.1.2 *Staining for RS Virus Antigen*

Control slides which are positive and negative for RS virus antigen should be examined simultaneously with any unknown test slides.

(i) Reconstitute and dilute commercial RS virus antiserum in PBS to the recommended working strength.
(ii) Place one drop of anti-RS virus on the cell preparation using a Pasteur pipette.
(iii) Place the slide in a humid container (sandwich box or Petri dish containing moistened paper tissue).
(iv) Incubate for 30 min at 37°C.
(v) Gently wash the anti-RS virus off the slides with PBS in a wash bottle.
(vi) Immerse the slides in PBS for three 10 min periods, changing the PBS between each period of washing.
(vii) Drain the slides, wipe off excess PBS with paper tissue (being careful not to touch the cell preparation) and air dry.
(viii) Reconstitute and dilute the fluorescein-conjugated anti-globulin to the recommended working strength.
(ix) Place one drop of conjugated anti-globulin on the cell preparation.
(x) Incubate for 30 min at 37°C.
(xi) Wash three times in PBS as before.
(xii) Wash for 2 min in distilled water.
(xiii) With care, wipe off the excess distilled water and air dry.
(xiv) Examine the slides by oil immersion with a fluorescence microscope. Coverslip and mountant are not required.

3.1.3 *Interpretation*

The presence of RS virus antigen will be shown by areas of apple green fluorescence in the cytoplasm of some cells.

4. CELL CULTURE

A major advance in the development of clinical virology was the ability to culture cells which would support virus replication in monolayers on glass or plastic surfaces. With the advent of antibiotics to suppress bacterial and fungal contamination the technique became established in routine practice. It is beyond the scope of this chapter to give full details of the preparation and maintenance of monolayers. For this aspect the reader is referred to Paul (5) and for details specifically relevant to clinical virology to Lennette and Schmidt (1).

It is not cost-effective for a single diagnostic laboratory to maintain a full range of cell lines capable of isolating all known human viruses. Rather the range of cell lines used by an individual laboratory will depend upon factors such as access to foetal tissue, particular interest in certain viruses and local preference. In general a satisfactory range consists of an epithelial cell line used as a primary or secondary culture (e.g., human embryo kidney or baboon kidney), a semi-continuous or continuous fibroblast cell-line (human embryo lung or MRC 5 cells) and a continuous epithelial cell line such as HEp2 or HeLa cells.

The isolation of viruses in clinical practice consists of two steps, first the recognition of the presence of a virus and second the specific identification of any virus present. The presence of a virus is most frequently detected by the development of morphological changes (cytopathic effect or cpe) in the cell monolayer. For some viruses the cpe is sufficiently characteristic to allow presumptive identification of the virus particularly when considered in relation to the clinical specimen. For example, most laboratories on seeing a characteristic cpe in fibroblast cells inoculated with material from a genital swab will report along the lines 'cpe characteristic of herpes simplex virus seen' and proceed no further with formal identification. In other instances the cpe may be characteristic of a large group of viruses such as enteroviruses with at least 67 types. Here specific identification must depend upon the neutralisation of the infectivity with specific antisera (the neutralisation test). The latter is often time-consuming, and an interim report such as 'Enterovirus isolated; identification to follow' may be issued.

Some viruses such as myxo- and paramyxoviruses do not regularly produce an obvious cpe but alter the surface of the culture cells so that they will bind red blood cells. This haemadsorption is usually detected by flooding the monolayer with a suspension of guinea pig or human group O red blood cells after a period of incubation with the clinical material and looking for evidence of haemadsorption after careful washing. Identification is then performed using a haemadsorption-inhibition test incorporating specific antisera. Some of these viruses also produce a soluble haemagglutinin and identification can be achieved by haemagglutination-inhibition.

Other viruses may be identified by spotting infected cells onto glass slides,

Table 2. Suitable Cell Culture Lines and Detection and Identification of Viruses most Frequently Encountered in Clinical Virology.

Virus	*Suitable cell lines for isolation*	*Detection and identification*
Adenovirus	HEK	cpe, NT
Coxsackievirus A	BK[a]	cpe, NT
Coxsackievirus B	BK	cpe, NT
Cytomegalovirus	HEL	cpe
Echovirus	BK	cpe, NT
Herpes simplex	HEL	cpe, NT, IF
Influenza A and B	BK	HAD, HADI, IF
Measles	BK, HEK	cpe, NT, IF
Mumps	BK	HAD, HADI
Parainfluenza	BK	HAD, HADI, IF
Poliovirus	BK	cpe, NT
Respiratory syncytial virus	HEp2	cpe, NT, IF
Rhinovirus	HEL	cpe, NT
Rubella	RK 13	cpe, NT, IF
Varicella/zoster	HEL	cpe

[a]Some types only.
HEL: human embryo lung fibroblasts, BK: baboon kidney, HEK: human embryo kidney, RK: rabbit kidney, HEp2: continuous human epithelial line, cpe: cytopathic effect, NT: neutralisation test, IF: immunofluorescence, HAD: haemadsorption, HADI: haemadsorption inhibition.

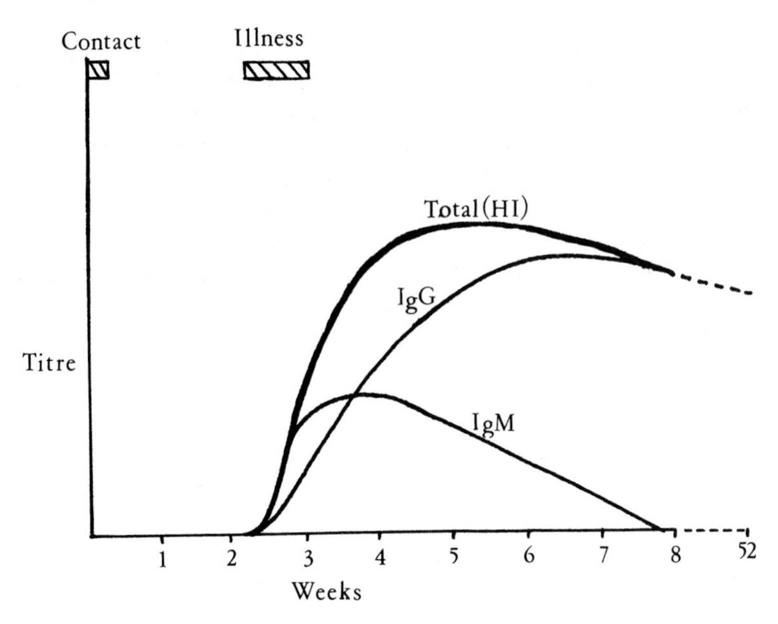

Figure 3. Serological response in primary rubella.

staining sequentially with an anti-viral antibody and a fluorescein-labelled anti-globulin and examining under u.v. light. Such a procedure can be used for RS virus (see Section 3.1.2). Finally, if the clinical details and cpe do not permit conclusions concerning the likely group of viruses to which the isolate belongs, a cell culture homogenate should be examined by electron microscopy (see Section 2). This procedure is also useful to check from time to time that adventitious viruses are not present in the cell-culture lines being used.

In *Table 2* we summarise the main groups of clinically important viruses, appropriate cell cultures for their isolation and the means of recognition and identification. The details of the tests involved are given by Lennette and Schmidt (1).

5. SEROLOGY

Numerous techniques are available for the detection and quantification of virus-specific antibodies. They may detect virus-specific IgM and IgG (e.g., haemagglutination-inhibition), virus-specific IgG only (e.g., radial haemolysis) or virus-specific IgM only (e.g., M-antibody capture assays). The complement fixation test detects only specific IgG for most virus antigens but for other microbial antigens may detect both specific IgM and IgG.

Serology is used both for diagnosing a current, recent or congenital virus infection, and for screening for antibody, which indicates previous infection and therefore immunity to a further primary infection. There is considerable individual variation in the quantitative aspects of a specific anti-viral antibody response, but a typical example of the serological response in a primary rubella infection is shown in *Figure 3*. It can be seen that diagnosis of recent rubella may be made by demonstrating the development of specific antibody (seroconversion), by demonstrating a significant (i.e., greater than 4-fold) rise in titre of total antibody or by the demonstration of specific IgM. Intrauterine infection can also be diagnosed by demonstrating specific IgM in neonatal blood since maternal IgM does not cross the placenta although maternal IgG does. Specific IgG usually persists for life after the primary infection and the detection of this class of antibody can be taken to indicate immunity to primary infection.

The examples of serological techniques which follow are all for the detection of rubella-specific antibody, since such techniques have reached a high degree of refinement for both diagnosis, especially in relation to pregnancy, and antibody screening to ascertain whether rubella immunisation is required. Further details of the techniques applicable to the diagnosis of rubella can be found in recent reviews by Pattison (6) and Morgan-Capner (7).

5.1 **Haemagglutination-inhibition (HI) Test**

Rubella virus is able to haemagglutinate the red blood cells (rbc) of a wide variety of species, those of day-old chicks being most frequently used for the HI test. Rubella haemagglutinating (HA) antigens consist of cell-culture-grown virus treated with Tween 80/ether to increase the HA titre.

5.1.1 *Standardisation of Rubella HA*

(i) Wash 1 ml of an approximately 50% suspension of chick rbc three times in dextrose gelatin veronal (DGV) buffer (*Table 3*) by centrifugation and resuspension in a 15 ml graduated centrifugation tube. Finally reconstitute to 30% in DGV.

Table 3. Dextrose Gelatin Veronal (DGV) Buffer.

(i)	DGV buffer solution:	
	Complement fixation test (CFT) buffer tablets	20
	Barbitone sodium	400 mg
	Gelatin	1200 mg
	Distilled water	2000 ml
	Dissolve CFT tablets in distilled water. Add barbitone sodium and gelatin. Place in 56°C water bath until gelatin is dissolved. Distribute in bottles in 100 ml volumes. Sterilise by autoclaving and store at 4°C.	
(ii)	25% Bovine plasma albumin (BSA):	
	Bovine plasma albumin fraction V	25 g
	Sterile distilled water	100 ml
	Sterilise by filtration and distribute aseptically in 1 ml volumes. Store at −20°C.	
(iii)	10% Glucose:	
	Glucose	10 g
	Distilled water	100 ml
	Dispense in 1 ml volumes. Autoclave and store at 4°C.	
(iv)	For use:	
	DGV buffer solution	100 ml
	25% BSA	0.8 ml
	10% glucose	1.0 ml
	Store at 4°C.	

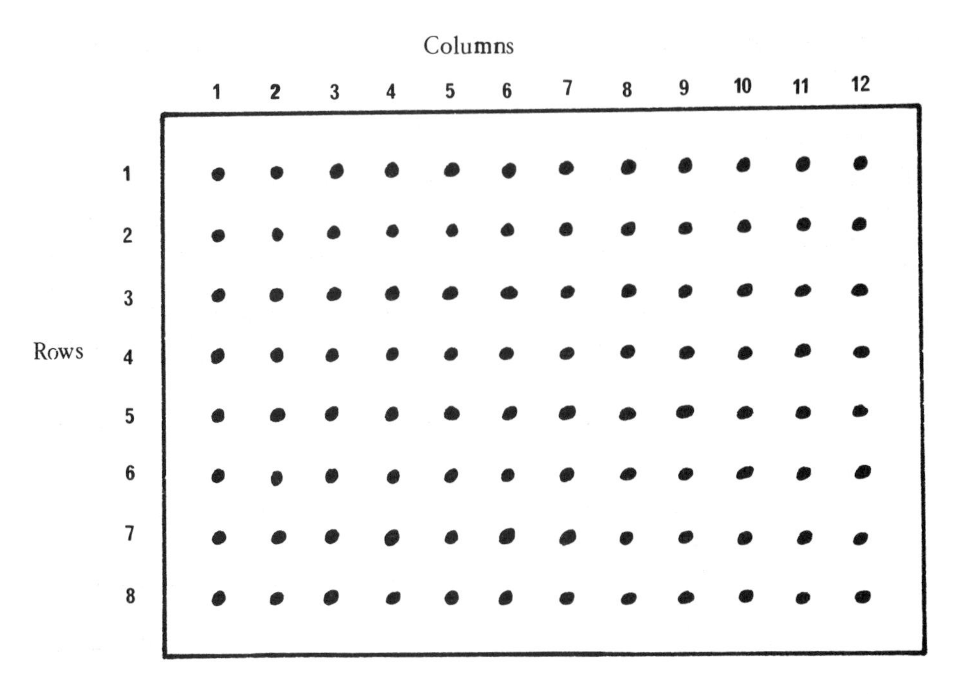

Figure 4. Polystyrene microtitre plate.

(ii) Reconstitute the freeze-dried rubella HA with the recommended volume of distilled water.

(iii) Add one volume (0.025 ml) of DGV to each of eight wells in two adjacent rows of a polystyrene 96-well microtitre plate with U-shaped wells (*Figure 4*, rows 1 and 2, columns 1–8).

(iv) To the two wells in column 1 add one volume of rubella HA.

(v) Make serial 2-fold dilutions of rubella HA from this initial 1 in 2 dilution to 1 in 256 using a 0.025 ml microdiluter. Before use, the microdiluter head must be heated in a bunsen to red heat, allowed to cool for a few seconds and then placed in distilled water. Blot the 'wetted' microdiluter on absorbent filter paper and it is then ready for use.

(vi) Add a further volume of DGV to each well used and two volumes of DGV to the wells in column 12, rows 1 and 2, to act as cell control.

(vii) Dilute the washed chick rbc 1 in 100 in DGV to make a 0.03% suspension.

(viii) Add two volumes of 0.03% chick rbc to all 18 wells, cover the plate with an unused plate and place at 4°C for 1 h.

(ix) The cell control wells should show a tight button of rbc in the bottom of the wells so excluding autoagglutination of the chick rbc. Agglutinated rbc will form a carpet of cells on the bottom of the well and the HA titre of the antigen is the reciprocal of the highest dilution of rubella HA which shows complete agglutination. This dilution contains 1 HA unit.

For testing sera, the rubella HA is used at 4 HA units. Thus, if the titre of the HA antigen is 32 the reconstituted rubella HA should be diluted 1 in 8 in DGV for the detection of rubella antibody. This diluted antigen is stable at 4°C for 24 – 48 h.

5.1.2 *Pre-treatment of Sera*

All sera contain non-specific inhibitors of haemagglutination which need to be removed prior to testing for rubella-specific antibody. These inhibitors are predominantly lipoproteins and may be removed by treatment with kaolin (see below). Non-specific agglutinins of chick rbc may occasionally be present in human sera but it is not necessary to routinely absorb the sera with chick rbc prior to testing. Any non-specific haemagglutinins present in a serum will be revealed in the serum control wells of the test.

(i) Pipette 0.2 ml of serum into an appropriately labelled 75 mm x 12 mm glass tube. Known positive and negative sera must be included in each batch of tests.
(ii) To each tube add 0.8 ml of a 25% suspension of acid-washed kaolin in borate-saline buffer pH 9.0.
(iii) Mix by tapping the tubes and stand the tubes at room temperature for 20 min.
(iv) Centrifuge at 2000 r.p.m. (~800 *g*) for 20 min in a bench centrifuge to deposit the kaolin.
(v) Remove supernatant into clean, labelled tubes. This gives a 1 in 4 dilution of serum ready for the HI test and can be stored overnight at 4°C.

5.1.3 *Rubella HI Antibody Test*

(i) For each serum to be tested add one volume of DGV to a row of wells (columns 1 – 10 and 12).
(ii) To the first and last well (columns 1 and 12) add one volume of the 1 in 4 dilution of serum.
(iii) With the microdiluter perform doubling dilutions from column 1 (1 in 8) to column 10 (1 in 4096).
(iv) To columns 1 to 10 add one volume of rubella HA antigen at 4 HA units. Column 12 is the serum control with no rubella HA present.
(v) On another microtitre plate add one volume of DGV to columns 1 to 8 in rows 1 and 2. Add one volume of rubella HA at the working dilution to column 1 and doubly dilute from column 1 to column 8. Add one volume of DGV to these 16 wells. This 'back-titration' checks the titre of the rubella HA antigen. Add 2 volumes of DGV to column 12, rows 1 and 2. This is the cell control.
(vi) After covering the plates, leave at room temperature for 1 h or at 4°C overnight.
(vii) To every well used add two volumes of 0.03% chick rbc and leave to settle for 1 – 1.5 h (or overnight) at 4°C.
(viii) Check that the rubella HA antigen was at 2 – 8 HA units by examining the rubella HA antigen control titration and check that the rbc did not auto-agglutinate by examining the cell control wells.

(ix) Examine the serum control wells for haemagglutination. If any serum has haemagglutinated the rbc in the absence of rubella HA, the 1 in 4 dilution of serum should be re-tested after absorbing with one drop of 30% chick rbc for 1 h at room temperature and sedimenting the rbc by centrifugation.

(x) The titre of rubella-specific antibody in a serum is the reciprocal of that dilution of serum which fully inhibits haemagglutination.

5.1.4 *Interpretation*

Titres of eight are difficult to interpret as they may be due to residual non-specific inhibitors. Titres of 16 or above are considered to represent rubella-specific antibody.

5.2 **Complement Fixation Antibody Test**

The principle of the complement fixation test (CFT) is that the reaction of an antibody with a viral antigen will result in activation, and therefore utilisation, of any complement present. Any active complement remaining is detected by the addition of sheep rbc which have been sensitised by incubation with antibody to sheep rbc – usually called haemolysin and produced by immunisation of rabbits. If sufficient active complement remains, the rbc will be lysed by activation of complement by the antigen/antibody complex at the rbc surface. If the complement has been activated by a specific antigen/antibody reaction in the first stage of the assay, the sensitised sheep rbc will not be lysed.

The amount of antibody in a serum is therefore determined by testing a dilution series with a known microbial antigen in the presence of complement, with the subsequent addition of sensitised rbcs.

To obtain reproducible, reliable results it is essential that the components of the test are used in the correct proportions. Therefore, initially the appropriate concentrations of complement, haemolytic serum and antigen have to be determined by chessboard titration. The following example is the testing of sera for antibodies to rubella CF antigen.

5.2.1 *Titration of Complement and Haemolytic Serum*

As in the HI test, microtitre polystyrene plates with U-shaped wells, 0.025 ml volumes and microdiluters are used.

5.2.1 *Setting up the Titration*

(i) Reconstitute freeze-dried complement with distilled water as directed.

(ii) Label eight 75 mm x 12 mm glass tubes from 1 to 8.

(iii) Make a dilution of 1 in 60 of complement in tube 1 by adding 0.1 ml of reconstituted complement, 0.7 ml of distilled water and 4.0 ml of veronal-buffered saline (VBS) (20 CFT tablets in 2000 ml distilled water). The freeze-dried complement contains preservative so that addition of 0.7 ml distilled water to 0.1 ml of reconstituted complement gives a 1 in 10 dilution.

(iv) Place 0.4 ml of VBS into tubes 2 to 8.

(v) Transfer 1.6 ml from tube 1 to tube 2, mix, transfer 1.6 ml to tube 3, and so on to tube 8. Wash the pipette with VBS between each dilution. This procedure prepares a dilution series of complement with a 20% difference in concentration between each, i.e., 1 in 60, 75, 94, 118, 148, 184, 230 and 288.

(vi) Label a microtitre plate as in *Figure 3*.

(vii) Place 2 volumes of VBS in columns 1–8, rows 1–7.

(viii) Place 3 volumes of VBS in the haemolytic serum control wells, column 9, rows 1–6.

(ix) Add 1 volume of each dilution of complement to each of the seven wells in the appropriate column.

(x) Cover and leave overnight at 4°C.

(xi) The following morning, prepare a series of doubling dilutions of haemolytic serum from 1 in 25 to 1 in 800 and a further tube containing 1 ml of VBS for the complement control. In tube 1 place 2.4 ml of VBS and 0.1 ml of haemolytic serum. In tubes 2–6 place 1 ml of VBS. Transfer 1 ml from tube 1 to tube 2 and so on to tube 6, and wash the pipette after each dilution. Discard 0.5 ml from tube 1 and 1 ml from tube 6 to leave a volume of 1 ml in each tube.

(xii) Wash sheep rbc three times in VBS.

(xiii) Discard the supernatant after the last wash and re-suspend the packed cells.

(xiv) Measure the packed cell volume of rbc and dilute in VBS to 4%.

(xv) Put 1 ml of rbc suspension into seven appropriately labelled tubes next to the row of haemolytic serum dilutions and the control tube.

(xvi) Add haemolysin dilutions to the rbc suspensions, mix, cover the tubes and incubate in a 37°C water bath for 20 min.

(xvii) Simultaneously warm microtitre plate at 37°C in an incubator or hot room.

(xviii) Mix tubes by tapping and put 1 volume of each dilution into the appropriate row of wells.

(xix) Tap the plate to suspend the cells and place at 37°C for 30 min, re-suspending at 15 min.

(xx) Leave the plate at 4°C for approximately 90 min before reading.

Reading and interpretation. A typical result is shown in *Figure 5*. No haemolysis in the complement and haemolysin controls demonstrates the absence of non-specific lysis.

The optimum sensitising concentration (OSC) of haemolytic serum is that dilution which gives most lysis with the highest dilution of complement. In *Figure 5* the OSC is 1 in 100 and haemolytic serum is used at this dilution.

One unit of complement (HC_{50}) is in the dilution which gives 50% lysis (reading 2) at the OSC of the haemolytic serum. For testing sera, three units of complement are used (i.e., three times the HC_{50}). In *Figure 5* the HC_{50} is in the 1 in 184 dilution and thus the complement should be used at 1 in 60.

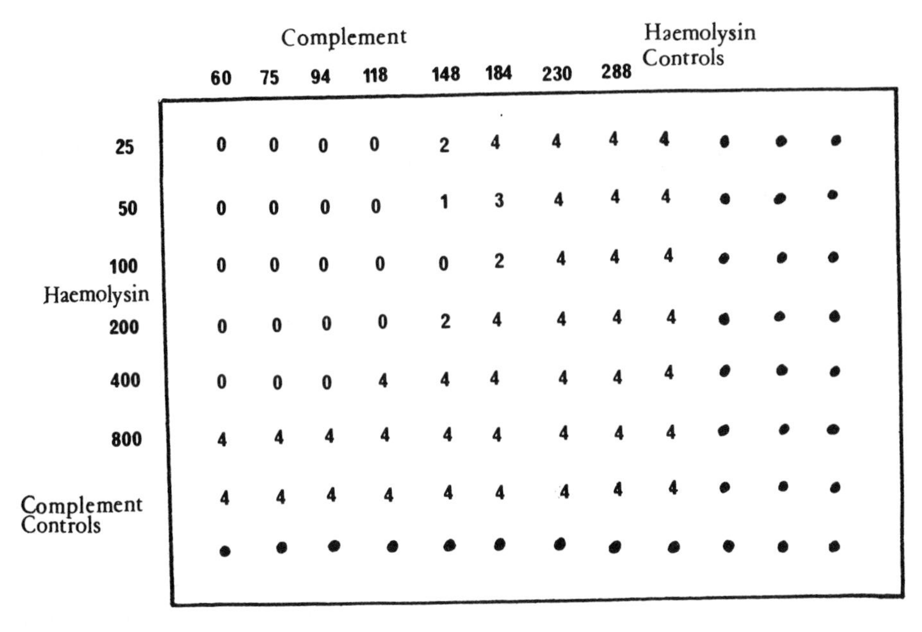

Figure 5. Complement/haemolytic serum chessboard titration. 0, complete lysis; 1, 25% cells remaining; 2, 50% cells remaining; 3, 75% cells remaining; 4, 100% cells remaining.

5.2.2 *Titration of Rubella CF Antigen and Positive Control Standard Antiserum*

Setting up the titration

(i) Reconstitute freeze-dried rubella CF antigen with distilled water as directed.
(ii) Label six 75 mm x 12 mm tubes 1 to 6.
(iii) To tube 1 add 0.2 ml of rubella antigen and 0.8 ml of VBS.
(iv) To tubes 2 – 6 add 0.5 ml of VBS.
(v) Transfer 0.5 ml from tube 1 and so on to tube 6 to produce a dilution series of antigen 1 in 5, 10, 20, 40, 80 and 160.
(vi) Reconstitute control freeze-dried, positive antiserum with distilled water.
(vii) Dilute serum 1 in 40 by adding 0.1 ml to 3.9 ml VBS.
(viii) Doubly dilute 0.5 ml volumes in VBS to give antiserum concentrations of 1 in 40, 80, 160, 320 and 640.
(ix) Label a microtitre polystyrene plate as in *Figure 6*.
(x) Add 1 volume of VBS to each well of the antiserum control column and to each well of the antigen control row.
(xi) Add 1 volume of each of the serum dilutions to each well of the appropriate row.
(xii) Add 1 volume of each of the antigen dilutions to each well of the appropriate column.
(xiii) Add 1 volume of complement at 3 HC_{50} to each well.
(xiv) To ensure that the complement is at the correct concentration a series of four complement control wells is made with the complement at concentra-

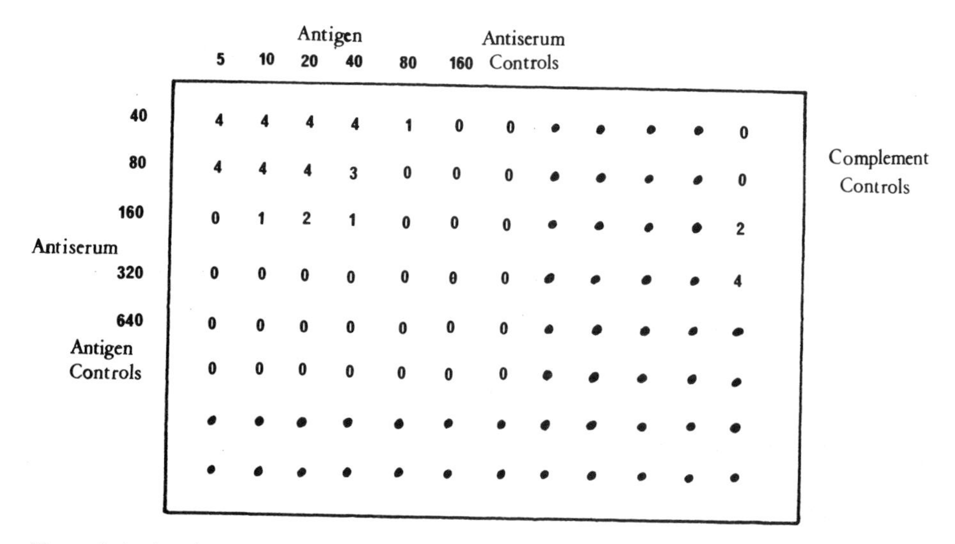

Figure 6. Antigen/antibody chessboard titration.

tions of 6, 3, 1.5 and 0.75 HC_{50}. These are made by the following procedures in the four wells, rows 1–4, column 12:

(a) Add 1 volume of VBS to the well in rows 1, 3 and 4 and add 2 volumes of VBS to the well in row 2.

(b) To the well in row 1 add 2 volumes of complement; to the well in rows 2 and 3 add 1 volume of complement.

(c) Transfer, with a microdiluter, 0.025 ml from the well in row 3 to that in row 4.

(d) Discard 0.025 ml with the microdiluter from the well in row 4.

(e) Add 2 volumes of VBS to the well in rows 3 and 4.

(xv) Leave the plate overnight at 4°C.

(xvi) Place the plate at 37°C for 30 min.

(xvii) Make a 4% suspension of sheep rbc and mix 2.5 ml with an equal volume of haemolysin diluted in VBS to its OSC. Mix by adding the haemolysin to the rbc and inverting 10–12 times. Incubate sensitised cells for 30 min in a 37°C waterbath.

(xviii) Add 1 volume of sensitised rbc to each well in the plate that has been used.

(xix) Tap the plate to suspend the cells, cover and incubate at 37°C for 30 min with tapping at 15 min.

(xx) Stand at 4°C for approximately 90 min before reading.

Reading and interpretation. A typical result is shown in *Figure 6*. The complement control shows that it was used at the correct concentration and total haemolysis in the antigen and antisera controls shows that they were not intrinsically anti-complementary.

The optimum dilution of the rubella CF antigen is that dilution which gives most fixation with the highest dilution of serum. In the example (*Figure 6*) this dilution is 1 in 20 and is the dilution to be used in tests with unknown sera.

In the example given, the titre of the positive serum is 160, i.e., the reciprocal of that dilution which gives a reading of two with the optional dilution of rubella CF antigen.

5.2.3 *Testing of Unknown Sera*

Titration of sera

(i) Dilute test, control positive and control negative sera 1 in 4 by adding 0.1 ml of serum to 0.3 ml of VBS in glass tubes.
(ii) Place in a water bath at 56°C for 30 min to inactivate the intrinsic complement in the sera.
(iii) Label a microtitre plate as in *Figure 7*.
(iv) Add 1 volume of VBS to each well in columns 1 – 8 and 10, the number of rows used depending on the number of sera to be tested.
(v) Add 1 volume of the 1 in 4 serum dilution to wells 1 and 10.
(vi) Using the microdiluter, doubly dilute the serum from well 1 to well 8 so giving a dilution series from 1 in 8 to 1 in 1024. Using the microdiluter, discard 0.025 ml from wells column 8.
(vii) To the antigen control well add 1 volume of VBS and to the cell control add 3 volumes of VBS.
(viii) Add 1 volume of rubella CF antigen to all wells used except the serum controls (column 10), the cell control and the complement control.
(ix) Add 1 volume of complement at a concentration of 3 HC_{50} to every well except the cell control and complement control wells.
(x) Prepare the complement control wells as described previously.
(xi) Place the covered plate at 4°C overnight.
(xii) Warm the plates at 37°C for 30 min and add sensitised rbc to every well used as described previously.

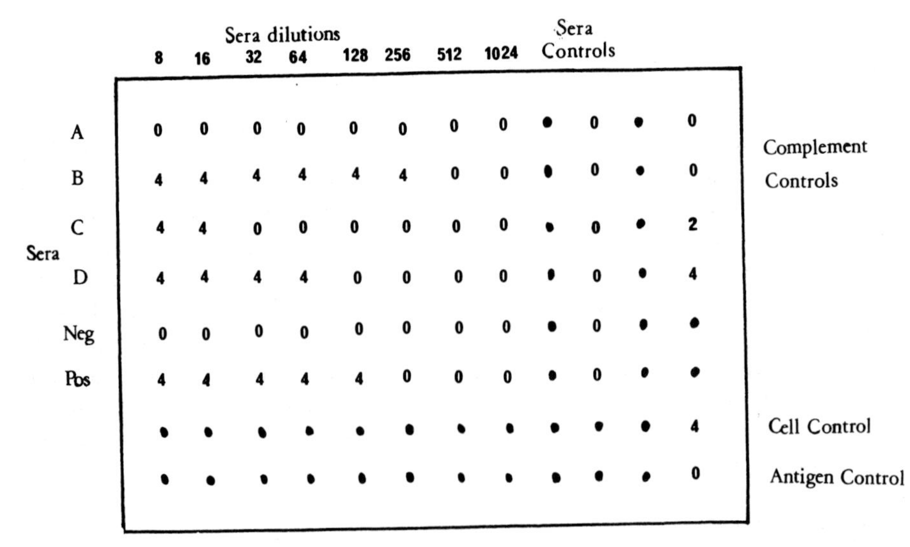

Figure 7. Evaluation of sera by complement fixation test.

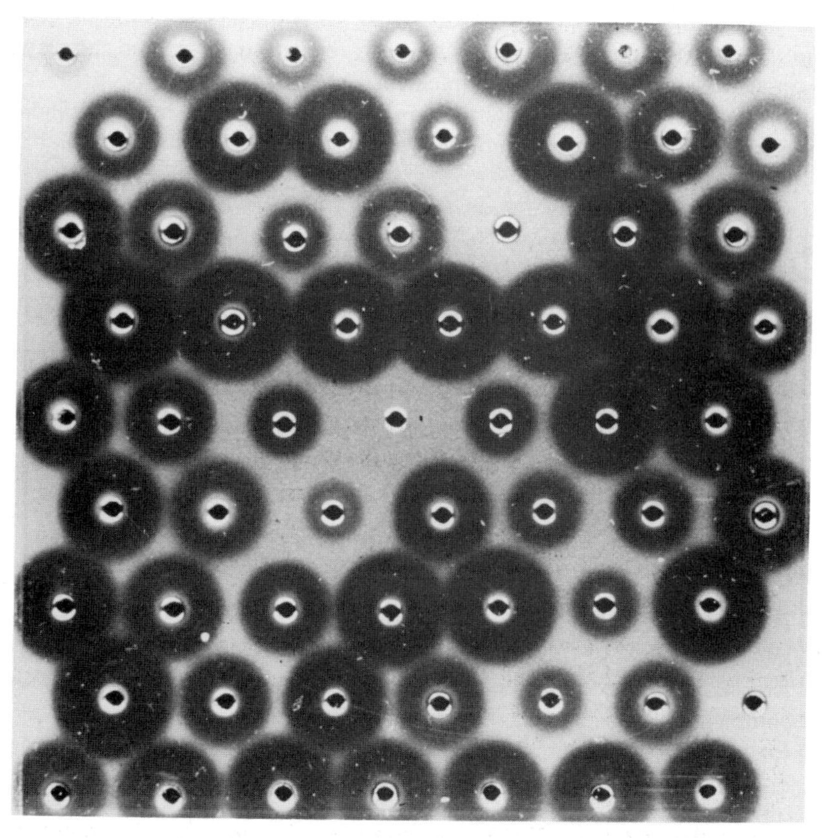

Figure 8. Radial haemolysis for rubella-specific antibody. A positive control of low rubella-specific IgG content (15 international units) is in wells 3 and 5 of row 5 and the negative control serum is between the two.

(xiii) Incubate at 37°C as described and allow to settle at 4°C for approximately 90 min.

Reading and interpretation. An example of a plate showing the testing of four unknown sera is given in *Figure 7.* The complement titration control shows that the complement was at 3 HC_{50}, the cell control demonstrates the absence of spontaneous lysis of the sensitised cells and the total lysis in the antigen control shows that the antigen was not anti-complementary. The serum control wells should also show complete lysis, so indicating that the sera are not intrinsically anti-complementary – a problem which may result if the sera contain immune complexes or are bacterially infected. The negative control serum shows no detectable antibody and the positive control serum shows rubella CF antibody present at a 1 in 128 dilution but not at 1 in 256. Thus, the titre of the positive control serum is 128. The titres of the test sera A to D are respectively, 8, 256, 16 and 64.

5.3 Radial Haemolysis

Radial haemolysis is an ideal assay to detect rubella-specific IgG (8) as many sera can be examined simultaneously by this sensitive and specific technique (*Figure 8*). Rubella-specific IgG is detected by lysis of rubella antigen-coated rbc suspended in an agarose gel containing complement.

The presence in a serum of IgG to the rbc used is detected by simultaneously testing in control gels containing complement and rbc which have not been coated with rubella antigen.

5.3.1 *Preparation of Radial Haemolysis Gels*

(i) Melt two 15 ml volumes of 1% agarose in CF buffer in glass universals by steaming and then hold at 43°C in a water bath.
(ii) Wash sheep rbc three times in CF buffer by centrifugation and resuspension.
(iii) After the final wash, re-suspend to 15% in CF buffer and pipette 0.3 ml into two glass universals, one labelled 'test' and the other 'control'.
(iv) Reconstitute the rubella HA and mix 0.3 ml with the rbc in the 'test' universal.
(v) Leave at room temperature for 30 min.
(vi) Fill both universals with CF buffer and centrifuge to deposit the rbc.
(vii) Discard supernatants and re-suspend rbc in 0.5 ml of CF buffer containing kanamycin (100 μg/ml).
(viii) Reconstitute freeze-dried complement and pipette 0.5 ml of the undiluted complement into the cell suspension in the 'test' universal.
(ix) Place the 'test' universal in the 43°C water bath for 10 sec and then pour one universal (15 ml) of the molten agarose into it. Quickly screw the universal cap into place and invert a few times to suspend the rbc evenly.
(x) Immediately pour the contents of the universal into a 100 mm x 100 mm square plastic Petri dish (labelled 'test') on a level surface.
(xi) By gentle rocking ensure that the agarose is evenly distributed in the Petri dish.
(xii) Repeat steps (vii) – (ix) for the control universal, labelling the Petri dish 'control'.
(xiii) Leave for 20 min to set. The plates can be stored for up to 4 days at 4°C prior to use.

5.3.2 *Testing of Sera*

(i) Heat-inactivate the complement in sera to be tested in a 60°C water bath for 20 min.
(ii) Cut 3 mm wells in the test and control gels over a template (*Figure 9*) using a steel borer attached to a vacuum line and trap.
(iii) Add sera to corresponding wells in test and control gels. Each batch should include a negative and a low positive serum (in diagnostic use a serum containing 15 international units of rubella-specific antibody is used).
(iv) Place gels in a humid box overnight at 37°C.

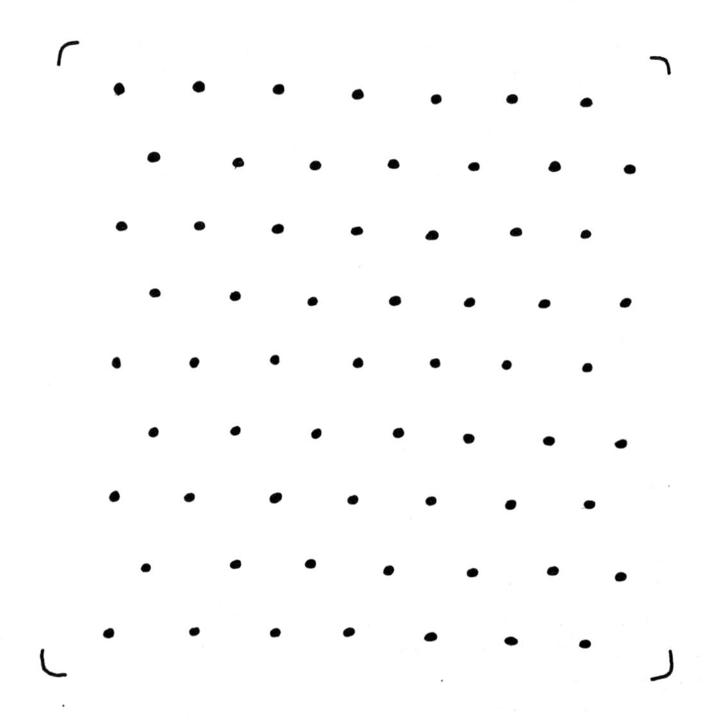

Figure 9. Template for evaluation of 63 sera (including controls) by radial haemolysis.

5.3.3 *Reading and Interpretation*

Examine the plates for zones of haemolysis around the wells with trans-illumination against a dark background. Sera that give a larger zone of haemolysis on the test plate than on the control plate are interpreted as containing rubella-specific IgG. Those sera which give no zones of haemolysis or have zones of equal diameter on the test plate and control plate are interpreted as having no rubella-specific IgG and therefore the donor of the serum is susceptible to primary infection with rubella.

5.4 Detection of Rubella-specific IgM by an Antibody-capture Technique

The detection of rubella-specific IgM is of importance for the diagnosis of both recent rubella infection and congenital rubella. Numerous methods are available. The long-established methods depended on the separation of IgG from IgM by serum fractionation. This was usually by gel filtration or sucrose density gradient centrifugation with HI tests being performed on the immunoglobulin-containing fractions. These techniques have been replaced by immunological methods which do not require such serum fractionation. They are of two main varieties. The first is the solid-phase antigen assay where a plastic surface (e.g., microtitre well) is coated with rubella antigen and sequentially incubated with the patient's serum and a labelled antibody to human IgM prepared in another species. If careful washing takes place after each of these steps, detection of label remaining in the wells indicates the presence of rubella-specific IgM.

An alternative is the antibody capture assay in which the solid phase is coated with an anti-human IgM. Incubation with serum will result in IgM being retained on the solid phase after washing. Incubation with rubella antigen will select any rubella-specific IgM and, after washing, bound rubella antigen can be detected by incubation with a labelled anti-rubella antibody.

The final detector antibody for this test may be labelled with radioactive ^{125}I or with an enzyme which will produce a colour change in the presence of a suitable substrate. We describe here the radioimmunoassay since this has been fully evaluated and is in use in a number of laboratories. The labelled anti-rubella antibody now used is a mouse monoclonal antibody but unfortunately this is not generally available. When setting up this test it is advisable to liaise with a local virology laboratory since they may be able to supply a high titre monospecific anti-rubella serum. Otherwise such a serum will have to be raised by conventional techniques (e.g., in a rabbit immunised with rubella virus grown in RK 13 cells) or by the hybridoma technique according to Tedder *et al.* (8). These antibodies are conveniently labelled by the iodogen method as previously described (6). Currently the fashion is to change to enzyme labels and the commercially available Rubenz-M (Northumbria Biologicals) is an antibody capture assay with an enzyme label which is suitable for the detection of rubella-specific IgM. Moreover, the enzyme labelled monoclonal anti-rubella antibody can be obtained as a separate item.

5.4.1 *Coating the Solid Phase*

(i) Place 6.4 mm diameter polystyrene beads (Northumbria Biologicals Ltd.) in a container and cover with rabbit anti-human μ chain (Dako Ltd.) diluted 1 in 500 in 1 N HCl.

(ii) Shake the beads for 1 h at room temperature, then hold at 4°C for at least 48 h before use.

(iii) For small-scale assays the beads can be placed in 12 mm diameter polystyrene tubes and washed using a wash bottle, the wash fluid being aspirated through a Pasteur pipette attached to a vacuum line with a trap. However, the handling and washing of beads is greatly facilitated by using Abbot trays and an Abbott pentawash system (Abbott Diagnostics Ltd.).

5.4.2 *Test Procedure*

(i) Remove the required number of beads from stock and aspirate the residual HCl. Cover the beads with PBSA with 1% bovine serum albumin and leave for approximately 3 h at room temperature.

(ii) In the wells or tubes used to hold the beads, place 200 μl PBSA plus 0.05% Tween 20 (PBST) and add 5 μl of test or control sera (each should be tested in duplicate). A pool of 4–8 early convalescent sera from cases of rubella should be designated as containing 100 arbitrary units (a.u.) of rubella-specific IgM. Positive control sera containing 40, 10, 3.3 and 1.0 a.u. are prepared by diluting in rubella antibody-negative human sera. These are in-

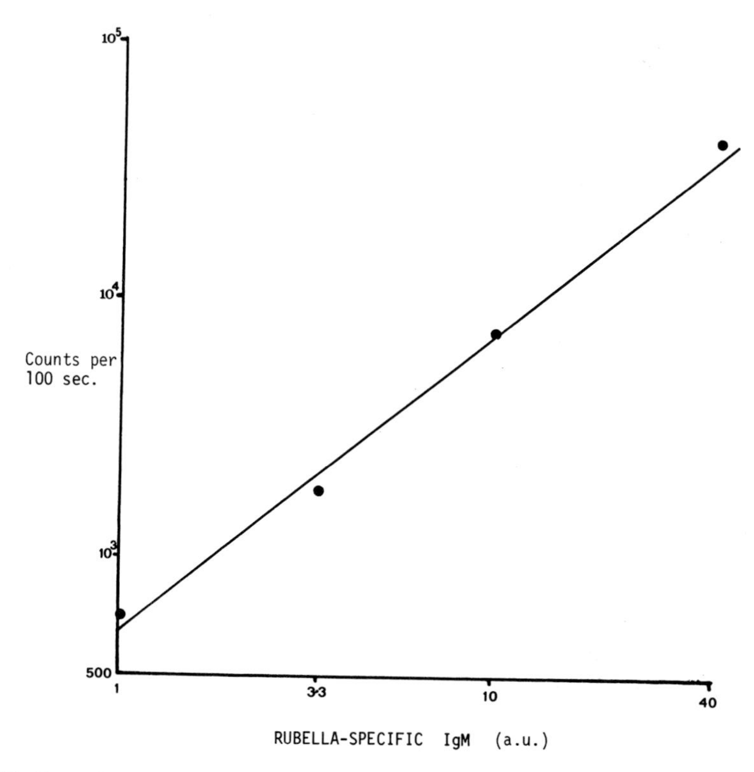

Figure 10. Example of a standard curve for IgM antibody capture radioimmunoassay for rubella-specific IgM.

cluded in each test as a standard curve. A negative control serum should also be included.

(iii) Add an anti-μ coated bead to each well or tube containing the 1 in 40 dilution of sera. Hold at 37°C for 3 h.

(iv) Wash the beads three times in PBST and add 200 μl of rubella HA antigen (titre 128) diluted 1 in 10 in PBST. Hold at 4°C for 18–24 h. Wash the beads three times in PBST.

(v) Add 200 μl of ^{125}I-labelled detector antibody which has been diluted in PBS containing 0.2% Tween 20, 20% heat-inactivated rabbit serum and 10% heat-inactivated human rubella antibody-negative serum to give counts of approximately 50 000/min/200 μl. Hold at 37°C for 3–4 h.

(vi) Wash the beads four times with PBST and measure the radioactivity bound to the bead with a gamma counter. Average duplicate readings, subtract background readings and construct a standard curve by plotting results for control sera against arbitrary unitage on log/log paper (*Figure 10*). Determine the unitage of test sera by comparison with the standard curve.

5.4.3 *Interpretation*

Test sera which contain 3.3 or more a.u. of rubella-IgM are positive and indicate recent or congenital rubella. Those with less than 1.0 a.u. are negative. Values between 1.0 and 3.3 a.u. are regarded as equivocal and must be interpreted in relation to the clinical details of the patient from whom the sera came. These equivocal values are usually considered as negative but they may indicate infection many weeks previously.

6. ACKNOWLEDGEMENTS

We would like to thank Julian Hodgson and the staff of Virology, King's College Hospital, for their help in establishing the techniques described and Phyllis Lovett for typing the manuscript.

7. REFERENCES

1. Lennette,E.H. and Schmidt,N.J., eds. (1979) *Diagnostic Procedures for Viral, Rickettsial and Chlamydial Infections,* 5th Edition, published by American Public Health Association, Washington.
2. Grist,N.R., Bell,E.J., Follett,E.A.C. and Urquhart,G.E.D. (1979) *Diagnostic Methods in Clinical Virology,* 3rd Edition, published by Blackwell Scientific Publications, Oxford, London, Edinburgh and Melbourne.
3. Field,A.M. (1982) *Adv. Virus Res.,* **27**, 1.
4. Gardner,P.S. and McQuillan,J. (1980) *Rapid Virus Diagnosis,* 2nd Edition, published by Butterworths, London.
5. Paul,J. (1975) *Cell and Tissue Culture,* 5th Edition, published by Churchill Livingstone, Edinburgh, London and New York.
6. Pattison,J.R., ed. (1982) *Laboratory Investigation of Rubella,* Public Health Laboratory Service Monograph Series No. 16. Her Majesty's Stationery Office, London.
7. Morgan-Capner,P. (1983) *Public Health Lab. Serv. Microbiol. Digest,* **1**, 6.
8. Tedder,R.S., Yao,J.L. and Anderson,M.J. (1982) *J. Hyg.,* **88**, 335.

INDEX